An Anthology of Theories and Models of Design

Amaresh Chakrabarti · Lucienne T. M. Blessing
Editors

An Anthology of Theories and Models of Design

Philosophy, Approaches and Empirical Explorations

Springer

Editors
Amaresh Chakrabarti
Centre for Product Design and Manufacturing
Indian Institute of Science
Bangalore
India

Lucienne T. M. Blessing
Université du Luxembourg, Campus Limpertsberg
Luxembourg
Luxembourg

ISBN 978-1-4471-6337-4 ISBN 978-1-4471-6338-1 (eBook)
DOI 10.1007/978-1-4471-6338-1
Springer London Heidelberg New York Dordrecht

Library of Congress Control Number: 2013958140

Printed on acid-free paper

Springer is part of Springer Science+Business Media (www.springer.com)

Preface

Background

Developing models and theories of design is one of the major, growing activities in design research. Yet, many of these theories and models are not widely known. It was, therefore, felt worthwhile to bring together, in a book, an anthology of as many as possible of the major models and theories that have emerged in this relatively young discipline. The other goal of the book was to present the highlights of the discussions that took place during the International Workshop on Models and Theories of Design (IWMT 2013) held at the Centre for Product Design and Manufacturing, Indian Institute of Science, Bangalore, India, during 3–5 January 2013. The workshop was organized to support intensive discussion around the theories and models, identify progress, and seek future directions.

This book is intended to provide a ready reference to a comprehensive collection of theories and models in design research, so that these can act as catalysts for further research that is informed by, and based on a better understanding of past effort. The book is meant primarily for young researchers in the area of design theory and methodology.

The editors have a long background in this area. Both have been involved in developing various theoretical and empirical aspects of design theories and models, and conducted as co-chairs several workshops in the past in this area (e.g., 1st Cambridge General Design Theory Workshop 1998, 2nd Cambridge General Design Theory Workshop 1999, and 1st Cambridge Design Synthesis Workshop 1999, all held at Churchill College, Cambridge, UK). Besides, the first author was involved in initiating a series of "Newness of Designs" workshops in Japan in 1997 that were precursors to the 1st Cambridge Design Synthesis Workshop in 1999.

Overview of the Book

The contributions in this book cover three related aspects of research into theories and models of design—philosophical, theoretical, and empirical. The book contains 21 chapters. The editorial chapter summarizes the findings in the book, a review of some of the major theories and models not covered by the authors in the book, and the major findings from the workshop. The other chapters are written by eminent authors from 15 universities in 11 countries. The book has three parts: Part I—Philosophical Contributions—contains 6 chapters; Part II—Theoretical Contributions—contains 9 chapters; and Part III—Empirical Contributions—contains 5 chapters.

Apart from showcasing a representative cross-section of major contributions in these three aspects, the contributions and discussions attempt to explore three, related (sets of) questions:

- What is a theory or model of design? What is its purpose: what should it describe, explain, or predict?
- What are the criteria it must satisfy to be considered a design theory or model?
- How should a theory or model of design be evaluated or validated?

Even though by no means complete, the contributions and the workshop outcomes showcase the rich and varied tapestry of thoughts, concepts, and results that have emerged in this area. At the same time, they highlight the effort still required to establish a sound theoretical and empirical basis for further research into design.

Acknowledgments

We are thankful, first of all, to all the authors for writing their chapters for the book. We further thank those who were able to participate in the workshop, take part in its discussions, and share their insights, ideas, and contributions: Marine Agogué, Claudia Eckert, John Gero, Gabriela Goldschmidt, Tom Howard, Lauri Koskela, Udo Lindemann, BSC Ranjan, Eike Wintergerst, Neeraj Sonalkar, Eṣhwaran Subrahmanian, Toshiharu Taura, and SrinivasanVenkataraman. We also thank the Rapporteurs (John Gero, Lauri Koskela, and Udo Lindemann), the Scribes (Sonal Keshwani, S. Harivardhini, Praveen Uchil and Boris Eisenbart), and the videographers (Kashyap Krishna, Anusha Raki, Praveen Uchil and Shakuntala Acharya Nair) for their help in capturing as much as possible of the knowledge produced during the workshop. We are particularly thankful to Ms. Sonal Keshwani for her valuable coordination in capturing the overall proceedings of the workshop.

Special thanks go to Mr. Ranjan who has been the mainstay of the workshop and of subsequent proceedings with the book, including his continuous engagement with our publisher, Springer-Verlag. He has been ably supported by Biplab Sarkar, Ujjwal Pal, Praveen Uchil, S. Harivardhini, and Sonal Keshwani in organizing the various aspects of the workshop. We also thank Ms. Chaitra, the then Secretarial Assistant at IdeasLab, Centre for Product Design and Manufacturing (CPDM) at Indian Institute of Science (IISc), and Ms. Suma, the current Secretarial Assistant, for their help in secretarial matters related to the workshop and the book.

We are thankful to Springer-Verlag, especially Oliver Jackson, for accepting our proposal for this book, and to Charlotte Cross for keeping us on track regarding deadlines. We are also thankful to Prof. Anindya Deb, the current Chairman of CPDM, for lending his support to, and for providing the venue for the workshop. Thanks are due also to Mr. Jagadesh, Ms. Elizabeth, and Mr. Yogananda for providing logistic support.

Finally, we are thankful to our families, Anuradha and Apala, as well as Peter, Koen, and Saskia, for being patient with us during the preparation of the workshop and the book.

We hope that this book will help further theoretical progress by bringing together a wide range of thoughts, approaches, assumptions, concepts, scopes, and foci developed in our research community, and in doing so inspire readers and provide them with a broader basis for their own research.

Amaresh Chakrabarti
Lucienne T. M. Blessing

Contents

Chapter 1
Theories and Models of Design: A Summary of Findings

Amaresh Chakrabarti and Lucienne T. M. Blessing

1.1 Introduction

The goal of this book is to bring together an anthology of some of the major theories and models of design that have emerged in the last 50 years of the relatively young discipline of design research. Another goal is to bring together the highlights of the discussions that took place during a workshop that was organised around the theories and models—The International Workshop on Models and Theories of Design (IWMT 2013) held at the Centre for Product Design and Manufacturing, Indian Institute of Science, Bangalore, India, during 3–5 January 2013.

The contributions in this book cover three related, but distinct aspects of research into theories and models of design—philosophical, theoretical, and empirical. Even though by no means complete, taken together the contributions and the workshop outcomes showcase the rich but varied tapestry of thoughts, concepts and results. At the same time, they highlight the effort still required to establish a sound theoretical and empirical basis for further research into design.

1.1.1 Contributions

The chapters in this book are grouped according to their *main* area of contribution, i.e. philosophical, theoretical or empirical.

Part I: Philosophical contributions: This part commences with two chapters presenting a discussion about research into design theories and models (Vermaas

A. Chakrabarti
Centre for Product Design and Manufacturing, Indian Institute of Science, Bangalore, India
e-mail: ac123@cpdm.iisc.ernet.in

L. T. M. Blessing (✉)
FSTC, Université du Luxembourg, Luxembourg, Luxembourg
e-mail: Lucienne.blessing@uni.lu

A. Chakrabarti and L. T. M. Blessing (eds.), *An Anthology of Theories and Models of Design*, DOI: 10.1007/978-1-4471-6338-1_1,

and Sonalkar et al.). This is followed by two chapters emphasising the need to move the boundaries of design research and thus the coverage of theories and models (Taura and Horváth). The last two chapters focus on models and modelling (Lindemann and Maier et al.).

2. Vermaas, on the scientific status of design research with respect to design theories, models and their testing.
3. Sonalkar et al., on a two-dimensional structure for design theory allowing scientific rigour as well as practical usefulness.
4. Taura, on considering Pre-Design and Post-Design by including the motive of design.
5. Horváth, on the theoretical challenges imposed by social-cyber-physical systems.
6. Lindemann, on the systematic development and the desirable characteristics of models.
7. Maier et al., on using a cybernetic perspective to explain modelling in design.

Part II: Theoretical contributions: The chapters in this part have their main contribution in the theoretical development of the field. To understand design, it is necessary to address both the artefact and the process. Design theories and models tend to cover both, but with a clear difference in focus. The core can be strongly product-focused, strongly process-focused, or intentionally focused on both in equal measure. It has to be noted, however, that as theories and models evolve, the core may change.

The theoretical contributions are grouped according to this core: Chaps. 8–10 are largely product-focused (Albers and Sadowski, Andreasen et al., and Eder), Chaps. 11–15 are largely process-focused (Agogué and Kazakçi, Cavallucci, Gero and Kannengiesser, and Koskela et al.), Chaps. 16 and 17 focus equally on product and process (Ranjan et al., Weber et al.).

8. Albers and Wintergerst, on the Contact and Channel Approach to integrate functional descriptions into a product's physical structure model.
9. Andreasen et al., on the Domain Theory as a systems approach for the analysis and synthesis of products.
10. Eder, on the role of theory, models and methods in engineering design, with emphasis on the Theory of Technical Systems.
11. Agogué and Kazakçi, on the mathematical foundations of C–K theory, its development and its impacts in design research and practice as well as in other fields.
12. Cavallucci, on the Inventive Design Method (IDM) to guide inventive practices based on and enhancing the theory of inventive problem solving (TRIZ).
13. Gero and Kannengiesser, on the development of their Function-Behaviour-Structure (FBS) ontology and framework to represent regularities in design and designing.
14. Koskela et al., on the Aristotelian proto-theory of design as a possible design theory.

15. Ranjan et al., on the development of the Extended-Integrated Model of Designing (E-IMoD) to describe and explain the design process.
16. Weber, on the CPM/PDD approach to model products and processes based on characteristics and properties.

Part III: Empirical contributions: The final five chapters describe empirical contributions that inform theoretical developments and their verification.

17. Badke-Schaub and Eris, on the exploration of the role intuitive processes play in thinking and acting of designers, as a precursor to the development of a theory of design intuition.
18. Culley, on the reinterpretation of the engineering design process as a process of generating information objects.
19. Eckert and Stacey, on identifying the major causal drivers of design and their effects as first steps in incremental design theory development.
20. Goel and Helms, on the development and application of knowledge models using the example of biologically inspired design.
21. Goldschmidt, on a cognitive model of sketching in the early design phases.

1.1.2 Questions Addressed

Three general questions were asked to all authors. For *philosophical* contributions, they constituted the main questions:

- What, according to you, is a theory or model of design, e.g. what is its purpose, i.e. what is it expected to describe, explain or predict?
- What, according to you, are criteria it must satisfy to be considered a design theory or model?
- How should a theory or model of design be evaluated or validated?

Authors of *theoretical* contributions were additionally asked to address the following questions:

- What is your design theory or model, what is its purpose and which criteria does it satisfy?
- What studies have you undertaken to develop and validate your theory or model, i.e. to what extent does your theory or model satisfy its purpose?

Authors of *empirical* contributions were additionally asked to address the following questions:

- What empirical findings in your area of research are the most significant for the development or validation of theories and models of design?
- What are the consequences of these empirical findings for the development or validation of theories and models of design?

This editorial chapter attempts to bring together the views of the authors, as expressed in their chapters and during the workshop, with our own views.

1.1.3 Workshop

Nearly all contributions in this book were presented during the aforementioned workshop. The final sessions of each day were dedicated to group discussions. The participants were divided into three groups to address the following questions:

1. What should a theory or model for design be?
    a. what is its *purpose*, i.e. what is it expected to describe, explain or predict?
    b. what are the *criteria* it must satisfy to be considered a theory or model of design?
2. How should a theory or model of design be evaluated or *validated*?
3. Considering the current state of research:
    a. what are the gaps between theoretical and empirical results?
    b. what should be the directions of future research into theories and models of design?

One of the group members was assigned as rapporteur, who was supported by one or two PhD students as scribe to capture the discussions and produce a summary. The summaries were presented on the last day of the workshop and followed by a closing discussion involving all participants.

The results of the discussion sessions are brought together in Appendix A of this editorial chapter.

1.2 Theoretical Developments

1.2.1 Phases of Development

Design research can be considered to have passed through three overlapping phases: the Experiential, Intellectual, and Experimental [83]. Notable attempts to develop theories and related comprehensive models during that time are ARIZ/TRIZ [3, 4], Theory of Technical Systems [43, 44], Domain Theory [5], General Design Theory [86] and Extended General Design Theory [79], Function-Behaviour-Structure Ontology [32], Logic of Design [63]. Some of these theories and models were regularly cited, but the majority never really became established (or widely accepted) as a fundamental basis for further research, at least not during this period, which has been referred to as pre-theoretical, pre-paradigmatic [19] or pre-hypothesis [41].

The situation changed rather quickly, shortly before the turn of the millennium. A new phase in design research seemed to have started, the Theoretical Phase [13]. Several new theories of a very different nature were proposed at almost the same point in time: Mathematical Theory of Design [17], Universal Design Theory [37, 53], $K^LD^E{}_0$—Theory [69, 70], Axiomatic Design [74, 75], and Theory of Synthesis [76, 80]. These were soon followed by C-K Theory [38], Infused Design [66, 67], Domain Independent Design Theory [50, 51], GEMS of SAPPhIRE Model now called Integrated Model of Designing [58, 71], CPM/PPD framework [84], and the systematised theory for concept generation [78]. At the same time, earlier work was subject to considerable further development, such as Gero's Function-Behaviour-Structure Framework [33], Chap. 13 in this book,[1] Andreasen's Domain Theory [6], Chap. 9, and Altschuller's ARIZ [21], Chap. 12.

Most of these theories and models have been covered by the chapters in this book. Some of the major theories and models could not be included as the authors were not able to attend the workshop. As they are well worth mentioning and for the purpose of completeness, they are briefly introduced in Appendix B. Historical overviews can also be found in Blessing [9, 11], Lossack [51], Pahl and Beitz [57], Heymann [40] and Weber [85], Chap. 16.

1.2.2 *Differences*

The developments in the Theoretical Phase clearly distinguish themselves from the theoretical developments in the earlier phases. Firstly, the new theories and models received much more attention and have become more widely known. Importantly, they have done so in a much shorter period of time. The increased number of publications (due to the pressure to publish), the increased accessibility of publications due to the internet and open access policies, as well as a larger and more established design research community are certainly factors that contributed to the speed of dissemination, but they cannot fully explain this visibility. We think that dissatisfaction with the state of design research, as expressed in various publications (such as [8, 10, 12–15, 41, 60–62]) has fuelled interest in theoretical developments as a much needed foundation for the growing research community to build upon. Such a foundation is required not only for further development of a theoretical basis, but also to allow theory-based analysis, e.g. to explain differences between methods [47, 67].

Second, the developments in the Theoretical Phase differ from earlier ones in that they increasingly build on each other, rather than being developed largely independently from each other. Furthermore, they are accompanied by more fundamental discussions about design research and design science, gradually

[1] Hereafter, any reference to "Chap." refers to a chapter in this book.

allowing comparisons of research results, the identification of research paradigms, and discussions about quality and rigour (e.g. [13, 15]).

Third, there is now an explicit focus on validating theories and models using empirical data using observational studies or historical cases. This is fuelled on one hand by the increased demand for rigour in the discipline, and on the other hand by the increasing availability of empirical studies.

Fourth, the newer theories and models are richer in nature, using more and different concepts compared to the earlier theories and models (see Appendix C). A likely reason is our increased understanding of design resulting from a growing number of empirical studies into design. In her investigation of existing empirical studies up to 1992, the second author could only find 74 publications describing a total of 47 studies [9]. In 1999, Cantamessa counted 90 studies in one conference alone (the International Conference on Engineering Design), even though this conference was not dedicated to empirical studies [20]. Since then, empirical studies have become an established part of design research. Most research groups employ such research, albeit to varying degrees, and special conferences and interest groups have emerged. Notwithstanding this progress, our understanding remains fragmented [16] and as Koskela et al. [46], Chap. 14 conclude: 'many design theories and methods seem to be based on descriptive but somewhat shallow knowledge on some aspect of the design process'.

Finally, design research has always focused on increasing understanding *and* supporting practice, but often as separate streams [8, 13]. It was only in the Theoretical Phase that research that was focused on theories and models paid *explicit* attention to applicability in practice. A possible reason is the widely expressed dissatisfaction with the lack of adequate demonstration of the impact of earlier attempts on practice.

1.2.3 Theory and Practice

Design research has largely adopted the scientific paradigm in which it is assumed that there are regularities that underlie phenomena and it is the role of research to discover and represent those regularities [33], Chap. 13. We would add that design research also assumes that many of the observed phenomena can be changed, i.e. design practice (and education) can be improved, and that design research has an additional role: to develop and evaluate ways of realizing these changes. The majority of authors in this book confirm this combination of developing understanding and support, i.e. of scientific and practical/societal goals, as the purpose of design theories and models (see Sect. 1.4).

Having this double aim strongly affects both the research process and its outcomes. Design research is 'pulled in two opposing directions—towards scientific rigour on one hand, and a greater relevance for professional practice on the other', resulting in 'formal design theories deriving from mathematical roots that rarely influence practice' and 'process models that serve as scaffolds for professional designing, but lack scientific validity' (Sonalkar et al. [72], Chap. 3). To resolve this dichotomy, Sonalkar et al. propose a two-dimensional structure for design theory that displays scientific rigour while being useful to professionals. Our own attempt to resolve the dichotomy has been to propose a research methodology, DRM, which explicitly addresses both aims [16].

1.2.4 Competing and Complementing Theories and Models

Some theories and models are further developments of a particular theory, such as Gero's situated FBS, some are developments based on existing theories in other domains, such as Hatchuel and Weil's C-K Theory, others are based on critical reflections on existing theories. Usually theories and models are based on a combination of sources. The result is a multitude of theories and models. The question is, whether this constitutes a problem. Some authors, such as Buchanan [18], consider the existence of different views a strength, while others are worried that this might prevent coherent theory development [82], Chap. 2 and—cause the Problem of Disintegration [31]. In our opinion, both views can be correct, depending on the relationship between the theories or models.

Overall, the existence of multiple theories within the same domain *over time* can be interpreted positively as a sign of work in progress, indicating that an area is alive and developing. The evolution of design theories can be interpreted as an attempt to increase their generative power without endangering their robustness [39].

Theories and models that exist *at the same time* can be competing (addressing the same phenomena) or complementary. The latter can be divided into those that address different phenomena in design, and those that address the same phenomena from a different perspective. Vermaas [82], Chap. 2 seems to focus on the former (competing theories and models) when he warns that we might be creating too many theories and models, which jeopardises the coherence of the discipline. The reason of the multitude, according to Vermaas, is that 'design research does not yet have means to test and refute design theories and models'. Maier et al. [54], Chap. 7 found that in general 'researchers consider too much heterogeneity of models problematic and that design research should aim towards rationalisation, consolidation and integration of the ideas'. For Goel and Helms [34], Chap. 20, having multiple theories and models is inherent to design research: 'Research on design adopts many perspectives ranging from anthropology to neurobiology to philosophy. The various research paradigms produce not only different theories and models of different aspects of design, but also different types of theories and models'. Cavallucci [21], Chap. 12 emphasises the need for fundamentally different theories of design that will engender fundamentally different methods and tools for design activity's framing. For Eckert and Stacey [28], Chap. 19 having multiple, complementary, what they call partial, theories is a transitional phase: 'Design is far too complex and too diverse for understanding the whole of design in one step, so we need an incremental approach to accumulating understanding'. Only after validating theory fragments, should they be connected into larger, more complete, partial theories covering more of the interlocking causal processes shaping how designing is done, by matching and merging the elements of different theory fragments. Weber supports this view: 'Developing/designing products is such a complex process that not one model alone can explain every aspect; several models may exist in parallel. However, an integrating framework would be beneficial' [85], Chap. 16. The latter is also emphasised in, and is a major driver for the work in Ranjan et al. [59], Chap. 15.

Earlier we wrote that discussions about what constitutes design research and how it is distinct from or similar to other disciplines are still very much on-going [13, 16]. We worried, however, about the lack of a common view as to what design research attempts to investigate, what its aims are, and how it should be investigated: many different aspects are investigated, many different aims pursued, and many different methods are applied. We quoted Samuel and Lewis [65], who stated that 'design research is highly fragmented and focused streams of activity are lacking', and Horváth [41] who found it 'not easy to see the trends of evolution, to identify landmarks of development, to judge the scientific significance of the various approaches, and to decide on the target fields for investments'.

In the last few years, the number of discussions about design research and theoretical developments has seen a further strong increase, in particular due to special sessions at the main conferences in the field, as well as through the workshops of the Design Theory SIG (Special Interest Group) of the Design Society. Nevertheless, the main issues (see e.g. [15]) have not been resolved yet, as the list of main difficulties for research on Design Theory suggests Le Masson et al. [48]:

- no self-evident unity of the design theory field,
- multiple paradigm shifts that threaten the specificity of design,
- the fragmentation of the design professions and,
- the limits of empirical research.

Le Masson et al. conclude that the renewal of design theory should lead today to a body of sustainable collective research, will help build a powerful discipline, a unified body of knowledge, should help to understand and support contemporary forms of collective action and might help to invent new forms of design action.

1.3 Definitions of Design Theories and Models

The authors of the chapters in this book were asked to describe what they considered a theory or model of design to be, what its purpose is, i.e. what it is expected to describe, explain or predict. In this section, we provide a structured overview of their definitions. Details can be found in the respective chapters.

1.3.1 Introduction

In literature, considerable variation exists in what a theory is, what a model is, and what the overlap is between these two. One reason is certainly the general use of the terms in everyday life which covers a spectrum of meanings as dictionary entries show: the definitions of *theory* range from 'belief', 'ideal or hypothetical set of facts' and 'an unproved assumption' to 'a plausible or scientifically accepted general principle or body of principles offered to explain a phenomena' (Merriam-Webster in Ranjan et al. [59], Chap. 15). Similarly, definitions of *model* include 'an example for imitation or emulation', 'a type or design of product', 'a description or analogy used to help visualise something that cannot be directly observed', 'a system of postulates, data, and inferences presented as a mathematical description of an entity or state of affairs; also: a computer simulation based on such a system' (Merriam-Webster online dictionary). Overviews are given by Ranjan et al. [59], Chap. 15, Lindemann [49], Chap. 6 and Vermaas [82], Chap. 2. In the following sections, we focus on definitions used by the authors in this book.

1.3.2 Theory

For Goel and Helms [34], Chap. 20 'A scientific theory is (i) based on testable hypotheses and makes falsifiable predictions, (ii) internally consistent and compatible with extant theories, (iii) supported by evidence, and (iv) modifiable as new evidence is collected'.

According to Ranjan et al. [59], Chap. 15 a theory consists of 'a set of constructs and their definitions; and a set of propositions, expressed as descriptive relationships among the constructs, as statements about designing'.

Badke-Schaub and Eris [7], Chap. 17 add a user perspective in their definition of design theory as 'a body of knowledge which *provides an understanding* of the principles, practices and procedures of design'.

The same with Vermaas [82], Chap. 2, who refers to the definition of theory given by Ruse [64]: 'A scientific theory is an attempt to bind together in a systematic fashion the knowledge that one has of some particular aspect of the world of experience. The aim is to achieve some form of understanding, where this is usually cashed out as explanatory power and predictive fertility'. Thus, in Vermaas' view, Design Theory 'is an attempt to systematically bind together the knowledge we have of experiences of design practices'.

According to Eder [29], Chap. 10, 'the theory should describe and provide a foundation for explaining and predicting 'the behaviour of the concept or (natural or artificial, process or tangible) object', as subject. The theory should answer the questions of 'why,' 'when,' 'where,' 'how' (with what means), 'who' (for whom and by whom), with sufficient precision'.

Sonalkar et al. ([72], Chap. 3) emphasise 'the importance to distinguish between bounding the phenomenon that a theory attempts to explain and the generality of that explanation'.

Gero and Kannengiesser [33], Chap. 13 are explicit about the boundary of the phenomenon, emphasising the need to include both foundational concepts of design and designing: 'a design theory should describe any instance of designing irrespectively of the specific domain of design or the specific methods used' and should 'account for the dynamics of the situation within which most instances of design occurs'. Weber [85], Chap. 16 provides a very similar description: 'the designs (as artefacts) and the designing (as a rationally captured process to create artefacts)' should be considered and they have to be 'situated, i.e. 'external influences have to be considered as they evolve'. The explicit inclusion of designs and designing can also be found in the definition of Andreasen et al. [6], Chap. 9, even though they use a far looser basis for the theory than other authors, when they refer to their own theory as 'the authors' imagination or mental model about the nature of artefacts and their design'. Ranjan et al. [59], Chap. 7 too take both designs and designing as part of a theory of designing, as they argue that 'These propositions are meant to be used to describe or explain' the 'various characteristics of the facets of designs and designing'. They, however, go beyond these as the goals of design theories, and extend these to 'relationships among the facets' and relationships among these and various characteristics of design success'.

Cavallucci [21], Chap. 12, includes the relevance of theory for practice. A theory or model of design 'should describe the world and its realities through a prism from which, when observed through, designers could envision useful insights as regarding their designing tasks. These useful insights could be provoked by an original description, a clear definition and allow designers to anticipate with artefacts design processes with some kind of robustness. The notion of robustness can only be reached if what the theory proposes matches with temporal realities'.

All above definitions refer to theory as a *description* of a phenomenon. Weber [85], Chap. 16 is one of the authors to include a prescriptive part, when he refer to 'collecting and systematising knowledge about 'what is' (descriptive part) as well as collecting and systematising knowledge about actions and skills that can change the present state into another, previously not existing state (prescriptive)'. This is very much in line with our own view [10]: 'A typical characteristic of design research is that it not only aims at understanding the phenomenon of design, but also at using this understanding in order to change the way the design process is carried out. The latter requires more than a theory of what is; it also requires a theory of what would be desirable and how the existing situation could be changed into the desired'.

1.3.3 Models

The phrase 'models of design' can be interpreted in two different ways: models that are used in designing, such as scale models, CAD models, sketches etc.—this is henceforth referred to as 'models *in* design'; and models that are used to describe or prescribe how design is or should be (carried out)—this is henceforth referred to as 'models *of* design'.

1.3.3.1 Models in Design

Maier et al. [54], Chap. 7 and Lindemann [49], Chap. 6 focus on models in design. Both provide a number of exemplars to illustrate the variety of models for processes as well as outcomes that are used in design. Lindemann describes a number of important characteristics for models, like transformation and reduction, purpose and subject. His discussion of quality and requirements for modelling is mainly based on these characteristics.

Albers and Wintergerst [2], Chap. 8, in referring to product models ('product models should refer to physical characteristics and the related functional properties of a system') seem to focus on models used by designers, rather than by researchers, although the borderline between the two is not always clear-cut.

1.3.3.2 Models of Design

Ranjan et al. [59], Chap. 15 refer to Anderson [1964], who uses the term 'model' to refer to any way of visualising or conceiving of a structure or a mechanism that can account for observable phenomena. As mentioned before, they do not distinguish between models of design and theories of design.

According to Vermaas [82], Chap. 2 'scientific models represent features of a target system in the world or a scientific theory'. Note that the former includes models *in* design. He introduces five categories of scientific models: *Physical objects*, *Fictional objects*, *Set-theoretical structures*, *Descriptions* or *Equations*. 'Models of design practices may also be differentiated as models with descriptive, demarcating and prescriptive aims, but now all types of models fit much better in the characterisation of models in science, since there is such a diversity of scientific models'.

In the definition of Goel and Helms [34], Chap. 20 'a scientific model is an interpretation of a target system, process or phenomenon that proposes or elaborates on the processes and mechanisms that underlie it..... models are abstractions of reality... models are cognitive tools for generating explanations'. They specify two kinds of models in design in which they are interested: a knowledge model in design provides an ontology for representing the knowledge and a structure for organizing the knowledge in a design domain, a computational model of design provides architectures, algorithms, and knowledge models for the theory'.

Goldschmidt [35], Chap. 21 defines a model as 'a simplified and schematic representation of the essence/skeleton of a theory', which is 'highly linked to the disciplinary approach within which the theory is embedded'.

Lindemann [49], Chap. 6 provides three model definitions showing an increasing scope (italics added to emphasise the scope change): 1. 'A model is a *representation* of an object, system or idea in some other form than itself' [68]; 2. A model is the image of a system or a process *'within another conceptual or representational system'* [25]; and 3. 'A model is the *simplified* reproduction of a *planned or an existing system* including its processes within another conceptual or representational system' [81]. He concludes that all definitions leave room for interpretation, but agrees with Stachowiak [73] that each model should have three important characteristics: transformation of the attributes of the original into the attributes of the model, reduction of the number of attributes from original to model, and the pragmatic characteristics purpose, users and time frame of usage.

Maier et al. [54], Chap. 7 define a model as 'a simplified and therefore to a certain extent a fictional or idealised representation'. They distinguish three types of models depending on the claimed relationship between a model and the real world: explanatory ('the workings of a model map directly onto, or truly explain, 'real-world' mechanisms that 'cause' observable behaviour'), predictive ('a model can predict phenomena, but it is acknowledged that underlying real-world mechanisms may not exist in the form the model suggests, or the issue is viewed as unimportant'), and synthetic ('a model is explicitly recognised to not represent a real situation, but rather to represent an idea and thus to bring a situation into being'). 'Most models in design fulfil a synthetic role'. 'In the cybernetic sense, a model must be a description or conception of a situation that is used to guide or influence the response to that situation'.

An important factor to realise in this context is that 'An understanding of a model is a cognitive construct rather than an inherent property of the model, and a shared understanding is constructed through social processes of discussion and clarification' [27].

1.3.4 Theory or Model

The difference and relation between theory and model is often discussed, but thus far no generally agreed upon definitions exist in our discipline.

Some authors do not make an explicit difference. For Agogué and Kazakçi [1], Chap. 11 'A design theory is a model of creative rationality'. For Albers and Wintergerst [2], Chap. 8 theories and models 'address a specific purpose and are intended to describe, explain or predict certain phenomena that pose an unsolved challenge both for the research community and for design practitioners'. Weber [85], Chap. 16 too refers in his definition to 'theories and models'. Ranjan et al. [59], Chap. 15 follow [30] in stating that 'a theory in its most basic form is a model'.

Vermaas [82], Chap. 2 does make a difference and describes three different ways in which scientific models are related to scientific theories: 1. models of theories are taken as providing rules for interpreting the terms and sentences of the theory they represent; 2. a scientific theory is seen as a set of models; 3. models are not taken as closely representing the content of theories, but seen as means to understand that content, which may imply that the models contain elements that are not part of these theories.

Goldschmidt [35], Chap. 21 explicitly states that 'a model is not a theory' and seems to refer to relations 1 and 2 as described by Vermaas when she writes: 'a model is both derived from a theory and it contributes to the development of the theory'. 'A model in design research specifies the main components of a design theory and the relationships among these components. It is often represented as a diagram or graph'.

Eckert and Stacey [28], Chap. 19 clearly refer to relation 2: 'Theory fragments comprise partial models' that 'represent the structure of real, if abstractly described, causal processes', that are 'networks of interlocking causal processes influenced by causal drivers'. And so do Andreasen et al. [6], Chap. 9 when they refer to their theory as a model based theory 'composed of concepts and models which explains certain design phenomena'.

The view of a theory as a series of models, rather than the so called received view of scientific theory, reflects changes in how philosophy of science perceives theories and models. As described in Sonalkar et al. ([72], Chap. 3) the common perspective of the design research community is the received view, which 'defines a three-part structure for scientific theory. The first part deals with logical formalism, the second part describes observable constructs and the third part describes theoretical constructs. The three parts are connected by rules of correspondence that hold the mathematical, observable and theoretical constructs together'. 'The rigidity and, hence, difficulty of developing such a theory has led to heavy criticism and rejection by most philosophers of science'. As a reaction, Craver [23] proposed the semantic or model view of scientific theory in which 'theories are abstract extra-linguistic structures quite removed from the phenomena in their domains. In this view, theories are not associated with any particular representation. Researchers have a much greater freedom than in the received view to describe their theory in terms of a series of models that explain a set of phenomenon through abstraction constructs that constitute the theory' (Sonalkar et al. [72], Chap. 3).

Sonalkar et al. follow the distinction made by Dörner [26] who succinctly describes a theory as 'a formulation that explains a phenomenon', and a model as 'an abstraction that simulates a phenomenon'. Simply put, models do things while theories explain things.

In our view, all theories are models, but not all models are theories.

Further to these views, the attendees addressed the definitions of, and the similarities and distinctions between the terms theory and model in the discussion sessions. Regarding the definitions of model and theory, the participants agreed on two main points.

First, it became clear that the term 'model' was used in two ways: models in design and models of design (see Sect. 1.1.3) and that confusion can arise if no clear distinction is made, even though several of the identified characteristics are valid for both.

Second, there is considerable overlap between the meanings of models of design and theories of design. A 'spectrum of meanings' emerged, starting from having 'no distinction in how these terms are currently used in our area', to where 'Theory defines a framework from which multiple models could be derived'. A consensus also emerged that there is a need to see 'theory as a spectrum', with terms such as taxonomies, models and theories having varying degrees of maturity in context, purpose and explanatory capacity.

It was agreed that for a discipline of research such as design, a clear understanding of these terms is crucial, since they form the basis for further research. Details of the discussions can be found in Appendix A.

1.3.5 Ontologies

Although the issue of ontology was not the focus of this book, it came up in several contributions and in the discussion session. Several authors emphasised the need for an ontology to provide accurate descriptions of the concepts they used in the frameworks, theories and models they propose Agogué and Kazakçi [1], Chap. 11, Albers and Sadowsky [2], Chap. 8, Andreasen et al. [6], Chap. 9, Cavallucci [21], Chap. 12, Goel and Helms [34], Chap. 20, Gero and Kannengiesser [33], Chap. 13, and Ranjan et al. [59], Chap. 15. An ontology or—as a minimum—a clearly defined set of concepts is considered not only an important basis for theoretical development but also an important aid in analysis of empirical data and in making a theory comprehensible and transferable to design practice and education.

Appendix C lists the sets of main concepts the authors in this book used or created for their theories and models. What becomes immediately apparent is the strong diversity in concepts. Looking at the theories and models this diversity can have three reasons. First, most theories and models describe different aspects of the design phenomena or describe the same phenomena at different levels of resolution. This implies that these theories and models are partial theories and models, and potentially complementary. Second, the main concepts within a theory or model are interdependent: the definition of one concept influences the definition of others. For example, the definition of conceptual stage influences the definitions of the preceding and subsequent stages. This implies that the same term(s) may represent different underlying concepts in different theories and models. Third, where a similar aspect of design is described, different theoretical origins cause differences in the concept set, the concept definitions, or the terms used for essentially the same concept.

In our view, in order to describe the design phenomenon in a more comprehensive way, the current theories and models have to be brought together. Given the interdependency of concepts, a redefinition of existing concepts, and a coherent terminology will be necessary to achieve consistency. The need for a common ontology or agreement about the main concepts in our field has been argued for since several decades (e.g. in [15, 22]) but is still lacking. This is also reflected in the sets of keywords proposed for papers in our domain: a total of 1049 keywords were proposed for 390 papers submitted to one conference in engineering design [55]. In our view, this issue needs urgent attention, as it can hamper a coherent and more comprehensive understanding of design (ontology as basis for analysis) and our theoretical developments (ontology as basis for bringing together partial theories and models).

1.4 Purpose of Theories and Models

Theories, according to Koskela et al. [46], Chap. 14 can be *descriptive* or *prescriptive*. Vermaas [82], Chap. 2 includes a third category, *demarcating* theories, and points out that not all design theories 'systematically bind together the knowledge we have of experiences of design practice' and can, hence, be called scientific theories. The difference lies in the aims or purposes of the theory [82], Chap. 2:

- *Descriptive design theories*. Its aims include describing design practices that are regularly taken as design. It should bind together our knowledge of these regular design practices, and arrive at understanding, explanation and prediction of and about them.
- *Demarcating design theories*. Its aims include fixing the borders of what is to be taken as design practices.
- *Prescriptive design theories*. Its aims include singling out particular types of existing or new design practices and positing favourable properties about these practices.

According to Vermaas, only those demarcating and prescriptive theories that include a descriptive aim can be considered scientific theories. Prescriptive design theories that single out *new* types of design practices and posit favourable properties, i.e. are not descriptive, are for Vermaas at most *hypothetical* scientific theories. He emphasises that design theories that are generated in design research typically are not *pure* theories but combine aims.

1.4.1 Demarcating Purpose

The work of Eckert and Stacey [28], Chap. 19, Koskela et al. [46], Chap. 14, Taura [77], Chap. 4, Horváth [42], Chap. 5, Gero and Kannengiesser [33], Chap. 13 and Badke-Schaub and Eris [7], Chap. 17 can be seen as contributing to the demarcation by questioning the current boundary of what is to be taken as design, and hence of what is to be covered by design theory.

Eckert and Stacey [28], Chap. 19, e.g., criticise existing theories of design that 'have aimed at understanding design as a unified phenomenon', but fail to 'explain or predict the differences and similarities that we observe when studying design processes across a range of products and domains'. Design theories are 'typically presented with insufficient consideration of how much of designing they actually cover'. Eckert and Stacey propose to use *constraints* and *drivers* as major elements in demarcating various design processes, and to use this to specify the scope of models and theories of design.

Koskela et al. [46], Chap. 14 emphasise that their proto-theory (as other theories) cannot cover the whole area of design: 'it has to be contented that there are aspects and stages in design that are best approached through rhetoric. The task of agreeing on the boundaries of the phenomenon of design seems still seem to be in front of us'.

Taura [77], Chap. 4 proposes a typology of designing consisting of pre-design, design, and post-design stages in order to include the 'motive of design', thereby proposing a demarcating theory of possible design practices. He argues that discussions on particular aspects of design that have not been considered yet have 'the potential to extend existing methods and to develop products that will be more readily acceptable to society'. The motive of design is discussed in terms of the fundamental issues faced in designing highly advanced products. Specifically, Taura proposes the conception of a social motive that is created and contained in society in contrast to the so-called motive of the individual, which can be referred to as personal motive.

The argumentation of Horváth [42], Chap. 5 is quite similar. He describes how the shift to developing socio-cyber-physical systems raises major design challenges, since such systems cover the broadest possible range of phenomena as the focus of design, and hence would lead to development of design theories that are robust enough to address any subset of such systems, e.g. physical, social, cognitive, socio-physical, socio-cyber, or cyber-physical systems. He argues that multi-disciplinary research is needed to successfully address these challenges: 'new design theories and principles and system design methodologies are needed to be developed'. Although implicit, he considers the borders of what is taken as design practice in current theories no longer valid. 'A unified design theory and methodology that facilitates addressing of the issues of both worlds (cyber and physical)' is required. This considerably expands the scope of theories and models of design.

Gero and Kannengiesser [33], Chap. 13 also extend what is to be taken as design: 'a design theory should describe any instance of designing irrespectively of the specific domain of design or the specific methods use' and should 'account for the dynamics of the situation within which most instances of design occurs'.

Badke-Schaub and Eris [7], Chap. 17 point out that 'rational decision making and its influence on design performance has been (and should be) a major source of empirical studies for the purposes of developing theories and models of design', but that the design phenomena is broader: 'design theories need to be able to also explain the need of and the processes for the unconscious such as intuition in design', as there is 'rich empirical evidence highlighting unconscious and mainly inaccessible processes that support the designer in making pragmatic and useful decisions that do not offer explicit rationale'. They note that although 'researchers seem to acknowledge that designers 'use' intuition on a daily basis, there is hardly any targeted empirical work which tries to understand whether intuition works in designing and if so, how'. Their research on intuition aims to fill this gap.

In our view, demarcating theories are still very relevant for design research as an area with ill-defined boundaries. Defining the boundaries, which may be very wide, will also contribute to the earlier mentioned need for a common ontology or agreed set of main concepts.

1.4.2 Descriptive and Prescriptive Purposes

1.4.2.1 Theories

As any other theory, the purpose of a design theory is to describe, explain and predict. In addition, the majority of authors emphasises that the ultimate purpose is to create support to improve practice, based on the understanding obtained. Note that this does not automatically imply the development of a prescriptive theory: descriptive theories and models are used to obtain understanding that can be used to develop improvement measures. As the following paragraphs show, the characteristics of design to be described, explained and predicted can vary, but tend to be fairly wide.

A typical example is Ranjan et al. [59], Chap. 15, following Blessing and Chakrabarti [16]: 'a model or a theory of designing should be able to describe or explain characteristics of one or more facets of design and designing, including relationships among the facets involved (at one or more stages of designing, including the transitions from one stage to another, of a design process) and the relationships among these and various characteristics of design success. Furthermore, a model or theory of designing should be used as a basis to identify the positive and negative characteristics influencing design. Further, design models or theories can be used as a basis to improve the design process'.

Similarly, Badke-Schaub and Eris [7], Chap. 17 view design theory as 'a body of knowledge which provides an understanding of the principles, practices and procedures of design. That knowledge leads to hypotheses on how designers should work, and such hypotheses provide the basis for the prescriptive part of design methodology'.

Eckert and Stacey [28], Chap. 19 argue that a theory of design should explain and predict the behaviour of real processes and should be useful for understanding and improving design processes in industry. 'We are primarily interested in why design processes are as they are, and how they could be made to work better, to produce better products, to increase the profitability of companies or produce products faster and with less effort, or involve happier, less stressed, more fulfilled participants'. Taura [77], Chap. 4, Koskela et al. [46], Chap. 14 and Weber [85], Chap. 16 make similar statements. Taura expects a theory or model of design 'to extract the essences of phenomena within the real design process' but also 'to predict and lead future new design methods'. According to Koskela et al. a theory should provide better 'explanation, prediction, direction (for further progress) and testing' and 'provide tools for decision and control, communication, learning and transfer (to other settings)'. For Weber a model or theory should 'explain and predict observations in its field.' The framework he proposes should 'integrate many existing approaches and to deliver some explanations of phenomena in product development/design that have been insufficiently understood so far'.

Eder [29], Chap. 10 follows the above, but does extend the purpose to include the various life-cycle phases. The theory should describe and provide a foundation for explaining and predicting 'the behaviour of the concept or (natural or artificial, process or tangible) object', as subject. […] The theory should support the utilised methods, i.e. 'how' (procedure), 'to what' (object), for the operating subject (the process or tangible object) or the subject being operated, and for planning, designing, manufacturing, marketing, distributing, operating, liquidating (etc.) the subject'.

Albers and Wintergerst [2], Chap. 8 include the designers as a target audience. A design theory should be 'explaining, or predicting certain phenomena', but also 'facilitating designers to analyse design problems and to create appropriate solutions'. Referring to the latter, they specify that 'theories and product models provide a framework for making information accessible (analysis) as well as for expressing design concepts and decisions (synthesis). They serve designers to capture, to focus, to structure, to make explicit and to simplify the complex relationships of a system's properties and characteristics. Thus, they serve as a means to overview, explore, understand and communicate such relationships at a systems level'.

For Cavallucci [21], Chap. 12 the main purpose is practical use: 'a theory or model of design is supposed to provide designers with answers to their everyday professional difficulties. Along each tasks assumed by designers, a relevant Theory of Design should provide first theoretical roots, scientifically proven, then a methodological declination of it for appropriate use and practice'.

Andreasen et al. [6], Chap. 9 look in particular at the concepts used in a theory or model by specifying the purpose of a design theory as 'the creation of a collection of concepts related to design phenomena, which can support design work and to form elements of designers' mindsets and thereby their practice'.

Some of the authors mention additional purposes that extend the role of design theory for the design research community and beyond.

Goel and Helms [34], Chap. 20 add that 'An important cognitive feature of a scientific theory is that it suggests a process or method for building, evaluating, revising, and accepting (or abandoning) a theory'. They, e.g., used their knowledge model, which specifies the ontology and the schema for representing and organizing knowledge of design problems (the aspect of design they considered) as a coding scheme for their research into design processes, as a pedagogical technique to help students in formulating design problems, as support for designers, and to structure a knowledge base to help facilitate search. Gero and Kannengiesser [33], Chap. 13 refer to a similar aim: the use of their model (or ontology) as a project-independent scheme to code data from the protocols.

Eckert and Stacey [28], Chap. 19 see the possibility to use the set of drivers they identified (i.e. the elements of their model) to categorise a design, and to be better able to inform practice what kind of design processes are and should be followed for such design.

Agogué and Kazakçi [1], Chap. 11 contribute with a description and purpose of each step involved in developing a theory. First, it aims at *revitalizing the knowledge* accumulated in engineering design. Then, deepening the formal aspects of a design theory helps to both *unveil and explain* the surprises, the paradoxes, the oddness of design reasoning that goes beyond classic rationality and logics. Moreover, a design theory being a model of creative rationality, it can circulate and *become a framework for disciplines outside of design*, where there is a need for innovation and for building understanding on creative reasoning. 'A design theory enables a dialogue that either benefits from or contributes to other disciplines'.

1.4.2.2 Models

As mentioned earlier, several authors do not distinguish between theory and model and, hence, consider a model to have the same descriptive, explanatory and predictive purposes as a theory (see Sect. 1.4.2). In this section we focus on those authors that explicitly discussed the purpose of models.

Lindemann [49], Chap. 6 points out that models are developed for a multitude of purposes. Some examples he mentions are specification and demonstration models, experimental models, geometry models, theoretical models, i.e., these are models in design. The purpose determines which attributes of the original are selected and how they are transformed, but also puts 'limits to the validity of a model'. He accepts 'the reality of having a large and ever increasing number of models', but emphatically expresses the need for providing the pragmatic characteristics of a model (purpose, users and time frame of usage): He particularly stresses the importance of *usefulness* of the model in satisfying a purpose (its purpose) as the main criterion for its use'.

The purposes mentioned by Maier et al. [54], Chap. 7 are: 'explaining or predicting behaviour, or articulating and realizing something new'. The former overlap with earlier definition of descriptive theories, the latter is of a more predictive nature.

Goel and Helms [34], Chap. 20 are more specific: the purpose of a model is to 'productively constrain reasoning by simplifying complex problems and thus suggest a course of analysis' and 'serve as tools both for specifying and organizing the current understanding of a system and for using that understanding for explanation and communication'. Vermaas [82], Chap. 2 adds that 'Scientific models also have epistemic value: their creation, analysis and development allow scientist to understand the target systems and the theories represented'.

This is in line with the purpose mentioned by Goldschmidt [35], Chap. 21: 'to facilitate the disjunction of a theory into constituent parts and to lay down relationships among components, for further investigation and/or proof. Likewise, vice versa, a model displays the integration of distinct parts into a whole—'the larger picture'. In design research the purpose of a model is to explicate the process of designing or elements thereof from one or another standpoint'.

The discussions in the workshop highlighted a lack of clarity concerning theories and models. A major agreement emerged: it was felt that any proposal for a model or theory should be accompanied with its purpose (what it does) and context (where it applies)—its 'system boundary'.

1.5 Criteria to Satisfy to be Considered a Design Theory or Model

The authors in this book largely agree about the criteria that a theory or model should satisfy in order to be called a design theory or model of design.

A theory should 'refer to actual and existing phenomena' [29], Chap. 10, to 'real design processes at a level that is not trivially true for all processes' [28], Chap. 19, and 'contain a set of propositions to describe or explain some characteristics of (one or more facets of) designing (and design success)' [59], Chap. 15.

Its coverage should be broad: 'It must account for both the similarities and the differences between them, across products, companies and industries' [28], Chap. 19, 'provide a *broader* set of aspects of designing [...] explaining communication in design as an activity by many individuals covering various possible *types* of *reasoning* in design (e.g. plausible reasoning), making sense of the *never complete* particular starting point of design, and providing *aesthetical* considerations in design' [46], Chap. 14, be as complete as possible [29], Chap. 10, have 'generativity, that is, the capacity to model creative reasoning and to relate to innovative engineering in all its aspects' and 'generality, i.e. 'the capacity to propose a common language on the design reasoning and design processes' [1], Chap. 11. Sonalkar et al. add that the 'perception–action dimension needs to be an integral part'. The perception–action dimension 'does not explain, but rather gives reflection of the theoretical constructs in situations relevant to practice'. This dimension 'accounts for the human agency in design' and lets 'the theory be rooted in situations relevant for professional practice'. This results in a 'much higher coupling between logical relationships and the situational relationships of constructs that design theory uses to explain phenomenon'.

As discussed in Sect. 1.4, a theory should be able to fulfil its purposes, that is, being able to describe, explain, predict. A theory should be as complete and logically consistent as possible [29], Chap. 10, empirically accurate [82], Chap. 2, based on testable hypotheses [34], Chap. 20, have clarity of explanation [46], Chap. 14, and be accessible by and meaningful to both researchers and practitioners (Sonalkar et al. [72], Chap. 3). For example, 'theories and models should be tools for practice' [46], Chap. 14, that lead to 'hypotheses on how designers should work [that] are the basis of the prescriptive part of design methodology' [7], Chap. 17, and indicate how design processes in industry can be influenced [28], Chap. 19. Weber [85], Chap. 16 points out that the usefulness of theories and models depends on the stakeholder: 'there may be different 'stakeholders' who pose requirements on models and theories of designs and designing', such as 'scientists, designers in practice, students, and tool/software developers'.

Finally, Koskela et al. [46], Chap. 14 add that a design theory should provide directions for further research. Some of the questions posed are: What is the core of a design theory? What is the scope of the phenomena of design? What are the main constructs that describe design?

Goldschmidt [35], Chap. 21 focuses on the criteria for a model: 'The criteria to be satisfied by a model include the presence of all essential components and links in the modelled process (or other phenomenon) and the possibility to extract any portion of it and develop it in more detail. Contraction and expansion must not undermine the integrity of the model, and the expectations from each level of detailing must be clearly defined'.

According to Maier et al. [54], Chap. 7 a good model should make it 'appropriate to enable design cognition and collaboration'. Note that they focus on models for use in design. 'The specific issues in determining the goodness of a model depends on the perspective: explanatory models should be able to accurately explain underlying mechanisms, predictive models should accurately predict patterns in observations. For a synthetic model it is 'not so much the goodness of fit, but rather the degree to which it enables decision-making that turns out to add value given a certain purpose and context'.

Lindemann [49], Chap. 6 includes models in design and models of design. He lists three important *characteristics* of models [73]: reduction (the model contains less attributes than the original), transformation (some attributes may have been modified or may have been additionally added, such as a coordinate system in CAD), and pragmatism (addressing purpose, users and time frame of usage), which influences reduction and transformation. He further refers to *conventions* to be considered during modelling and provides a first set of requirements for a model from [45]: *accuracy* (correspondence between original and model), *clarity* (how clear the purpose and limits are to the user), *relevance* (where is it relevant), *comparability* (can it be compared with original or with other models), *profitability* (what are the benefits of using the model), systematic *settings* (how to set up the model for using it).

From the discussions in the workshop, a strong consensus emerged across the teams in the criteria to be considered a theory or model of design: theories should be testable and refutable (i.e. falsifiable).

1.6 How Should a Theory or Model be Evaluated or Validated?

A design theory or model not only has to meet 'the usual criteria of a descriptive science (e.g. truth, completeness, level of detail) but also the criteria of usefulness and timeliness' [85], Chap. 16 'Usefulness needs testing' [82], Chap. 2 and 'should be the focus of the validation of methods, models and theories in design [as validation] is a process of building confidence in their usefulness' Gero and Kannengiesser [33], Chap. 13. For Lindemann [49], Chap. 6 too, purpose plays the most significant role in validation of theories and models, but at the same time the purpose limits validity. Validity depends on stakeholders [85], Chap. 16.

Andreasen et al. [6], Chap. 9 see 'two dimensions in a theory's goodness, namely its range and productivity. Range is the breadth of related phenomena that the theory is able to describe based upon a shared set of concepts. The productivity of a theory shall be found in its suitability for teaching its applicability for designers' practice and its utility for researchers to understand and analyse the phenomena of design'. Albers and Sadoswki [2], Chap. 8 also mention these criteria: the variety of problems and domains that can be addressed in industry and research, and the impact on education. Eckert and Stacey [28], Chap. 19 stress the importance to indicate where a theory applies when validating these: 'Theories about the nature of design or how designing is done are typically presented with insufficient consideration of how much of designing they actually cover'.

For Andreasen et al. [6], Chap. 9 it is important 'whether the theory lead to new theories or to new models and methods that can support design'. They see rigour 'in the efforts to link a theory to design practice'. Similarly, Eckert and Stacey [28], Chap. 19 emphasise the role of validation in supporting the development of theory fragments into a more coherent theory of design by 'comparing pieces of theory with the reality of particular design processes, and explaining failures to observe the phenomena the theory fragments predict either in terms of the falsification of the theory, or by elaborating the theory fragments to cover a wider range of causal factors and distinct situations'. That is, 'developing design theory involves constructing pieces of theory, assessing their validity, assessing their limits of applicability, and progressively stitching them together to make a larger coherent whole'. Badke-Schaub and Eris [7], Chap. 17 add that evaluation and validation can extend the theoretical considerations or show that 'existing theories in other domains (that were considered generic) did not always apply in the design domain'.

Referring to models, Lindemann [49], Chap. 6 distinguishes verification and validation: 'Verification has to guarantee that all requirements are fulfilled in a correct way, and validation has to show that the purpose of the model will be fulfilled. Usability checks should ensure that the subject (the user of the model) will be able to use the model in a correct way'.

Vermaas [82], Chap. 2 argues for falsification rather than validation to address 'two deficiencies that lower the scientific status of design research': 'the lack of generally accepted and efficient research methods for testing design theories and models', and a 'fragmentation in separate research strands'. He suggests naive Popperian falsification as a swifter way of testing, and sophisticated falsification as described by Lakatos to compare rival design theories and models'. The need to focus on falsification is mentioned by several other authors: a design theory should make falsifiable predictions [34], Goel and Helms, Chap. 20, a model or theory should be falsifiable rather than verifiable [85], Weber, Chap. 16, 'researchers need to infer hypotheses that test the theory by being amenable to falsification' (Sonalkar et al. [72], Chap. 3), and 'theory development should involve deliberate falsification of arguments' [28], Chap. 19. The development of the E-IMoD—the model of designing proposed by Ranjan et al. [59], Chap. 15 is a case of Lakatosian falsification, where extension of the scope of the model beyond conceptual design leads to the need for further elements in the model.

Ranjan et al. [59], Chap. 15 propose two ways to test propositions: first, using empirical data, and second, using 'logical consistency with other theories or models, that are already validated'. Vermaas [82], Chap. 2 emphasises that testing cannot be done independently of rival design theories and models.

Examples of *testing using empirical data* are given by various authors. Ranjan et al. [59], Chap. 15 use protocol analysis of existing protocols to identify whether all constructs of the model are present. Badke-Schaub and Eris [7], Chap. 17 could confirm and extend their theoretical considerations based on a qualitative analysis of the data gathered by interviews of professional designers from different disciplines. Goel and Helms [34], Chap. 20 mapped data from a large number of cases to an initial coding scheme from an earlier knowledge model and added new conceptual categories as they emerged from the data. Based on additional sets of data, the new model was refined and relationships added. This model was validated using a third data set. Gero and Kannengiesser [33], Chap. 13 validated the utility of their ontology both conceptually and empirically by using it to code hundreds of design protocols in various design disciplines and for various tasks, allowing comparison 'across protocols independent of the designers, the design task and all aspects of the design environment' and thus 'provide insight into designing'. The results imply 'that the FBS ontology provides a robust foundation for the development of a generic coding scheme. Cavallucci [21], Chap. 12 verified his IDM framework through case studies in industry in which he moderated the use of the framework by company experts. Albers and Wintergerst [2], Chap. 8 analysed the results of design projects of students who had received training in the approach as well as the results of the application of the approach in a variety of problems and domains.

Agogué and Kazakçi [1], Chap. 11 focus on *logical consistency with other theories and models* that are already validated when they speak about 'relatedness to contemporary knowledge and science (i.e. the capacity to relate to advances in all fields even when they seem far from the design community, such as mathematics or cognitive psychology: a design theory enables a dialogue that either benefits from or contributes to other disciplines)'. They compare data of CK-theory with a similar approach on the developments of axiomatic theory. They propose further ways of validating a theory: looking at the impact in practice, both in the own field and in other fields; using a theory to interpret or lead to a deeper understanding of existing models and methods, and as a framework to model very diverse issues. Koskela et al. [46], Chap. 14 evaluated the validity of the aristotelian proto-theory as a theory of design by looking whether its explicit and implicit features can be found in modern, corresponding ideas, concepts and methods. They also verified whether it provides an explanation of design. Weber [85], Chap. 16 confronted his own approach 'with a multitude of questions in order to fathom its limits or even find at least one falsification'.

From the discussions in the workshop, validation was found to have a spectrum of meanings, from checking for internal consistency, through truth, to utility. Testing the limits of a theory or a model was considered important and lead to a strong consensus on falsification as an approach.

Several challenges to validation were also identified: difficulty or lack of repeatability of phenomena, the large number of factors blurring clear and identifiably strong influences, difficulty of finding statistically large number of appropriate subjects or cases, and difficulty of generating reliable data about the phenomena under investigation. Furthermore, the lack of clarity of purpose and intended context of many theories and models (see Sect. 1.4.2) is considered a hindrance for proper validation.

1.7 Future Work

The various tasks ahead that were formulated by the authors clearly show that design research is still a rapidly developing field. Apart from tasks related to their own research programme, the authors in this book also propose more fundamental tasks for the research community that should contribute to the maturity of our field. These are:

1.7.1 Coverage

- Agreeing on the boundaries of the phenomenon of design [46], Chap. 14.
- Acknowledging that engineering design is distinct from other forms of designing [29], Chap. 10.
- Learning from history as a fertile legacy for understanding design [46], Chap. 14.
- Developing genuine system adaptation, evolution, and reproduction theories [42], Chap. 5.
- Developing new system abstraction, modelling, prototyping, and testing theories [42], Chap. 5.

1.7.2 Concepts

- Clarifying the terminological problems [46], Chap. 14.
- Developing an ontology of key concepts to enable a clear distinction of concepts and how they can vary [28], Chap. 19.
- Developing irreducible foundational concepts of design and designing and ontologies as frameworks for the knowledge in the field of designing [33], Chap. 13.
- Compiling a common conceptual and theoretical core for the various design and production sciences, and develop associated ways of contextualizing it to specific situations [46], Chap. 14.
- Fusing heterogeneous bodies of disciplinary knowledge into a holistic body of trans-disciplinary knowledge (Horváth [42], Chap. 5).
- Linking different theories and models to cover multiple domains [85], Chap. 16.

1.7.3 Multiplicity

- Using different paradigms to provide different perspectives on design [34], Chap. 20.
- Explaining or predicting the differences and similarities that we observe when studying design processes across a range of product and domains [28], Chap. 19.
- Developing fundamentally different theories of design to engender fundamentally different methods and tools for design [21], Chap. 12.
- Development of a tradition to let design theories and models compete to avoid proliferation of theories and models [82], Chap. 2.
- Rationalising, consolidating and integrating the ideas behind the heterogeneity of models and methods [54], Chap. 7.
- Reducing the large number of different types of models and languages and to have them meet the requirements of usability and purpose orientation [49], Chap. 6.
- Developing an integrating framework of the several models that exist in parallel, each explaining certain aspects [85], Chap. 16.
- Reducing the fragmentation in separate research strands [82], Chap. 2.

1.7.4 Validation

- Differentiating between descriptive, prescriptive and demarcating aims of design theories [82], Chap. 2.
- Developing design theories that display scientific rigour while being useful to professionals [72], Chap. 3.
- Developing generally accepted and efficient research methods for testing design theories and models [82], Chap. 2.

- Testing design theories and models by naïve and sophisticated falsification for effective testing and for coherence of design theories and models, respectively [82], Chap. 2.

1.7.5 Impact

- Ensuring impact in academia and in empirical contexts by fulfilling three criteria: generality, generativity and relatedness [1], Chap. 11.
- Development of theories rooted in the pragmatics of professional practice by including a perception–action dimension in addition to the event-relationship dimension [72], Chap. 3.
- Addressing transfer to industry to reduce effort and risk of full implementation [85], Chap. 16.
- Developing methods or process models that allow guidance for different situations [54], Chap. 7.
- Using design theory as a framework for disciplines outside of design, whenever there is a need to model and understand creative reasoning [1], Chap. 11.

1.7.6 Presentation

- Presenting models explaining design with a clear statement of their purpose, their subject, and the time frame to help recognise the limits of its validity [49], Chap. 6.
- Presenting theories with sufficient consideration of how much of designing they actually cover [28], Chap. 19.

Many of the issues raised in the individual chapters, as reflected in the individual statements above, coalesced during the workshop into a number of major, common issues. One of these is the general lack of a common understanding that can act as the underlying basis for the discipline of design research. A need for an overview, or even consolidation, of research carried out so far has been strongly emphasised. As a discipline, we need good 'demarcating theories' that provide a clearer understanding of what constitutes (and what does not constitute) part of the phenomena of designing (e.g. designing is demarcated by intentionality), the different types of designs and designing that form our discipline; and position the models and theories with respect to these.

This base, it was suggested, might be initiated by including the following (see details in Appendix):

- The philosophies of the discipline, including what design means, and what the 'phenomena of designing constitute'. 'We need a philosophy of design, like a philosophy of science'.

- A list of 'demarcating theories' that provide an understanding of the different types of designs and designing that form our discipline.
- A list of models and theories of design, along with their context and purpose.
- A list of agreed upon concepts that are used within the discipline, including theory and model, along with their contexts and purpose.
- A list of agreed upon research methodologies and methods for use within the discipline, along with their contexts and purpose.
- A list of empirical results, along with their context and purpose.
- A list of influences of results of design research on practice.

Another major issue raised was the need to clarify the common purpose of design research, and to identify what the pressing, concrete questions are that the discipline needs to address. Also emphasised was the need for investigating the specific characteristics, benefits and complementarities across the various theories and models, rather than discussing only about which one might be superior.

Towards addressing the above, several suggestions were made in the workshop (see details in Appendix A):

- Have more events at various levels, e.g. students, researchers, educators, etc., to discuss these issues. Getting together is the first step to 'form the discipline'. Developers of theories and empirical results should interact more with one another.
- Like in other disciplines, teach the common understanding reached to those (intending to be) in this discipline. This knowledge should be taught in a context-specific manner, i.e. 'make explicit what is applicable in which specific situation'.
- Interact with other disciplines with similar goals, such as management, and learn from their perspectives.
- Carry out more empirical studies that are unbiased, of high value, high-quality, and are clearly explained, as we still do not understand in sufficient depth why design processes happen the way they do.
- Have 'grand debates' where specific models are discussed and contrasted together.
- Work more on developing research methods that are appropriate for serving the specific needs of design research. A major issue is: how to develop and validate testable, refutable theories and models of adequate accuracy within the constraints of complexity of the phenomena observed and within the limited availability of appropriate cases and subjects? A starting point can be to form Special Interest Groups (SIG) to work on these, e.g. on research methodology.

1.8 Conclusions

With each theoretical development new concepts and/or relationships between concepts were introduced, earlier ones revived, and existing definitions refined or modified so as to become coherent with the set of concepts covered by the new theory or model. This introduced new perspectives on design, allowed increased understanding, and resulted in richer models and theories of design, and of models and theories for design. These developments were fuelled by an increase in results from empirical studies, a desire to better understand and/or support design, an openness to look into existing theories in other fields, and the need to do so in the light of an (perceived) increased complexity of both the product and the process. The increasing complexity is a combination of reality and, foremost, of our perception: the richer models with their increased number of concepts and relationships allow us to see more (depth), and/or consider more (width). The latter has also been fuelled by a change of perception as to what influences design and what is influenced by design (e.g. taking into account users (user-centred design), environment (eco-design), services (product-service systems) and society (socio-technical systems)). Theories, models and their concepts co-evolve with our understanding of design (and with the development of design support), i.e. theoretical and empirical (and applied) research should go hand-in-hand.

Intensive debates and dialogues, increased, richer sets of empirical studies as a basis, testing using established means, as well as endeavours to develop new, appropriate research methods as enablers are required to ensure a gradual movement towards an established set of core concepts and their definitions (which may change over time as understanding progresses) and to 'progressively stitching them (the pieces of theory) together to make a larger coherent whole' [28], Chap. 19. Whether we are working on the same puzzle or multiple puzzles remains to be seen.

The chapters in this book show that the development of theories and models may in name be linked to one person, the 'originator', but is in fact a joint effort taking many years of generating and evaluating, of discussion and comparison, of modification and refinement, of creating and rejecting concepts and relationships, of criticism and support, and of including concepts and relationships of other theories also outside one's own field. Even though we did not manage to obtain a contribution from all researchers who developed a theory, we hope this book can further theoretical progress by bringing together a wide range of thoughts, approaches, assumptions, concepts, scopes and foci developed in our research community, and in doing so inspire readers and provide them with a broader basis for their own research.

Appendix A: Summary of Discussions from the International Workshop on Models and Theories of Design

Discussions in the workshop, carried out primarily in three, parallel breakout sessions that continued through the days of the workshop, and culminated in a subsequent, common, final discussion session on the last day, focused on the four questions discussed below. This appendix provides a summary of the outcomes from these discussion sessions, which, we hope, will add to the richness of the knowledge already encapsulated in the individual chapters. As will be seen, while it is far from being conclusive, some major similarities in (lack of) understanding about theories and models, their purposes and criteria, and as to how they should be validated, have already began to emerge, and a number of common directions for further activity in this area have been proposed.

1. What is the distinction between a theory and a model?

Team 1: Rapporteur: John Gero; Scribe: Sonal Keshwani. The team took a broad approach of decomposition, and looked at the elements that constituted a model. A model was taken as a representation (i.e. away in which a language is used to describe something) of some observable phenomena. It had been noted that some phenomena may not be observable, and observation of phenomena may sometimes change the phenomena themselves. It was noted that the point of view of the observer plays an important role in what will be observed and how it will be interpreted: 'what you come up with is always limited by how you see the world and your output is evaluated by how the world looks at it'. All representations, it was felt, are limited, ideally by the purpose of the representation; hence, all models are also purposively limited. Models have generality and causality. Models project or predict, and can be used to explain. The team defined a theory to be an abstract representation of a generalisation of phenomena; a theory may have axioms that explain how a world behaves. Three views on the distinction between a model and a theory emerged: (i) a theory may be composed of multiple models; (ii) a model may be more concrete and specialised in its context than a theory, which is more abstract and general; (iii) a model may embed explanation of phenomena, while a theory may allow for such explanation. A theory may be represented by different models. There may be theory-driven and phenomena-driven models.

Overall, the team summarised its findings as follows. A model is a representation of some phenomena and relationships among these phenomena. With features that are operationalisable, a model provides some generality with respect to the phenomena, which can be causal, speculative and dynamic, and independent from theory. A theory is an abstract generalisation of phenomena, which can be modelled in multiple ways. Models, but not theories, can change with time. Phenomena are things that have regularity and are directly or indirectly observable, and are interpretable. A representation is an externalisation of a description of phenomena. Any representation leads to a reduction in some aspects of the phenomena and its granularity. What is represented is limited by the purpose or intention of the representation.

Team 2: Rapporteur: Udo Lindemann; Scribe: S Harivardhini, Praveen Uchil. The team distinguished between two types of models: research-based (driven by truth) and practice-based (driven by utility). The team raised the question: should models and theories in design be able to explain only (as in natural sciences) or should they also be useful, since the purpose of design research is to improve knowledge to improve design practice? The team also discussed what constituted goodness of a model, and argued that the goodness of a model depends on understanding of its system boundary, i.e. the context and purpose of the model. The team felt that there is an overlap in meaning between models and theories. A model may simulate a part of the world, but does not necessarily explain it. A model could be a subset of a theory, in that a theory provides explanation at a higher level than a model does.

Team 3: Rapporteur: Lauri Koskela; Scribe: Boris Eisenbart. The team distinguished between two types of models: models of design (i.e. of outcomes of design activity), and models of designing (of design activity). The latter is often used synonymously to theories of design. The team distinguished between a model and theory in the following. A model is an abstraction of reality created for a specific purpose, and the purpose includes representation of a theory; a model is helpful: it may serve multiple purposes and may be applied in multiple ways. A theory, on the other hand, may involve a number of hypotheses, each of which should be possible to be falsified. They recognised that describing something as a theory is sometimes a cultural issue; for instance, in some fields of research, less comprehensive approaches, frameworks etc. are called theories for the only reason that the term 'theory' added some kind of value to the proposition. The team recognised that while taxonomies are typically not considered theories in natural sciences, design research should consider theories as a spectrum with various levels of maturity in its context and purpose of use.

Overall, the team felt that a model and a theory have several aspects in common: both models and theories serve a (set of) specific purpose(s) that are useful for researchers and/or practitioners; both are explanatory in character which facilitates prediction and prescription. A goal of theories that is distinct from those of models is to provide an explanation of what design and designing mean within the context of use of the theory.

2. What is a model or a theory expected to describe, explain or predict? What criteria must it satisfy?

Team 1: Rapporteur: John Gero; Scribe: Sonal Keshwani. The purpose of a model is to transform something (e.g. produce an output given an input, which can form a prediction), to explain something. Explanatory power of the model comes from the result produced when using the model. A theory is a set of beliefs that are proposed as a generalisation of some phenomena, which are intended to give an explanation for the phenomena. Models have to be useful; theories have to be falsifiable. A model may help in prediction or exploration. A theory has to be testable/refutable. A model has to be usable in design, if this is a model for design. A theory cannot be evaluated directly, but can be evaluated only after its implementation. Theories contain rules and principles which together form their explanatory framework; this characteristic (i.e. of being constituted of rules and principles) is one of the criteria that a theory should satisfy.

Team 2: Rapporteur: Udo Lindemann; Scribe: S Harivardhini, Praveen Uchil. The team argued that a major distinction in the nature of phenomena dealt with between natural sciences and design research is that, design research focuses on design processes that are unique and operate within incomplete information and uncertainty. It is important to distinguish between different models in terms of their system boundary (i.e. scope of application) and their purpose. The purpose can be truth (in research) or utility (in practice). For a model to be good for truth, it should be true at least with the scope of its application. Goodness criteria for models for utility include: usability, ease of use, how quickly it can be used, system boundary, and limits of the model. Many theories and models are not used well in practice because it is hard for practitioners to understand the terms used in these theories and models. A theory or a model should be able to provide insight. A theory must be falsifiable.

Team 3: Rapporteur: Lauri Koskela; Scribe: Boris Eisenbart. The team felt that theories need to be useful: they can be curiosity-driven where the goal is to understand the nature and characteristics of objects, entities and their relationships, or problem-driven where the goal is to support practitioners and provide utility, or to support education. Understanding is necessary for predicting an outcome, and eventually prescribing how to perform design to achieve an expected outcome. Theories in design may be more probability-driven rather than being strictly causal, given the large number of influences, and may take the form of narratives rather than strict propositions. The team asked for whom theories are to be developed, and felt that these would be primarily for researchers or managers. The team discussed what phenomena a theory should address. While it noticed there may not be a single phenomenon of designing, there might be something fundamental to designing that every designer or design team does or shares, e.g. similar activities, aspects etc. appear across different design projects and disciplines. Overall, it was agreed that there are similarities and differences across designing in different contexts, and a theory of design should explain both similarities and differences across the contexts. It was strongly felt that 'We do not have a thorough understanding of all the assertions we make about designing. We ought to have theories about how to differentiate between different types of design'.

The team felt that phenomena of designing essentially refer to 'how design works'; various aspects (e.g. people, process, product, knowledge etc.) play a role in this, and therefore, designing may look very different as these aspects change. There are also many partial activities within designing (e.g. the work of an FEM engineer), i.e. there is 'designing within designing', which theories currently do not capture. Design processes are seen as a major aspect, and therefore, need to be comprehensively understood. Since human reasoning is an essential part of the phenomena of design, and since there is a variety of different kinds of reasoning that exist in design (e.g. logical, informal etc.), a theory should account for these differences and their influences.

Overall, the team argued that the criteria which a theory should satisfy is its amenability to validation and testing, where correspondence between what can be concluded from the theory and the phenomena it tries to explain are assessed. Another criterion is that a theory helps prediction which is useful; this can also be in the form of justification in a historical context. Theories are evolutionary rather than stationary. All assumptions underlying a theory should be made explicit, and one should be aware, as a researcher, about the process by which is a theory is developed.

3. How should a theory or model be evaluated or validated?

Team 1: Rapporteur: John Gero; Scribe: Sonal Keshwani. The team felt that all theories have to be falsifiable. The team defined evaluation as assessment of usefulness, and validation as assessment of consistency. It noted that a model that has so far always given correct results can still give incorrect results: theories are never tested to be true, but with more evidence, confidence in the theory grows. A model has to be validated (checked for internal consistencies) followed by evaluation (checked for usefulness). A difference between models and theories is that, 'hypotheses are derived from theories, while hypotheses are derived from application of models'. A causal model is a network of hypotheses. In evaluating, one has to test each of these hypotheses. To evaluate a theory, one has to operationalise its hypotheses and test these.

Two aspects are critical to pay attention to, when discussing validation: the first is, what should be taken as true and false, and what the process of refutation is whereby truth and falsity should be adjudged. According to this team, validation involves application of the theory or model in design, checking for their internal and external consistencies, and checking them against other, already validated theories or models.

Team 2: Rapporteur: Udo Lindemann; Scribe: S Harivardhini, Praveen Uchil. Validation, the team argues, is about finding the limits of a theory. A major difficulty in validating theories and models of design is that, unlike much of natural sciences, being able to carry out repeatable experiments is hard to impossible. The team proposes that one way of validating a model or theory would be in terms of the level of reliability of the model or theory to achieve its purpose. The team proposed several ways of validation e.g. by comparative studies, by comparing and reducing gaps between research and practice models, by comparing multiple practice based models, or by referring to an existing theory which is already validated.

Team 3: Rapporteur: Lauri Koskela; Scribe: Boris Eisenbart. No design is ever repeatable; however for many areas of natural sciences too. There are various levels of variation across so called repeatable phenomena (e.g. the breaking stress of no two samples of the same material is exactly the same, the effect of the same medicine on no two people is exactly the same, etc.). If the discipline looks into a vast number of design projects in various fields, it might find the phenomena at some level of repeatability (as both material science and medical science already do by taking a statistically large set of samples or subjects). However, two distinct challenges for our discipline are: (i) comparable data in our discipline is currently missing, and (ii) such data is hard to generate. For instance, designers may not be aware of what they do during designing, or may distort certain aspects of their work (e.g. to hide failure, due to miscommunication, post facto rationalisation, forgetting, etc.).

A major issue in validation is that, while some researchers develop theories and others develop empirical results, the two rarely discuss their results with one another to bootstrap their work. A platform to support such discussion is necessary. Another issue is that, many empirical studies are carried out with students only; as a consequence, what can be learnt from these about design in practice is relatively limited. In these studies, and even more so for studies of practice, sample sizes are small due to lack of availability of subjects and constraints on time for detailed analyses. There is a strong need for developing appropriate design research methods to tackle these issues. Another issue is the lack of information of the contexts in which a theory of design is applicable. Given the complexity and variety of designing, it may be too ambitious to develop one theory of design; the community needs to develop many theories, each of which applies in a particular context for a particular purpose. These may then form the basis for developing more comprehensive theories. Another challenge is the difficulty of validating prescriptive theories in practice, e.g. asking practicing designers to change their thinking or process of designing may be hard. Validation need not be done only via practice, but also via teaching, training budding designers into preferred ways of thinking and processes of designing. A possible, new direction for validating theories is theory-driven prediction of new, hitherto non-existing, types of design or design fields.

Overall conclusions about these three questions

Regarding the definition of models and theories, two main points emerged. One is that the term 'model' has multiple meanings. In one meaning, models are used as a means to carry out design, e.g. a digital model of the product; we may call these models for design. In the other meaning, models describe, explain or predict how designs and designing are, and how aspects of these are related to various criteria that are of importance to practice, e.g. how designing relate to costs of designs. We may call these models of design.

The second point is that there is considerable overlap between the meanings of models of design and theories of design. A spectrum of meanings emerged, starting from having 'no distinction in how these terms are currently used in our area', to one where 'Theory defines a framework from which multiple models could be derived'. A consensus emerged that there is need to understand 'theory as a spectrum', with terms such as taxonomies, models and theories having varying degrees of maturity in context, purpose and explanatory capacity.

The purpose of the need for understanding these terms was also discussed. It was felt that for a practitioner, it made no difference as to what these terms meant. However, for a discipline of research such as design, understanding of these terms is crucial, since this forms the basis for research. Overall, it was agreed that a clear understanding of the terms model and theory in the context of design research is necessary. It is also felt that any proposal for a model or theory should be accompanied with its purpose and context.

A strong consensus arrived at across the teams is in the criteria to be considered a theory: theories should be testable and refutable (i.e. falsifiable), and this should be possible to be carried out within the context and purpose of the theories, i.e. where it applies, and how well.

Validation was seen to be testing the limits of a theory or a model. Validation, too, emerged to have a spectrum of meanings, from testing for internal consistency, to truth and usefulness, in terms of providing explanation or insight in the form of predictions or post-dictions.

Several challenges to validation were identified: difficulty or lack of repeatability of phenomena, the large number of factors blurring clear and identifiably strong influences, difficulty of finding statistically large number of appropriate subjects or cases, and difficulty of generating reliable data about the phenomena under investigation.

4. What are Gaps in our Current Understanding and What are the Directions for Further Research?

Several directions emerged.

One major issue identified in the discussions is the general lack of a common understanding that can act as the underlying basis for the discipline. One symptom or a possible cause of this lack is the poor citing of each other's work in the discipline. A need for an overview, or even consolidation, of research carried out so far was strongly emphasised. As a discipline, we need good 'demarcating theories' that provide a clearer understanding of what constitutes (and what does not constitute) part of the phenomena of designing (e.g. designing is demarcated by intentionality), the different types of designs and designing that form our discipline; and position the models and theories with respect to these.

This base, it was suggested, might be initiated by including these:

- The philosophies of the discipline, including what design means, and what the 'phenomena of designing constitute'. 'We need a philosophy of design, like a philosophy of science'.
- A list of 'demarcating theories' that provide an understanding of the different types of designs and designing that form our discipline.
- A list of terms that are used within the discipline, including theory and model, along with their contexts and purpose.
- A list of research methodologies and methods within the discipline, along with their contexts and purpose.
- A list of empirical results, along with their context and purpose.
- A list of models and theories of design, along with their context and purpose.
- A list of influences of results of design research on practice.

Another major point was the need to clarify the common purpose of design research, and identify what the pressing, concrete questions are that the discipline needs to address. Also emphasised was the need for investigating the specific characteristics, benefits and complementarities across the various theories and models, rather than discussing only about which one among these.

A further major point was the challenge of validating theories of models of phenomena of design, which pointed to the need to develop research methods that are appropriate for scientific studies within the constraints and expectations of design research: how to develop and validate testable, refutable theories and models of adequate accuracy within the constraints of complexity of the phenomena observed and within the low availability of appropriate cases and subjects?

Towards addressing the above directions, several suggestions were made:

- Have more discussion events at various levels, e.g. students, researchers, educators, etc., to discuss these issues. Getting together is the first step to 'form the discipline'. Developers of theories and empirical results should interact more with one another.
- Like in other disciplines, teach the common understanding to those (intending to be) in this discipline. This knowledge should be taught in a context-specific manner, i.e. 'make explicit what is applicable in which specific situation'.
- Interact with other disciplines with similar goals, such as management, and learn from their perspectives.
- Carry out more empirical studies that are unbiased, of high value, high-quality, and are clearly explained, as we still do not understand in sufficient depth why design processes happen the way they do.
- Have 'grand debates' where specific models are discussed and contrasted together.
- Work more on developing research methods that are appropriate for serving the specific needs of design research. A starting point can be to propose Special Interest Groups (SIG) to work on these, e.g. on research methodology.

Appendix B: Major Theories and Models not Contained in this Book

This appendix provides a summary of some of the major theories not contained in this book, but are necessary to point to for the sake of completeness. The summaries are not meant to be comprehensive, but only as a pointer to more detailed sources.

General Design Theory (GDT) was proposed by Yoshikawa [86] and later expanded by Tomiyama and Yoshikawa [79]. It is one of the first design theories at the knowledge level—a concept originally proposed by Newell [56] in the context of computational theories. GDT describes design as a transformation between two spaces—function and attribute, and discusses the nature of this transformation in relation to availability of complete and incomplete knowledge.

Axiomatic Design Theory was proposed by Suh and colleagues [74, 75]. It describes design as a transformation between functions and parameters, and argues that good designs can be described by two axioms: axiom of independence and axiom of information content. According to Axiomatic Design Theory, the less coupled the functions are in a design and the less information content the design has, the better it is.

Another Knowledge Level theory—$K^LD_0^E$—was proposed by Smithers [69, 70]. This theory was tested by the author on design of a new font that the author himself designed. $K^LD_0^E$ distinguishes six types of knowledge needed in design: 1. knowledge needed to form *requirements*, knowledge of the requirements descriptions actually developed, and their associated justifications; 2. knowledge of how to develop *well-formed problem descriptions* and knowledge of the well-formed problem descriptions developed and their justifications; knowledge needed to *solve well-formed problems*, and the knowledge of the solutions and justifications actually formed; 4. knowledge needed to analyse and evaluate problem solutions, knowledge of the *analyses and evaluations* actually performed together with their justifications; 5. knowledge needed to *form design descriptions*, and the knowledge of the actual design descriptions and justifications; 6. knowledge needed to *construct design presentations*, and the knowledge of the presentations actually formed and their justifications.

A quest for a Universal Design Theory (UDT) was made by Grabowski et al. [37, 53]. UDT is attempted to be a design theory containing findings and knowledge about design from different engineering disciplines in a consistent, coherent and compact form [52]. It is aimed at serving as a scientific basis for rationalizing interdisciplinary product development. The aim of UDT is to provide models of explanation and prediction of artefacts and away of designing them. The theory takes the 'process of design as the mapping of a set of requirements onto a set of design parameters' that constitute a design solution. The process is proposed to be carried out in by transition through four linked, abstraction levels: modelling requirements, modelling functions, modelling effective geometry, and embodiment design. A design solution is a specification of information sets associated with levels of functions, effective geometry, and embodiment. UDT proposes three axioms: the first states that there is a finite number of levels of abstraction; the second axiom states that the 'the set of well-known basic elements on each level of abstraction is finite at a certain point of time'; the third axiom states that 'the number of transitions between the different levels of abstraction is also finite'. Based on these axioms, the authors considered that 'Elements of a design theory…can only include the components currently known to us whereas the invention of new effects etc. has to be the concern of research work'. In line with this, they hypothesised the following: 'The invention of a product is always a new combination of known basic elements', and that 'Discovery, achieved through research, is defined as the finding of new basic elements'. In this sense, the scope the universal design theory is limited to those types of design where new designs can be seen only as a combination of old basic elements.

Based on the methodological framework used for the development of Grabowski's universal design theory [52], Lossack [50, 51] proposes the foundations of a Domain Independent Design Theory. The theory describes design knowledge, design process knowledge and system theoretical approaches for processing this knowledge system. The underlying concept consists of three elements: object patterns, process patterns and design working-spaces. Lossack emphasises that 'design is not a workflow [...] workflows represent processes in a deterministic manner, whereas design is intrinsically indeterministic'. He therefore proposes an approach based on solution patterns to support indeterministic design processes, which include solution finding processes and creativity. A solution pattern is an aggregation of an object and a process pattern, although an object pattern can be used without process patterns. Object and process patterns describe design knowledge with which a mapping between properties of the design stages is defined. To define the design context, design working-spaces are introduced [36]. A design working space is a system (with elements, relationships and boundaries) which builds a framework to support the solution finding processes with object and process patterns. The approach is regarded to be general enough to support designing in mechanical, electrical and software engineering.

The theory of synthesis by Takeda et al. [76] focuses on the properties that the synthesis process should have as a thought process and propose a theory for synthesis. Knowledge for synthesis in design, they argue, 'needs physicality, unlikeness, and desirability'. Physicality ensures possibility of existence, while unlikeness and desirability ensure newness and value. The theory is based on the assumptions that a design process is an iterative logical process of abduction and deduction on design solutions, their properties and behaviours, and knowledge of objects. The synthesis theory for design is defined as a process of reconstruction of design experiences, where each experience contains a logical design process having three steps: 'collecting design experiences, building a model that includes the collected design experiences, and minimizing an element that designers want to find newness'.

Infused design [66] is an approach for 'establishing effective collaboration between designers from different engineering fields'. Infused design provides representation of the design problem at a mathematical meta-level that is common to all engineering disciplines. The problem solving is carried out by using mathematical terminology and tools that, due to generality, are common across design disciplines. The meta-level proposed consists of general discrete mathematical models termed combinatorial representations (CR). In particular, Infused design demonstrates 'how methods and solutions could be generated systematically from corresponding methods and solutions in other disciplines', and 'guarantees the correctness of results by relying on general ontology of systems that is embedded in the different representations'. Taura and Nagai [78], in their systematised theory of creative concept generation in design, proposed a theory on the thinking process at the 'very early stage of design', they define as the phase that 'includes the time just prior to or the precise beginning of the so-called conceptual design'. They segregate concept generation into two phases—the problem-driven phase and the inner sense-driven phase. They found that the concept generation process could be categorised into two types: first-order concept generation, which is related to the problem-driven phase, and high-order concept generation, which is related to the inner sense-driven phase.

Appendix C: Overview of Theories, Models and Key Concepts Proposed by the Authors

As discussed in Sect. 1.3.5 some authors have proposed ontologies for the development of their theories and models, others have defined their main concepts but not yet put these together into an ontology. In this section, we summarise the proposed theories or models and the related key concepts. What is immediately visible is the differences in concepts used, as well as the difference in their number. Some overlap in key concepts exists. As expected, this is the case where a theory or model has been built on other theories and models. The differences suggest that the phenomenon of design is (as yet) too large, or maybe its boundaries not fixed enough, to be treated as a whole, as also suggested by Eckert and Stacey [28], Chap. 19.

Agogué and Kazakçi [1], Chap. 11: Concept-Knowledge-theory of C–K theory, a theory of creative design reasoning.

Key concepts: K-space, C-space, logical status, properties, restrictive and expensive partitions, co-evolution of C- and K-spaces through operators (conjunction, disjunction, expansion by partition/inclusion, expansion by deduction/ experiments), d-ontologies, generic expansion, object revision, preservation of meaning, K-reordering.

Albers and Wintergerst [2], Chap. 8: Contact and Channel (C&C) Model and Approach to integrate functions and physical structure of a product in a shared representation using product models that are widely spread in practice.

Key concepts: Channel and support structures, working surface pairs, connectors, Wirk-Net, Wirk-structure, operation mode, input parameter characteristic, environmental conditions system state property.

Andreasen et al. [6], Chap. 9: Domain Theory as a systems approach for the analysis and synthesis of products.

Key concepts: Activity, organ, part, structure, elements, behaviour and function, state, property, characteristic, technical activity, need, operands, effects, surroundings, use function, wirk function transformation.

Badke-Schaub and Eris [7], Chap. 17: Understanding the role intuitive processes play in the thinking and acting of designers, to inform their Human Behaviour in Design (HBiD) framework which aims to understand the complex interplay between the designer, the design process, design output, and the related patterns and networks of influencing variable.

Key concepts: Intuition (physical, emotional, mental and spiritual), un/subconsciousness, reasoning.

Cavallucci [21], Chap. 12: Inventive Design Method based on and an extension of TRIZ theory, to rapidly arrive at a reasonable number of inventive solution concepts to evolve a complex initial situation that is currently unsatisfactory.

Key concepts: Contradiction (administrative, technical, physical), problem, partial solution, action parameter, evaluation parameter.

Culley [24], Chap. 18: An information-driven, rather than task-driven, design process to manage and control design activity.

Key concepts: 'Information as thing', knowledge (embedded, encoded, encultured, embrained, embodied).

Eckert and Stacey [28], Chap. 19: Identifying the causal drivers of design behaviour as a first step to generate partial theories of design.

Key concepts: Constraints (problem, process, solutions and meeting constraints), causal drivers (characteristics of classes of products or processes, conditions in which they are created), and requirements.

Eder [29], Chap. 10: Theory of Technical Systems and an engineering design methodology based on this theory.

Key concepts: Transformation process (operands and related states, effects, operators, technology, assisting inputs, secondary inputs and secondary outputs, active and reactive environment) and Technical System (function, organ, organ connector, constructional parts and their relationships: functional structure, constructional structure), life cycle of a technical system (a sequence of transformation systems), properties of transformation processes and technical systems (observable, mediating, elemental) and their related states.

Gero and Kannengiesser [33], Chap. 13: The Function-behaviour-structure (FBS) ontology to describe all designed things, irrespective of design domain, the FBS and the situated FBS (sFBS) frameworks to represent the process of designing, and its situatedness, respectively, irrespectively of the specific domain or methods used.

Key concepts: Function, behaviour (expected, derived from structure), situatedness (interactions between external, expected and interpreted world), interaction (interpretation, focussing, action), function, requirements, structure, design description, transformation (formulation, synthesis, analysis, evaluation, documentation, reformulation types 1–3), comparison.

Goel and Helms [34], Chap. 20: A knowledge model of design problems called SR.BID, derived from the Structure-Behaviour-Function knowledge model, and grounded in empirical data about biologically inspired design practice to capture problem descriptions more deeply than with the SBF knowledge model.

Key concepts: Function, performance criteria, solution, deficiencies/benefits, constraints/specification, and operating environment, structure, behaviour and function.

Goldschmidt [35], Chap. 21: A model of the role of sketching in the early, search phase of design.

Key concepts: Problem, search space, internal and external representations, rapid sketch, cognitive benefits and affordances (time effective/fluent, minimal cognitive resources, minimally rule-bound, transformable/reversible, tolerant to incompletion, tolerant to inaccuracy/lack of scale, provides unexpected cues).

Koskelaet al. [46], Chap. 14: The first theory—proto-theory—of design proposed by Aristotle based on the claim that design is similar or analogous to geometric analysis.

Key concepts: Analysis (theoretical and problematical), synthesis, deliberation, science of production, causes (efficient, formal, material and final), types of reasoning (regressive, transformational, decompositional or configurational).

Lindemann [49], Chap. 6: Definition and nature of the variety of models used for design, discussion on quality and requirements for modelling based on important characteristics like transformation and reduction, purpose and subject, and nature of the process of modelling.

Key concepts: Transformation, reduction, pragmatism (purpose, users, time frame), modelling conventions (accuracy, clearness, profitability, relevance, comparability, systematic settings), process of modelling (intention, modelling, validation, usage).

Maier et al. [54], Chap. 7: A cybernetic systems perspective to understand designing as a self-regulated modelling system, i.e. to consider the synthetic role of models in designing.

Key concepts: Sensoring, actuating.

Ranjan et al. [59], Chap. 15: Integrated Model of Designing' (IMoD) for describing task clarification and conceptual design, and for explaining how various characteristics of these stages relate to one another, by combining different views (or models).

Key concepts: Activity view (generate, evaluate, modify, select), outcome view (phenomenon, state change, effect, input, action, organ, part, other), requirement-solution view (requirement, solution, associated-information), and system-environment view (relationships, elements, subsystem, system and environment).

Sonalkar et al. ([72], Chap. 3): Two-dimensional structure for design theory: describing the theoretical constructs and relationships between them, and providing the perceptual field and action repertoire that makes a theory relevant in situations of professional practice.

Key concepts: Perceptual field, action repertoire, event, relationship,

Taura [77], Chap. 4: A framework composed of the Pre-Design, Design, and Post-Design stages is introduced to allow the explicit capture of the *motive of design*, as an underlying reason for the design of highly advanced products, that links the Post-Design and Pre-Design stages.

Key concepts: Pre-Design, Design, Post-Design, deductive, inductive and abductive processes, personal/social motive, inner/outer motive, need, problem, personal inner sense, inner criteria, function (visible/latent), force of a product, standard, field (physical/scenic/semantic; visible/latent).

Weber [85], Chap. 16: The CPM/PDD approach to modelling products and product development based on characteristics and properties (CPM: Characteristics-Properties Modelling, PDD: Property-Driven Development).

Key concepts: Characteristics, properties (current, desired), relations, external conditions, analysis, synthesis, solution elements/patterns.

References

1. Agogué M, Kazakçi A (2013) 10 years of C-K theory: a survey on the academic and industrial impacts of a design theory. In: Chakrabarti A, Blessing LTM (eds) An anthology of theories and models of design: philosophy, approaches and empirical explorations, Springer, Switzerland, pp 215–232
2. Albers A, Wintergerst E (2013) The contact and channel approach (C&C2-A): relating a system's physical structure to its functionality. In: Chakrabarti A, Blessing LTM (eds) An anthology of theories and models of design: philosophy, approaches and empirical explorations, Springer, Switzerland, pp 149–170
3. Altshuller GS (1961) How to learn to invent Tambov. Tambovskoe knijnoe izdatelstvo (in Russian)
4. Altschuller G (1984) Erfinden: Wege zur Lösung technischer Probleme. VEB Verlag, Berlin
5. Andreasen MM (1980) Machine design methods based on a systematic approach—contribution to a design theory. PhD Thesis, Department of Machine Design. Lund Institute of Technology, Lund, Sweden (in Danish)
6. Andreasen MM, Howard TJ, Bruun HPL (2013) Domain Theory, its models and concepts. In: Chakrabarti A, Blessing LTM (eds) An anthology of theories and models of design: philosophy, approaches and empirical explorations, Springer, Switzerland, pp 171–192
7. Badke-Schaub P, Eris O (2013) A theory of design intuition: does design methodology need to account for processes of the unconscious such as intuition? In: Chakrabarti A, Blessing LTM (eds) An anthology of theories and models of design: philosophy, approaches and empirical explorations, Springer, Switzerland, p 351–368
8. Blessing LTM, Chakrabarti A, Wallace KM (1992) Some issues in engineering design research. In: Cross N (ed) OU/SERC design methods workshop. The Open University, Milton Keynes
9. Blessing LTM (1994) A process-based approach to computer-supported engineering design. PhD thesis, University of Twente, Black Bear Press, Cambridge, UK
10. Blessing LTM, Chakrabarti A, Wallace K (1995) A design research methodology. In: Hubka V (ed) International conference on engineering design (ICED'95). Heurista, Zürich, Prague, pp 502–507
11. Blessing LTM (1995) Comparison of design models proposed in prescriptive literature. In: Perrin V, Vinck D (eds) Proceedings of COST A3/COST, 'the role of design in the shaping of technology". Social sciences series, vol 5, pp 187–212, Lyon
12. Blessing LTM, Chakrabarti A, Wallace KM (1998) An overview of descriptive studies in relation to a general design research methodology. In: Frankenberger E, Badke-SchaubP (eds) Designers: the key to successful product development. Springer, Switzerland, pp 56–70
13. Blessing LTM (2002) What is this thing called 'design research'? In: Annals of the 2002 international CIRP design seminar. CIRP, Hong Kong
14. Blessing LTM, Chakrabarti A (2002) DRM: a design research methodology. In: Les Sciences de la Conception: l'enjeuscientifique du 21e siècle en hommage à Herbert Simon, Ed. J. Perrin, INSA, Lyon, France
15. Blessing LTM (2003) Future issues in design research. In: Lindemann U (ed) Human behaviour in design: individuals, teams, tools. Springer, Heidelberg, pp 298–303

16. Blessing LTM, Chakrabarti A (2009) DRM: a design research methodology. Springer, Heidelberg
17. Braha D, Maimon OZ (1998) A mathematical theory of design: foundations, algorithms and applications. Kluwer Academic Publishers, Norwell
18. Buchanan R (2004) Design as inquiry: the common, future and current ground of design. In: Future ground design research society international conference, Melbourne
19. Cantamessa M (2001) Design research in perspective—a meta-research on ICED'97 and ICED'99. In: Culley S et al (eds) International conference on engineering design (ICED'01). IMechE, Glasgow, pp 29–36
20. Cantamessa M (2003) An empirical perspective upon design research. J Eng Des 14(1):1–15
21. Cavallucci D (2013) Designing the inventive way in the innovation area. In: Chakrabarti A, Blessing LTM (eds) An anthology of theories and models of design: philosophy, approaches and empirical explorations, Springer, Switzerland, pp 233–258
22. Chakrabarti, A, Murdoch, TNS, and Wallace, KM (1995) Towards a framework for a glossary of engineering design terms. In: Proceedings of the international conference in engineering design, Prague, vol 1, pp 185–186
23. Craver CF (2002) Structures of scientific theories. The Blackwell guide, 55
24. Culley SJ (2013) Re-visiting design as an information processing activity. In: Chakrabarti A, Blessing LTM (eds) An anthology of theories and models of design: philosophy, approaches and empirical explorations, Springer, Switzerland, pp 369–392
25. DIN 19226 (1994) Control Engineering. Beuth, Berlin
26. Dörner D (1994) Heuristik der Theorienbildung. Enzyklopädie der Psychologie 1:343–388
27. Eckert CM, Stacey MK (2010) What is a process model? Reflections on the epistemology of process models. In Heisig P, Clarkson PJ, Vajna S (eds) Modelling and management of engineering processes. Springer, New York pp 3–14
28. Eckert CM, Stacey MK (2013) Constraints and conditions: drivers for design processes. In: Chakrabarti A, Blessing LTM (eds) An anthology of theories and models of design: philosophy, approaches and empirical explorations, Springer, Switzerland, pp 393–414
29. Eder WE (2013) Engineering design: role of theory, models and methods. In: Chakrabarti A, Blessing LTM (eds) An anthology of theories and models of design: philosophy, approaches and empirical explorations, Springer, Switzerland, pp 193–214
30. Friedman K (2003) Theory construction in design research: criteria: approaches, and methods. Des Stud 24:507–522
31. Galle P (2006) Worldviews for design theory. In: Wondergrounds. Design Research Society International Conference, Lisbon
32. Gero J (1990) Design prototypes: a knowledge representation schema for design. AI Mag 11:26–36
33. Gero JS, Kannengiesser U (2013) The function-behaviour-structure ontology of design. In: Chakrabarti A, Blessing LTM (eds) An anthology of theories and models of design: philosophy, approaches and empirical explorations, Springer, Switzerland, pp 259–280
34. Goel AK, Helms ME (2013) Theories, models, programs and tools of design: Views from artificial intelligence, cognitive science and human-centered computing. In: Chakrabarti A, Blessing LTM (eds) An anthology of theories and models of design: philosophy, approaches and empirical explorations, Springer, Switzerland, pp 415-430
35. Goldschmidt G (2013) Modeling the role of sketching in design idea generation. In: Chakrabarti A, Blessing LTM (eds) An anthology of theories and models of design: philosophy, approaches and empirical explorations, Springer, Switzerland, pp 431–448
36. Grabowski H, Lossack R-S, Weis C (1995) A design process model based on design working spaces. In: Tomiyama T, Mäntylä M, Finger S (eds) Knowledge intensive CAD 1; Proceedings of the first IFIP WG 5.2 workshop, Springer, New York
37. Grabowski H, Rude S, Grein G (eds) (1998) Universal design theory—proceedings of the workshop 'Universal Design Theory". Shaker, Aachen

38. Hatchuel A, Weil B (2003) A new approach of innovative design: an introduction to C-K theory. In: Norell M (ed) International conference on engineering design (ICED'03), Stockholm, Sweden
39. Hatchuel A, Le Masson P, Reich Y, Weil B (2011) A systematic approach of design theories using generativeness and robustness. In: DS 68-2: proceedings of the 18th international conference on engineering design (ICED 11). Impacting society through engineering design. Design theory and research methodology, vol 2. Lyngby/Copenhagen, Denmark, pp 87–97
40. Heymann M (2005) Kunst und Wissenschaft in der Technik des 20. Jahrhunderts—zur Geschichte der Konstruktionswissenschaft. Chronos, Zurich
41. Horváth I (2001) A Contemporary Survey of Scientific Research into Engineering Design. In: Culley S et al. (eds) Design research—theories, methodologies and product modelling. Proceedings of ICED2001, Glasgow, pp 13–20
42. Horváth T (2013) New design theoretical challenges of social-cyber-physical systems. In: Chakrabarti A, Blessing LTM (eds) An anthology of theories and models of design: philosophy, approaches and empirical explorations, Springer, Switzerland, pp 99–120
43. Hubka V (1974) Theorie der Maschinensysteme. Springer, Berlin (2 ed, revised to become Hubka V (1984)Theorie Technischer Systeme)
44. Hubka V, Eder E (1988) Theory of technical systems: a total concept theory for engineering design. Springer, Berlin
45. Kohn, A, Reif, J, Wolfenstetter, T, Kernschmidt, K, Goswami, S,Krcmar, H,Brodbeck, F, Vogel-Heuser, B; Lindemann, U, Maurer, M (2013)Improving common model understanding within collaborative engineering design research projects. In: A Chakrabarti and RV Prakash (eds) ICoRD'13, Lecture notes in mechanical engineering. Springer, Switzerland, pp 642–654. doi:10/1007/978-81-322-1050-4_51
46. Koskela L, Codinhoto R, Tzortzopoulos P, Kagioglou M (2013) The Aristotelian proto-theory of design. In: Chakrabarti A, Blessing LTM (eds) An anthology of theories and models of design: philosophy, approaches and empirical explorations, Springer, Switzerland, pp 281–300
47. Kroll E (2013) Design theory and conceptual design: contrasting functional decomposition and morphology with parameter analysis. Res Eng Des 24:165–183
48. Le Masson P, Dorst K, Subrahmanian E (2013) Design theory: history, state of the art and advancements. Editorial to a special issue on design theory. Res Eng Des 24:97–103
49. Lindemann U (2013) Models of design. In: Chakrabarti A, Blessing LTM (eds) An anthology of theories and models of design: philosophy, approaches and empirical explorations, Springer, Switzerland, pp 121–132
50. Lossack R-S (2002) Foundations for a domain independent design theory. In: Annals of 2002 international CIRP design seminar, 16–18 May 2002, Hong Kong
51. Lossack R-S (2006) Wissenschaftstheoretische Grundlagen für die rechnerunterstützte Konstruktion. Springer, Berlin
52. Lossack R-S,Grabowski H (2000) The axiomatic approach in the universal design theory. In: Tate D (ed) Proceedings of the first international conference of axiomatic design. Institute for Axiomatic Design, MIT, MA, USA
53. Grabowski H, Rude S, Grein G (eds) (1998) Universal design theory. Shaker-Verlag, Aachen
54. Maier AM, Wynn DC, Howard TJ, Andreasen MM (2013) Perceiving design as modelling: A cybernetic systems perspective. In: Chakrabarti A, Blessing LTM (eds) An anthology of theories and models of design: philosophy, approaches and empirical explorations, Springer, Switzerland, pp 133–148
55. McMahon C (2006) Design research challenges for the 21st century—or my life as mistakes. In: Rigi meeting of the Design Society, Crete, Greece (Unpublished)
56. Newell A (1981) The knowledge level, Artif Intell 18:87–127
57. Pahl G, Beitz W (2007) Engineering design. Springer, London (1st edn. 1983, 3rd edn. 2007; translated and edited by Wallace KM and Blessing LTM)

58. Ranjan, BSC, Srinivasan, V, Chakrabarti, A (2012) An extended, integrated model of designing, In: Horváth I, Albers A, Behrendt M, Rusák Z (eds) Proceedings of TMCE 2012. Karlsruhe, Germany, May 7–11
59. Ranjan BSC, Srinivasan V, Chakrabarti A (2013) Perspectives on design models and theories and development of an extended – integrated model of designing. In: Chakrabarti A, Blessing LTM (eds) An anthology of theories and models of design: philosophy, approaches and empirical explorations. Springer, Switzerland, pp 301–324
60. Reich Y (1994) Layered models of research methodologies. Artif Intell Eng Des Anal Manuf 8:263–274
61. Reich Y (1994) Annotated bibliography on research methodologies. Artif Intell Eng Des Anal Manuf 8:355–366
62. Reich Y (1995) The study of design research methodology. J Mech Des 177:211–214
63. Roozenburg NFM, Eekels J (1995) Product design: fundamentals and methods. Wiley, Chichester
64. Ruse M (1995) Theory. In: Honderich T (ed) The oxford companion to philosophy. Oxford University Press, Oxford, pp 870–871
65. Samuel A, Lewis W (2001) Curiosity-oriented research in engineering design. In: Culley S et al (eds) International conference on engineering design (ICED'01). IMechE, Glasgow, pp 37–44
66. Shai O, Reich Y (2004) Infused design: I theory. Res Eng Des 15:93–107
67. Shai O, Reich Y, Hatchuel A, Subrahmanian E (2013) Creativity and scientific discovery with infused design and its analysis with C-K theory. Res Eng Des 24:201–214
68. Shannon RE (1975) Systems simulation: The art and science. Prentice-Hall, Englewood
69. Smithers T (1998) Towards a knowledge level theory of design process. In: Gero J, Sudweeks F (eds) Artificial intelligence in design '98. Kluwer Academic Publishers, Norwell, pp 3–21
70. Smithers T (2000) Designing a font to test a theory. In: Gero J (ed) Artificial intelligence in design '00. Kluwer Academic Publishers, Norwell, pp 3–22
71. Srinivasan V, Chakrabarti A (2008) Design for novelty—a framework? In: Marjanovic D, Storga M, Pavkovic N, Bojcetic N (eds) International design conference (Design 2008). Dubrovnik, Croatia, pp 237–244
72. Sonalkar N, Jung M, Mabogunje A, Leifer L (2013) A Structure for Design Theory. In: Chakrabarti A, Blessing LTM (eds) An anthology of theories and models of design: philosophy, approaches and empirical explorations, Springer, Switzerland, pp 67–82
73. Stachowiak H (1973) Allgemeine Modelltheorie. Springer, Wien
74. Suh NP (1998) Axiomatic design as a basis for universal design theory. In: Grabowski H et al (eds) Universal design theory. Shaker, Aachen
75. Suh NP (2001) Axiomatic design. Oxford University Press, Oxford
76. Takeda H, Tsumaya A, Tomiyama T (1999) Synthesis thought processes in design. In: Kals H, van Houten F (eds) CIRP international design seminar. Springer, University of Twente, Enschede, The Netherlands, pp 249–268
77. Taura T (2013) Motive of design: roles of pre- and post-design in highly advanced products. In: Chakrabarti A, Blessing LTM (eds) An anthology of theories and models of design: philosophy, approaches and empirical explorations, Springer, Switzerland, pp 83–98
78. Taura T, Nagai Y (2012) Concept generation for design creativity—a systematized theory and methodology. Springer, London
79. Tomiyama T, Yoshikawa H (1987) Extended general design theory. In: Design theory for CAD. Proceedings from IFIP WG 5.2, Amsterdam
80. Tomiyama T, Yoshioka M, Tsumaya A (2002) A knowledge operation model of synthesis. In: Chakrabarti A (ed) Engineering design synthesis: understanding, approaches and tools. Springer, London, pp 67–90
81. VDI 3633 (2010) VDI Guidelines for modelling and simulation. Beuth, Berlin
82. Vermaas PE (2013) Design theories, models and their testing: on the scientific status of design research. In: Chakrabarti A, Blessing LTM (eds) An anthology of theories and models

of design: philosophy, approaches and empirical explorations, Springer, Switzerland, pp 67–82
83. Wallace KM, Blessing LTM (2000) Observations on some German contributions to engineering design in memory of Professor Wolfgang Beitz. Res Eng Des 12:2–7
84. Weber C (2005) CPM/PDD—an extended theoretical approach to modelling products and product development processes. In: Bley H et al (eds) 2nd German-Israeli symposium on advanced in methods and systems for development of products and Processes. Frauhofer-IRB-Verlag, TU Berlin/Fraunhofer-Institut für Produktionsanlagen und Konstruktiontechnik, pp 159–179
85. Weber C (2013) Modelling products and product development based on characteristics and properties. In: Chakrabarti A, Blessing LTM (eds) An anthology of theories and models of design: philosophy, approaches and empirical explorations. Springer, Switzerland, pp 325–350
86. Yoshikawa H (1980) General design theory and a CAD system. In: Sata T, Warman E (eds) IFIP WG 5.2-5.3 working conference. North-Holland, Tokyo, pp 35–57

Chapter 2
Design Theories, Models and Their Testing: On the Scientific Status of Design Research

Pieter E. Vermaas

2.1 Introduction

In design research there is a general concern about the scientific status of the discipline. Design research is about observing existing and created design practices, about formulating design theories and models for describing and improving design practices, and about evaluating these design theories and models. As such design research is just like any other scientific endeavour: it has a domain, aims and research methods and it has results. Yet there is this concern that design research does not live up to the standards of science: it is creating in a sense too many theories and models, which jeopardises the coherence of the discipline and which indicates that design research does not yet have the means to test and refute design theories and models.

In this chapter I address this concern about the scientific status of design research using the resources of philosophy of natural science. First, I describe in Sects. 2.2 and 2.3 how scientific theories and models are understood in philosophy. It is analysed to what extent design theories and models of design can be taken as scientific, using a typology that groups design theories and models by their descriptive, demarcating and prescriptive aims. Second, I focus in Sects. 2.4 to 2.6 on testing. Testing design theories and models is in design research generally taken as validation, and I argue that this perspective may be one of the roots of the lack of effective testing and of coherence in design research. Philosophy of science provides more means for testing theories and models. Falsification is another possibility and I argue, using Lakatos' work, that falsification may give design research means for arriving at effective testing and coherence of design theories and models.

P. E. Vermaas (✉)
Philosophy Department, Delft University of Technology, Jaffalaan 5, 2628 BX Delft, The Netherlands
e-mail: p.e.vermaas@tudelft.nl

A. Chakrabarti and L. T. M. Blessing (eds.), *An Anthology of Theories and Models of Design*, DOI: 10.1007/978-1-4471-6338-1_2,

Philosophy is a rich and multifaceted source for understanding scientific theories and models, and their testing. This chapter contains a selection of this material, primarily from earlier philosophical work on the natural sciences. This selection is motivated by my aim to address concerns about the scientific status of design research. I start from a fairly traditional characterisation of what a scientific theory is, for showing that design theories can be taken as scientific, even in this traditional sense. The chapter contains a more extensive survey of what scientific models are, for showing that also models of design practices can be taken as scientific. Finally it introduces work on falsification, specifically by Lakatos, for arguing that design research can arrive at effective research methods for testing design theories and models that also strengthen the coherence of the discipline. This perspective on design research is however not meant as a claim that design research should be similar to research in the natural sciences, or that design practices are natural phenomena. The differences are obvious, as is described in the next section: design theories can, for instance, be prescriptive and define novel design practices. Other perspectives on design research are possible as well, say approaching it as a social science or taking it a *sui generis* discipline. Yet, the upshot of this chapter is that already by a more traditional perspective, design research can meet the standards of science.

A second caveat is about the term design. It may refer to the practice of designing or to the outcome of that practice; design theories and models may therefore be about design practices, about the outcomes or about both. In this chapter I understand the term as referring to design practices, and consider only theories and models about these practices. These theories and models can also describe the outcomes of those practices—it may be hard to capture a practice while avoiding saying something about its outcomes—yet I do not consider theories and models about only the outcomes of design practices. Such theories and models may exist, but belong to the classification and metaphysics of technical products.

2.2 Scientific Theory and Design Theories

The characterisation of a scientific theory that I adopt is a handbook description given by Ruse [1] in the *Oxford Companion to Philosophy*:

> A scientific theory is an attempt to bind together in a systematic fashion the knowledge that one has of some particular aspect of the world of experience. The aim is to achieve some form of understanding, where this is usually cashed out as explanatory power and predictive fertility.

When this description is taken as part of a definition of what a scientific theory is, then a design theory, when taken as a scientific theory, is an attempt to systematically bind together the knowledge we have of experiences of design practices. There is an aspect of the world of experience that we take as consisting of design practices. We conceptualise and describe our experiences of these design practices in terms of, say, actions or reasoning processes by agents, called designers, where the actions or reasoning involve material objects, information, economic funds, technical and cognitive tools, and so on. We develop criteria to evaluate design practices, for instance, as successful or not, as efficient or not or as innovative or not. We have knowledge of design practices in the form of, say, observed regularities between the actions or reasoning of designers in design practices, and the success and efficiency of these practices. Finally we arrive at a design theory of design practices by attempting to systematically binding that knowledge together. We then understand that aspect of the world and can explain design practices and generate predictions about them. We can, for instance, categorise design practices in types, and predict that future design practices of some type are more successful or more innovative than future design practices of another type.

This way of describing design theories may cover some design theories, yet seems not to fully capture the aims of design theories as developed in design research. One reason for this is that in design research there is no full consensus about what is to be taken as design practices. Some design theories can therefore also be aimed at fixing what aspect of the world of experience is to be taken as design practices. Other design theories can advance as designs altogether new practices of which we do not have experiences yet. Especially the latter possibility is relevant to design research since new types of design practices may be advanced as more successful, more efficient or more innovative than existing types. For bringing also those design theories into focus, I introduce a typology of design theories by means of their aims.

Descriptive design theories. Call a design theory *descriptive* if its aims include describing design practices that are regularly taken as design, say engineering and architectural design practices of existing types. If a descriptive design theory binds together our knowledge of these regular design practices, and arrives at understanding, explanation and prediction of and about them, it is a scientific theory by the given definition.

Demarcating design theories. Call a design theory *demarcating* if its aims include fixing the borders of what is to be taken as design practices, say by also taking developing courses of action as designing [2], or by ruling out that searches among off-the-shelf solutions are design practices [3]. If defining what is to be taken as design practices is the sole aim, then a demarcating design theory is not a scientific theory as defined. But if a demarcating theory is also binding together our knowledge of our experiences with design as demarcated, and providing understanding, explanation and prediction of and about these design practices, then it is a scientific theory by the given definition.

Prescriptive design theories. Call, finally, a design theory *prescriptive* if its aims include singling out particular types of design practices and positing favourable properties about these practices. These particular types of design practices may be types that already exist and are regularly taken as design practices, say when design as described by Pahl and Beitz et al. [4] is proposed as efficient. But the prescribed types of design practices may also be new ones, say, types of design practices that did not yet exist but are proposed to improve sustainability (e.g., [5]). A prescriptive design theory is a scientific theory as defined if it singles out types of design practices that already exist and if it binds together our knowledge about these practices, arriving at understanding, explanation and prediction; in fact, the claim that the singled out design practices possess the favourable properties, may be taken to be knowledge or prediction about these design practice. If, however, a prescriptive design theory singles out new types of design practices, the assessment whether it may be a scientific theory becomes more involved. One may take such a prescriptive design theory as a scientific theory as defined by arguing that the favourable properties it posits are predictions about design practices. Yet they are not predictions about practices that already exist; they are predictions about designs of types that are defined by the theory and that are not yet part of the world of experience. At best one could take a prescriptive design theory that defines new types of designs and posits favourable properties for them as a *hypothetical* scientific theory: it puts forward claims of the form "*if* design practices of the new type were an aspect of the world of experience, *then* the claim that they have the posited favourable properties is part of the knowledge about that aspect of the world of experience."

The design theories that are generated in design research are typically not pure descriptive, demarcating or prescriptive theories; they are amalgams. First, a regular approach to arrive at a design theory is by analysing a specific set of actual design practices, *describe* the structure of the actions or reasoning in these practices, and they *prescribe* this structure for novel design practices. For instance, the way expert designers have proceeded in successful design projects is described, and then as a design method or strategy prescribed to other designers, the prediction being that the favourable properties of successful expert projects are by mimicking also realisable by non-expert designers (e.g., [6, 7]). Second, a design theory that prescribes new types of design is demarcating as well, since it advances those new types of design as design practices whereas they were not yet taken as design practices. Third, a prescriptive theory becomes also descriptive when the tool is adopted. Say QFD [8] may have been a pure prescriptive design theory at the time it was introduced, as it was proposing a new reasoning scheme for design and thus defining new design practices. By being adopted QFD describes today practices that are regularly taken as design, making it also a descriptive theory. Finally, a demarcating design theory can be descriptive as well. C-K theory [9], for instance, describes both practices that are regularly taken as design practices, and practices that are not regularly taken as cases of design.

Not all amalgams of description, demarcation and prescription are possible. For instance, a design theory that describes or prescribes a particular type of design practice but by its demarcation does not acknowledge this practice as design, is contradictory. In [10] a more systematic analysis of possible and impossible amalgams is given.

The distinctions between descriptive, demarcating and prescriptive design theories are relevant to the question of whether design theories are scientific theories as defined. Descriptive design theories are reasonably candidate scientific theories, since the step from describing design practices in terms of processes of actions and reasoning, to knowledge, understanding, explanation and prediction of and about these processes seems rather feasible. Demarcating design theories need not be scientific theories since they may merely define what is to be taken as design. Prescriptive design theories may be scientific theories but need not be so. Design methods and tools are, for instance, all prescriptive design theories, since they all single out particular types of design practices and posit favourable properties about them. From a practical point of view, this singling out of preferred design practices may be sufficient. Yet, for counting as a scientific theory, a design method should minimally add understanding and explanation to the prediction that the singled out design practices have the posited properties. In Sect. 2.5 I argue that these distinctions are also relevant for the testing of design theories. For instance, the possibility to test demarcating design theories depends on the character of the definition of design practices these theories are advancing.

2.3 Scientific Models and Models of Design

A general description of what a scientific model is and of what a model is aimed at, is not so easily found in philosophy. Scientific models come in different types and have various aims. Taking my cue from [11], the following types of entities can be scientific models.

Physical objects. Scientific models can be physical objects, such as technical scale models of ships and buildings, Eisinga's orrery of the solar system and the sticks-and-balls model of DNA by Watson and Crick.

Fictional objects. Scientific models can be fictional objects in the mind, such as frictionless pendulums, the Carnot model of a heat-engine and the model of a rational agent in economics.

Set-theoretical structures. Scientific models can be set-theoretical structures consisting of a set of entities, operations on the entities and relations between the entities, where these entities can represent, for instance, numbers, probabilities or data as in database design.

Descriptions. Scientific models can be linguistic descriptions of objects, of systems, of practices and of experimental phenomena.

Equations. Scientific models can consist of equations, such as $F_x = -kx$ for the harmonic oscillator.

Models in science are generally taken to have the aim of representing something else, where there can be differences between what is represented and how strict the representation is.

First, a model is taken as representing features of a target system in the world. The scale models of ships or buildings are representing those ships and buildings, and these representations concern some but typically not all features of the target system: the dimensions of the ships and buildings are represented, but not, say, their material constitution. Models can represent target systems moreover in an idealised way. Orreries represent the movements of the planets around the Sun, and do so in an idealised way, say, by ignoring the gravitational influences of the various moons in the solar system. The idealisation part of models can even be such that the target system is at best an imaginary system. The Carnot model of the heat-engine [12] and the model of the rational agent in economics are not models that represent systems in the world, but are at best non-existing idealisations of them.

Second, and in addition to representing target systems, a model can represent a scientific theory. Orreries represent the movements of the planets in the solar system *and* represent Newtonian physics by being cases that comply with Newtonian physics; they are models of the solar system and of Newtonian physics.

The value of scientific models is not only that they represent target systems or scientific theories; they also have epistemic value. The creation of models, their analysis and their development, allow scientists to understand the target systems and the theories represented. With physical models such as scale models, knowledge about the target systems and theories can be collected by carrying out tests on the models in laboratories or in the field. With models consisting of fictional objects or of equations, such knowledge can be collected by derivation or computer simulation.

In philosophy there are two competing positions about the relationship between the aims of scientific models to represent and to have epistemic value, and here I take my cue from [12]. By the *semantic position* on models, the representation of target systems and of theories is the immediate aim of scientific models, and are the epistemic advantages secondary and due to this representation. By the *pragmatist position* on models, having epistemic value is an immediate aim of scientific models as well, and is the way in which models represent target systems and theories dependent on their epistemic value. The epistemic values of models may even compete with their aim to represent. The idealisations made in scientific models such as the Carnot model of the heat-engine and the rational agent model are useful for acquiring knowledge about thermodynamics and economics, yet are detrimental to the accuracy by which these models are representing real target systems in the world.

There are three general views in philosophy about how scientific models are related to scientific theories. In the first *syntactic view* scientific theories are taken as formal and axiomatised sets of sentences, and models of theories are then taken as providing rules for interpreting the terms and sentences of the theory they represent. An orrery is by this view not a model of Newtonian physics, since as a physical object it does not explicitly give an interpretation of terms of Newtonian physics as physical properties of the Sun and the planets. In the second *semantic view* a scientific theory is seen as a set of models, in particular of the more abstract types such as set-theoretic models. All theoretical models of Newtonian physics define together the contents of Newtonian physics, making the formulation of a formal and axiomatised set of formal sentences capturing all these models a possibly interesting but eventually unnecessary affair. Orreries as physical objects are by this second view probably still not models of Newtonian physics, but only so if it is added how they represent the physical properties of the Sun and the planets. In the third view, associated with [13], the link between models and theories is weakened. In this view models are not taken as closely representing the content of theories, but seen as means to understand that content. And in line with the pragmatist position that models have epistemic values as well, models allow this understanding of scientific theories by introducing elements that are not part of these theories. Models add, for instance, real and imaginary cases to theories, and introduce idealisations. Orreries are by this final view models that are not derivable from Newtonian physics but still means that enable us to understand this theory.

In design models are used for various reasons. Models are used to represent the investigated or final outcomes of design practices, to represent the objects, theories and data employed in finding these outcomes, and to represent the design practices themselves. These models in design typically count as scientific by being of the types of scientific models described above, as is illustrated by scale models and the Carnot model of a heat-engine. Given all the types of scientific models discerned, and given the different philosophical positions about their aims and relation to theory, a full discussion of design models will easily diverge into extensive classification and commentary. Yet when considering only models used in design research for representing design practices, the discussion becomes again focussed and of use to an analysis of the testing of the results of design research (see [14], Part IV) for broader discussions of models in engineering and design).

A discussion of the aims of models of design practices in design research reiterates much of the above discussion of the aims of scientific models.

First, abstracted descriptions of actual design as given in design research (e.g., [15]) may be taken as models that primarily aim at representing design practices in the world, that is, as models that are relatively independent of design theory. Diagrams of design practices as they are often given in design research, ranging from the VDI [16] flowcharts for activities in engineering design, to the reasoning schemes of design thinking (e.g., [17]), are not straightforwardly models. Diagrams are in philosophy not seen as models themselves but as means to express models ([18], p. 2). Accepting this point, diagrams of design practices are minimally expressions of models aimed at representing these design practices.

Second, diagrams and descriptions of design practices as advanced in design theories (e.g., [9, 19]) may be taken as (expressions of) models aimed at representing design practices and at representing the design theories concerned.

Finally, diagrams of design practices as advanced in more prescriptive design theories (e.g., [6, 17]) may be taken as expressing models that also have the epistemic value of making clear to designers how to improve on their design practices. And given the often made observation that actual design practices are typically more complex than the prescribed flow charts and reasoning schemes, these diagrams are typically expressions of models representing idealised design practices. So, in line with the pragmatist position on models, the epistemic value of design models of making clear how to improve on design practices may overrule their aim of representing design practices in the world.

There are design theories that strive toward formalisation and axiomatisation (e.g., [19, 20]), yet typically design theories are not meeting the logico-philosophical ideal of being formulated as axiomatised sets of formal sentences. The syntactic view on the relation between models and theory seems therefore not plausible for most design theories; at best the semantic view can be adopted, meaning that the models of design practices as advanced by a design theory together define the content of the design theory.

In the previous section, I took distance from understanding all types of design theories as scientific theories. Design theories can have descriptive, demarcating and prescriptive aims, and specifically demarcating and prescriptive design theories did not straightforwardly fit the characterisation of scientific theories. Models of design practices may also be differentiated as models with descriptive, demarcating and prescriptive aims, but now all types of models fit much better in the characterisation of models in science, since there is such a diversity of scientific models.

Models representing actual design practices and models representing a descriptive design theory may be taken as *descriptive models* of design, meaning that they are models with the aim of representing practices that are regularly taken as design. The abstracted descriptions of actual design practices as given in, e.g., [15] may be taken as such descriptive models, and they are scientific models of the type *descriptions*. Design diagrams aimed at capturing actual design practices are expressions of descriptive models, which are also scientific; the diagrams are expressions of scientific models of the types *fictional objects*, *descriptions*, or maybe even *set-theoretical structures*.

Models that represent an aspect of the world that is to be understood as design practices and models that represent a demarcating design theory, may be taken as *demarcating models* of design, meaning that they have the aim of representing practices that are to be taken as design practices. The diagrams and characterisations of design practices as advanced in C-K theory [9] may be seen as, in part, (expressions of) demarcating models, and they are scientific models of the types *fictional objects*, *descriptions*, or *set-theoretical structures*.

Finally, models representing real or imaginary design practices with favourable properties and models representing a prescriptive design theory, may be taken as *prescriptive models*, meaning that they are models for singling out design practices for having those favourable properties. The diagrams of design practices as advanced by Lawson and Dorst [6] are expressing such prescriptive models, and these models are scientific models of the types *fictional objects* or *descriptions*.

Again, models of design practices can be amalgams. The models of C-K theory [9] are both demarcating and descriptive, and the models by Lawson and Dorst [6] are both descriptive of particular expert design practices and prescriptive to other design practices. And, again, not all amalgams are possible. Models can represent cognitive processes as they take place in the mind of designers, and models can represent more rationalised cognitive processes of thinking in design practices [21]. A prescriptive model may by its pragmatic aim of characterising favourable design practices represent the thinking of designers in this more rationalised way, which blocks the possibility that this model can also have the aim to represent the way actual designers think during design practices.

2.4 Testing Design Theories and Models by Validation

Design theories and models of design advance a diversity of claims about design practices: they describe and demarcate design practices, and they prescribe design practices as having favourable properties. For evaluating design theories and model these claims are to be tested for determining the empirical accuracy and practical usefulness of the design theories and models. For instance, descriptive design theories aim at describing actual design practices, and are to be tested on whether they give accurate descriptions of these practices. And prescriptive models aim at representing design practices that have specific favourable properties, and are to be tested to determine whether these practices indeed have the posited properties. (In the next section it is more systematically discussed how to test descriptive, demarcating and prescriptive design theories and models.) In philosophy of science different approaches to testing theories have been advanced and criticised, and for the discussion in this chapter it is relevant to distinguish two rather opposite approaches. The first is testing a theory by *falsification* and consists of tests of individual observational statements derived from the theory. If such a test yields that the statement does not hold, the theory is refuted. And if the test confirms the statement, the theory is corroborated and accepted provisionally, since a new test may in the future still refute the theory. The second approach is testing a theory by *validation* and consists of tests of all (main) observational statements derivable from the theory. If one such test yields that a particular statement does not hold, the theory is refuted. If the tests confirm all statements, the theory is accepted. And as long as the tests are not yet all done, it remains open whether the theory can be accepted. In design research, as will be illustrated in this section, testing is typically taken as validation, whereas, as I will argue, testing by falsification may improve the scientific status of design research.

Design research may have known periods in which more theoretical work was done and testing was less prominent. But certainly today it is saturated by empirical studies in which design theories and models are discussed in close connection to analyses of actual design cases, to outcomes of design experiments and to statistical data gathered about design experiences in industry and academia. Yet, despite this ongoing empirical focus there is in design research a general concern about the quality of the testing of design theories and models. In work reflecting on the results that design research has produced, it is complained that generally accepted and effective research methods for testing design theories and models are lacking in design research, and that the discipline is fragmented in separate research strands (e.g., [22–25]). Hence, efforts are made in design research for strengthening testing (e.g., [23, 26, 27]), for arriving at coherence, and for thus improving the scientific rigour of the discipline.

When considering design research from a philosophical perspective it can be argued that the two complaints of weak testing methodology and fragmentation may have a common root in the very assumption that testing of design theories and models is taken as validation. First, by understanding this testing as validation it is assumed that all claims of design theories and models should be confirmed by empirical observation or other means. This assumption makes testing quite open-ended and rather difficult (e.g., [24, 26]). Prescriptive design theories, for instance, are ultimately meant to improve design practices in industry. Testing of design theories that posit that particular design practices lead to innovative products (e.g., [7, 28]), then requires years-long experiments in which sufficiently large numbers of engineering firms abandon their established design practices in favour of the prescribed practices, and in which relevant contextual factors like international economic growth and the behaviour of competitor firms are kept controlled. Given the unrealistic nature of such experiments, it seems not surprising that design research does not yet have effective research methods to carry them out.

Second, by understanding testing of a design theory or model as validation, it is suggested that this testing can be done independently of rival design theories and models. The claims of a design theory or model are to be confirmed as stand-alone claims or, in the case of a prescriptive theory or model, at best as claims relative to some design-practice benchmark; the claims of the tested design theory or model need, however, not be compared to the claims of rival design theories or models. By this suggestion design research arrives at series of stand-alone evaluations of individual design theories and models, supporting that design research indeed develops in separate research strands associated with these theories and models.

These arguments are not directed against testing design theories and models by validation; they are rather meant to provide room to taking this testing not only as validation. In the next section I drop the assumption that testing means confirming all claims of design theories and models, and consider swifter testing of design theories and models by falsification in a Popperian [29] manner. In Sect. 2.6 I also drop the suggestion that design theories and models should be assessed in isolation by discussing sophisticated forms of falsification as proposed by Lakatos [30] by which rival design theories and models are compared. Considering testing by falsification may seem a spurious return to what has already been discarded in philosophy of science. Yet, it can be shown that current work in design research on testing design theories and models by validation contains ample points of contact with also testing them by falsification.

A first example of a design research method for testing design theories and models is the validation research method developed and brought together by Blessing and Chakrabarti [23]. This research method, called DRM (Design Research Methodology) is aimed at helping design researchers at formulating and validating theories and models for understanding design and for supporting improvements. In terms of the terminology used in this chapter, DRM is a research

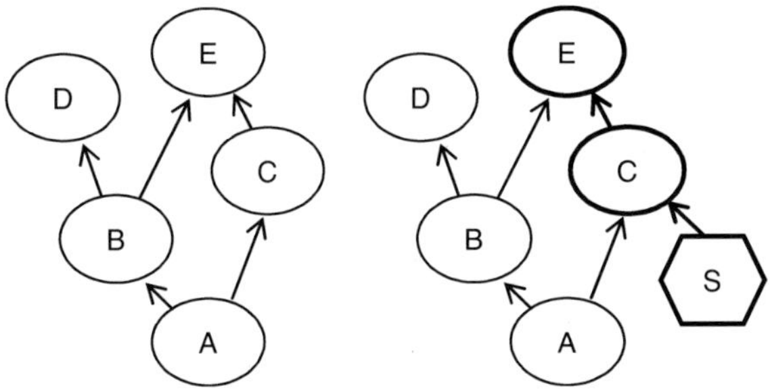

Fig. 2.1 Examples of a reference model (*left*) and an impact model (*right*) in DRM; the hexagonal node in the impact model represents the added support S and the *bold arrows* and nodes represent the causal chain leading to the improvement of E

method for formulating and validating descriptive and prescriptive design theories and models.

Abstracting from the many elements and subtleties, DRM consists of four stages: clarification, description of current design practices, formulation of the planned improvements, and testing of these improvements.[1] In these stages two models play a central role: the *reference model* and the *impact model*. Both these models are *networks of influencing factors*, containing nodes that represent factors, being aspects of design that influence other aspects of design, and containing arrows that represent how the factors causally influence each other (see Fig. 2.1). The reference model is representing the existing situation in design and acts as a benchmark to the improvements. The impact model represents the desired situation and adds the planned support for creating the improvements as an additional factor relative to the reference model.

The description of the causal influences in the reference model should be confirmed by a thorough literature study (which is one of the ways in which Blessing and Chakrabarti aim to overcome fragmentation; researchers are by DRM required to study the literature from all design research strands). And when information about influences between factors is lacking, this information should be created by empirical research. The supporting factor that is added to the impact model is meant to change one or more (downstream) factors in the reference model, and then to improve other (upstream) factors by means of the causal influences given in the reference model. This improvement may however not merely be assumed on the basis of the causal influences part of the reference model, but should be confirmed by separate empirical research.

A second example is the research method proposed by Seepersad et al. [27] for validating design methods by means of a *validation square*. By this proposal a design method is taken as constituted by *constructs*, which are loosely speaking the building blocks of the method, and as aimed at resolving design problems of a specific class. Yet, by the proposed research method a design method is not tested

[1] In DRM these stages are called, respectively, *Research Clarification*, *Descriptive Study I*, *Prescriptive Study* and *Descriptive Study II*. These stages comprise more tasks than described here; in the Descriptive Study II-stage, for instance, it is also evaluated whether designers can effectively manipulate the planned support for improving design practices [23].

for all the design problems in the class of problems the design method aims to resolve. Rather, a few *example problems* are introduced that have the characteristics of the class of design problems, and in that sense represent the design problems for which the design method is intended. Validation by means of the validation square consists then of four steps (these steps are graphically represented by the four quadrants of the evaluation square). First, the constructs of the design method are accepted separately and as an integrated whole. Second, the example problems are accepted as appropriately characterising the design problems the design method aims to address. Third, it has to be established that application of the design method to the example problems resolves them, and that this resolution is due to applying the method. Fourth, in a 'leap of faith' the usefulness of the design method should be accepted for all the design problems the example problems represent.

Finally, Frey and Dym [26] explore possibilities to validate design methods in a similar manner as medical treatments are evaluated. Frey and Dym are critical about taking the analogy too close; they observe, for instance, that it is hard to imagine what it would mean to let designers follow a 'placebo design method' without that the designers become aware of this. And the costs of letting a sufficiently large number of companies implement new design methods may be too high. Yet they propose to consider tests consisting of applying design methods on models of engineering design processes, in analogy with test of medical drugs on animal models, and to consider evaluating design methods by collecting data about the use of design methods in the 'field', in analogy with clinical trials.

What the first two validation research methods have in common is that they limit validation from confirmation of *all* claims of a design theory or model, to confirmation of only specific *key claims* of the design theory of model. For Blessing and Chakrabarti [23] these key claims are the causal influences of factors in their reference and impact models. For Seeparsad et al. [27] the key claims include that example problems are resolved. If these key claims of a design theory or model can be confirmed by research on design practices, the design theory or model is taken as validated. And if some key claims are not confirmed, then validation is absent, what may mean that these key claims can still be confirmed by further research on design practices, or that the key claims are in contradiction with design practices. In the latter case the conclusion should be that the design theory or model has to be rejected. The exploration of Frey and Dym [26], in turn, opens the possibility of a *step-wise* testing of design methods, first by applying them on models of engineering design processes, and then by data about their use in industry and academia, and possibly by other steps in analogy to the step-wise evaluation of medical treatments. This step-wise testing is already part of design research, when new design ideas and design theories are tested first in the design lab and then in the field, before being proposed in say product design (e.g., [31]). And again this step-wise testing may lead to rejection when a design theory or model does not pass one of the steps.

Hence, existing validation research methods for design theories and models are in addition to being useful for validation, also good starting points for falsification of the design theories and models. The validation research methods single out key claims of design theories and models, and testing of design theories and models may then consist of pilot studies in academia and industry in which these key claims are either confirmed or rejected. Hence, only a few design projects are to be carried out for testing a design theory or model, which is a lot more feasible also because they may be short projects that can be done under more controlled conditions.

2.5 Testing by (Naïve) Falsification: Towards Effectiveness

In the discussion of validation research methods the focus was on testing descriptive and prescriptive design theories and models. By the typology given in Sects. 2.2 and 2.3, design theories and models can also have the aim of demarcating design practices. So, let us consider design theories and models generally, and explore the possibilities to test them using Popperian falsification. The general advantage of falsification is that not all claims of design theories and models need to be considered, and not even all their key claims. Hence, testing more effective than validation is possible, be it testing that singles out provisionally acceptable design theories and models by elimination.

Descriptive design theories and models. A descriptive design theory is as a scientific theory that binds together knowledge of what we regularly take as design practices as they occur in the world, and that aims at understanding, explanation and prediction of and about these design practices. Descriptive models of design represent descriptive design theories, or define the content of these theories.

In principle it is clear how to falsify a descriptive design theory or model. If one practice that we regularly take as design does not fit the theory or model, or if one of these predictions is not correct, the design theory or model should be rejected as a descriptive one. This testing is possible already by considering more simple practices that are regularly taken as design; all other claims and predictions of a descriptive design theory or model about more elaborate design practices in industry can initially be ignored, thus avoiding waiting for the outcomes of years-long experiments before being able to say something about the acceptability of descriptive design theories and models. For instance, if one has for a descriptive design theory a reference model as defined by Blessing and Chakrabarti [23], key claims of that theory are represented as a network of influencing factors. A literature study or a simple empirical observation may then falsify one of these key claims and thus refute the descriptive design theory.

Demarcating design theories and models. For demarcating design theories and models the aim is to determine what practices can be taken as design, and falsification is in principle not possible. Demarcating design theories and models are similar to definitions and can as definitions only be rejected or falsified under specific further assumptions.

If it is assumed that a demarcating design theory or model singles out practices that are *essentially* design practices—a rather metaphysical assumption—then that theory or model is aimed at saying something about reality and can be true or false, just as an *essentialistic definition* can be true or false. Falsification is now possible: by finding one practice that is essentially a design practice but not identified as such by the demarcating design theory or model, or by finding one practice that is not essentially design, but identified as such by the theory or model, the demarcating design theory or model is refuted.

If it is assumed that a demarcating design theory or model singles out how design practices are identified in common or expert language, then that theory or model is aimed at saying something about how people give meaning to the term design. The theory of model can then again be true or false, as a *lexical definition* can be true or false. Falsification is possible: by finding one practice that is generally accepted as design but not identified as such by the demarcating design theory or model, or by finding one practice that is not accepted as design, but identified as such by the theory or model, the demarcating design theory of model is refuted.

Yet, if these assumptions are not made, then demarcating design theories and models have other overall goals. A demarcating theory or model may define design practices as part of a larger effort at analysing, say, engineering curricula or national economies. The goal of demarcation is then usefulness to that analysis, like the goal of a *stipulative definition*. Typically this usefulness implies that practices that are generally accepted as design practices are acknowledged as such by a demarcating theory or model. But if some borderline cases either fall inside or outside the demarcation, or if practices that typically are not taken as design become design, then the demarcating design theory or model is not refuted. Consider, for instance, Simon's [2] well-known characterisation of design:

> Everyone designs who devices courses of action aimed at changing existing situations into preferred ones. The intellectual activity that produced material artifacts is no different fundamentally from the one that prescribes remedies for a sick patient or the one that devises a new sales plan for a company or a social welfare policy for a state.

If this characterisation is meant as singling out the essence of design practices, metaphysicians may be called in for trying to falsify it. If the characterisation is meant as lexical, the second sentence of the citation provides ample reason to reject Simon's view. Yet if the characterisation is meant as stipulating what design is in the research program Simon envisaged, observing that prescribing medical treatment is typically not seen as a design practice, is beside the point; one may at most attempt to refute the characterisation of design by trying to argue that it is not particularly helpful to Simon's research program to take prescribing medical treatment as design.

Prescriptive design theories and models. Prescriptive design theories posit favourable properties of particular types of design, and can be falsified for that. Prescriptive models represent prescriptive design theories, or define the content of these theories.

Falsification of a prescriptive design theory or model is in principle possible. If there exists one design practice of the prescribed type that does not have the favourable properties posited, then a central prediction of the prescriptive design theory or model is not correct. One can then refute this theory or model as a prescriptive one. This testing is again possible already with more simple design practices in a controlled lab context; the predictions of a prescriptive design theory or model also concern long-lasting design practices in industry, yet with falsification the testing of prescriptive design theories and models is already possible without considering these claims. For instance, one can construct for a prescriptive design theory the example problems as defined by Seepersad et al. [27] and determine if designers following the prescription arrive at practices having the posited favourable properties. Or one can test a prescriptive design theory stepwise as envisaged by Frey and Dym [26] thus creating ample opportunities to test the theory by more controlled and simpler experiments. This testing by falsification of prescriptive design theories and models can again focus on their key claims, say when an impact model is constructed for it as defined by Blessing and Chakrabarti [23]. This impact model has the additional advantage for falsification by analysing the overall favourable properties of the prescribed design practices in terms of separate claims that supporting one factor of design has an effect on another effect. One can arrive at such separate claims also by spelling out in detail what types of design practices are prescribed by prescriptive design theories and models. For instance, if the methodological steps to be followed by design methods are formulated in a SMART manner, where SMART is the management acronym of *Specific*, *Measurable*, *Assignable*, *Realistic* and *Time-related* [32], then prescriptive design theories and models advance predictions about what each of these methodological steps has to result in [33]. These predictions can be tested separately, leading again to an effective way to determine the acceptability of prescriptive design theories and models.

2.6 Testing by Sophisticated Falsification: Towards Comparison

The position that evaluation of scientific theories takes place through only direct falsification has been abandoned in philosophy of science. The original position by Popper [29] was criticised by authors such as Lakatos [30], Kuhn [34] and Feyerabend [35], who considered more empirically how theories are accepted or rejected in science. The received view in philosophy is currently that science progresses by paradigm changes in which theories get replaced by other theories. These paradigm changes may still be motivated by empirical observations, yet theoretical argument and social processes play a decisive role as well, in part because observations are depending on theory and social considerations.

Given this abandoning of Popperian falsification, the analysis given in the previous section may seem spurious. Yet, as I argue in this section, the results of the philosophical critique of falsification may still be of interest to design research. Specifically Lakatos' criticism has resulted in a more sophisticated form of falsification, which opens up the possibility to design research to test design theories and models in comparison to each other, and thus to overcome fragmentation due to the validation of the theories and models in independent strands. Only if one holds that the critique has shown that theory evaluation is an entirely social affair in science, then it does not make sense to look further into what falsification could offer to design research. Yet, if this is the conclusion to be drawn from philosophy of science, also a larger part of design research work done on the validation of design theories and models becomes spurious. A social-constructivist view of evaluation undermines the very assumption that observations of design practices can validate design theories and models; on that view, it makes more sense to analyse validation sociologically as a negotiation process between design researchers. Hence, when current validation research methods in design research make sense, it equally makes sense to look at what sophisticated falsification as envisaged by Lakatos may mean for testing design theories and models.[2]

Lakatos [30] makes a distinction between *naïve* and *sophisticated falsification*. Naïve falsification comes close to the standard view of Popperian falsification: it concerns a single scientific theory T, which is falsified when one observational statement[3] is counterevidence to it. Lakatos argues that naïve falsification is typically circumvented in science: by adjusting auxiliary hypotheses in the *protective belt* of T, the *hard core* of T may be saved (later more about this protective belt and hard core).

What does happen in science, according to Lakatos, is what he calls sophisticated falsification. In sophisticated falsification one has two theories T and T', where the second is a rival to the first. Theory T is now falsified according to Lakatos if three conditions hold, which all refer to the rival theory T'. The conditions are that (1) T' has excess empirical content over T, (2) T' explains the previous success of T, and (3) some of the excess content of T' is corroborated

[2] This argument also holds if it is assumed that the observations that design researchers collect about design practices are interpreted as judgements of the designers carrying out the practices. The observations then concern social facts and are depending on social processes among designers. But these observations are not depending on social processes among the design researchers engaged in evaluating design theories and models. Hence, the observations can still be taken as objective facts, allowing an objective evaluation of design theories and models by only these observations, either by validation or falsification. On a social-constructivist view the judgements of designers about their design practices are also depending on the social processes taking place between the evaluating design researchers, undermining validation research methods as envisaged by, e.g., Blessing and Chakrabarti [23].

[3] I describe only a few elements of Lakatos' [30] analysis of falsification and ignore others. Lakatos, for instance, does not talk about observational statements simpliciter, but about 'observational statements', where the parentheses are reminders that observations in science typically depend on scientific theory.

[30]. Counterevidence to T still plays a central role in sophisticated falsification but this role is now twofold: it is part of the excess empirical content of T' over T, meaning that, first, T cannot describe the counterevidence whereas, second, the rival theory T' can. Counterevidence to T does thus not immediately falsify T; as long as there is no rival theory T' that can describe it, the counterevidence is merely set aside as *anomalies* to T.

With sophisticated falsification theory evaluation amounts to successions of accepted scientific theories and Lakatos captures this by talking about *scientific research programmes*. Such a programme consists of a series of scientific theories T, T', T'', T''' ..., and is characterised by a hard core of, e.g., basic assumptions and basic natural laws, and a protective belt of auxiliary hypotheses. Initial theories T in a scientific research programme typically are confronted by numerous counterevidence, which is all set aside as anomalies. Successor theories T' in the programme are meant to reproduce the empirical content of their predecessors T and to add novel content which includes some of these anomalies (this novel content is the excess empirical content referred to in the three conditions for sophisticated falsification). All theories in a scientific research programme share the hard core, and the auxiliary hypotheses are adjusted for letting successor theories have new empirical content that includes earlier anomalies. And by finding a successor theory T' with the right novel empirical content, the preceding theory T in the research programme is falsified without changing the hard core of the programme.

Scientific evaluation can also amount to the rejection of a scientific research programme as a whole, and again this takes place only when a rival scientific research programme is available. For this, both programmes should describe a common observational domain. If now a first scientific research programme after repeated attempts fails to generate theories that can describe counterevidence in the common domain, whereas the rival programme has theories available to do so, then the first is defeated (see [30], pp. 69–72).

In design research sophisticated falsification as described by Lakatos would mean doing tests in which design theories and models are taken as rivals and in which theories and models would be rejected relative to other more successful ones. This implication is probably sufficient to conclude that sophisticated falsification does not take place in design research. Design theories and models are proposed, studied, developed and tested relatively independently from each other, which leads to a rich variety of such theories and models (see [23] for a list) and an associated fragmentation of design research. Also there does not seem to exist a tradition in design research to let design theories and models compete. For instance, the question how to define the concept of function in design methods has been answered in different ways and an academic exchange at determining which answer is best is virtually absent [36], creating a development in design research in which theories and models proliferate.

Introducing sophisticated falsification into design research would create competition between design theories and models. It would direct testing towards collecting counterevidence to design theories and models, and it would direct research towards determining whether some of that counterevidence can falsify one theory or model in favour of another. It would raise the question of whether the current design theories and models can be ordered to belong to what may be called a single *design research programme* by sharing a hard core, and it would raise questions about how to evaluate such a design research programme as a whole.

Specifically, research on prescriptive design theories and models may benefit from such a development. There are design methods for engineering design that share common assumptions, say, first formulating design requirements as functional requirements, and then exploring design solutions that meet those functional requirements (e.g., [3, 4, 16]). Such design methods may be construed as prescriptive design theories and models that are part of the same design research programme, leading to the question of whether there is counterevidence in the design research literature that rejects one method in favour of another. Design methods that include the tool of reframing design problems (e.g., [37, 38]) may form another design research programme, and a third may consist of the emerging methods of design thinking (e.g., [7, 17, 28]), leading to the question of whether these programmes have a common domain of design practices by which it can be decided which of these programmes fare best in describing it.

With sophisticated falsification existing work on validating design theories and models can be used to bring design research to a more advanced level than merely separately testing all the claims advanced by the theories and models and then comparing the overall outcomes. With the example problems as defined by Seepersad et al. [27] and with the reference and impact models introduced by Blessing and Chakrabarti [23], already the key claims of, say, design methods can be used to determine the relative merits of these methods and associated design research programs. Hence, the comparison between methods can take place in a more fine-grained manner than the coarse-grained 'my method is better'-way alluded to by Reich [24].

The proposal to start testing design theories and models by sophisticated falsification clearly has a programmatic nature, since, as observed, it does not seem to take place yet. The benefits are clear, however, since doing it would mean a substantial step towards comparing design theories and models and towards overcoming the fragmentation that characterises current design research.

2.7 Conclusions

In this chapter I described how in philosophy scientific theories and scientific models are understood, and applied this understanding to design theories and models as generated in design research. A typology was introduced for distinguishing descriptive, demarcating and prescriptive design theories and models. All types of models of design fitted the philosophical understanding of scientific models, descriptive design theories may count as scientific theories, and demarcating and prescriptive design theories need not do so.

In this chapter I also considered research methods for testing design theories and models. In design research this testing is generally taken as validation, whereas philosophy of science also provides research methods for testing theories and models by falsification. Specific attention was given to sophisticated falsification as formulated by Lakatos to improve on (naïve) falsification associated to Popper.

The description of theories and models in design research and the survey of design research methods to testing them, was set against a discussion of the scientific status of design research. It was argued that using falsification to test design theories and models can help address two general concerns about this status, being a lack of effective research methods to test design theories and models, and a lack of coherence in design research. Popperian falsification gives design research means to more effectively test the claims of design theories and models. Lakatos' sophisticated falsification gives means to overcome fragmentation of design research by broadening testing to the comparison of design theories and models from separate research strands.

References

1. Ruse M (1995) Theory. In: Honderich T (ed) The oxford companion to philosophy. Oxford University Press, Oxford, pp 870–871
2. Simon HA (1996) The sciences of the artificial, 3rd edn. MIT Press, Cambridge
3. Gero JS (1990) Design prototypes: a knowledge representation schema for design. AI Mag 11(4):26–36
4. Pahl G, Beitz W, Feldhusen J, Grote KH (2007) Engineering design: a systematic approach, 3rd edn. Springer, London
5. McDonough W, Braungart M (2002) Cradle to cradle: remaking the way we make things. North Point Press, New York
6. Lawson B, Dorst K (2009) Design expertise. Architectural Press, Oxford
7. Verganti R (2009) Design driven innovation: changing the rules of competition by radically innovating what things mean. Harvard Business Press, Boston
8. Akao Y (ed) (1990) Quality function deployment: integrating customer requirements into product design. Productivity Press, New York
9. Hatchuel A, Weil B (2009) C-K design theory: an advanced formulation. Res Eng Des 19:181–192
10. Vermaas PE (2010) Beyond expert design thinking: on general, descriptive and prescriptive models. In: Dorst K, Stewart S, Staudinger I, Paton B, Dong A (eds) Proceedings of the 8th design thinking research symposium (DTRS8) Sydney, 19–20 Oct 2010. DAB documents, Sydney, pp 405–413
11. Frigg R, Hartmann S (2012) Models in science. In: Zalta EN (ed) The stanford encyclopedia of philosophy (Fall 2012 Edition). http://plato.stanford.edu/archives/fall2012/entries/models-science
12. Boon M, Knuuttila T (2009) Models as epistemic tools in engineering sciences. In: Meijers AWM (ed) Philosophy of technology and engineering sciences. Elsevier, Amsterdam, pp 693–726

13. Morgan M, Morrison M (1999) Models as mediators: perspectives on natural and social science. Cambridge University Press, Cambridge
14. Meijers AWM (ed) (2009) Philosophy of technology and engineering sciences. Elsevier, Amsterdam
15. Cross N, Christiaans H, Dorst K (eds) (1996) Analysing design activity. Wiley, West Sussex
16. VDI (1993) VDI Guideline 2221: Methodik zum Entwickeln und Konstruieren technischer Systeme und Produckte. Verein Deutscher Ingenieure, Düsseldorf
17. School D (2011) The d.school bootcamp bootleg. http://dschool.stanford.edu/wp-content/uploads/2011/03/BootcampBootleg2010v2SLIM.pdf
18. Bailer-Jones DM (2009) Scientific models in philosophy of science. University of Pittsburgh Press, Pittsburg
19. Hubka V, Eder WE (1988) Theory of technical systems: a total concept theory for engineering design. Springer, Berlin
20. Suh NP (1990) The principle of design. Oxford University Press, New York
21. Müller R (2009) The notion of model: a historical overview. In: Meijers AWM (ed) Philosophy of technology and engineering sciences. Elsevier, Amsterdam, pp 637–664
22. Birkhofer H (2011) Introduction. In: Birkhofer H (ed) The future of design methodology. Springer, London, pp 1–18
23. Blessing LTM, Chakrabarti A (2009) DRM: a design research methodology. Springer, London
24. Reich Y (2010) My method is better! Res Eng Des 21:137–142
25. Wallace K (2011) Transferring design methods into practice. In: Birkhofer H (ed) The future of design methodology. Springer, London, pp 239–248
26. Frey DD, Dym CL (2006) Validation of design methods: lessons from medicine. Res Eng Design 17:45–57
27. Seepersad CC, Pedersen K, Emblemsvåg J, Bailey R, Allen JK, Mistree F (2006) The validation square: how does one verify and validate a design method? In: Lewis KE, Chen W, Schmidt LC (eds) Decision making in engineering design. ASME, New York, pp 303–314
28. Brown T (2009) Change by design: how design thinking transforms organizations and inspires innovation. Harper Business, New York
29. Popper KR (1968) The logic of scientific discovery. Hutchinson, London
30. Lakatos I (1978) Falsification and the methodology of scientific research programmes. In: Lakatos I, Worrall J, Currie G (eds) The methodology of scientific research programmes. Cambridge University Press, Cambridge, pp 8–110
31. Koskinen I, Zimmerman J, Binder T, Redström J, Wensveen S (2011) Design research through practice: from the lab, field, and showroom. Morgan Kaufmann, Waltham
32. Doran GT (1981) There's a S.M.A.R.T. way to write management's goals and objectives. Manage Rev 70(11):35–36
33. Vermaas PE (2012) On managing Innovation by design: towards SMART methods. In: Hansen PK, Rasmussen J, Jørgensen KA, Tollestrup C (eds) Proceedings of the ninth norddesign conference, Aalborg University, paper no 67
34. Kuhn TS (1970) The structure of scientific revolutions. University of Chicago Press, Chicago
35. Feyerabend P (1975) Against method. Verso, London
36. Vermaas PE (2013) On the co-existence of engineering meanings of function: four responses and their methodological implications. Artif Intell Eng Des Anal Manuf 27:191–202
37. Cross N (2006) Designerly ways of knowing. Springer, London
38. Schön DA (1983) The reflective practitioner: how professionals think in action. Temple Smith, London

Chapter 3
A Structure for Design Theory

Neeraj Sonalkar, Malte Jung, Ade Mabogunje and Larry Leifer

3.1 Introduction: The Science-Practice Dichotomy

The field of engineering design research aims to study the activity of engineering design in order to improve it further. While engineering design practitioners could themselves work toward improving their practice through experience, the field of design research adds a component of systematic inquiry toward the development of robust, reliable tools and methods along with their underlying theories and models. This design research inquiry has been historically pulled in two opposing directions—toward scientific theories on one hand, and a greater relevance for professional practice on the other. These two directions are opposing because development of scientific theories drives research toward abstract conceptualization that has general validity while relevance to professional practice requires inquiry to be rooted in the pragmatics of particular situations.

Design researchers in the past have called for design research inquiry to be scientific. Over the decades, the perspective of what is scientific has included both the design activity itself and the inquiry on design. Researchers initially took the view that design activity itself could be scientific. Simon [26] in his influential work the "Sciences of the Artificial" proposed what he called "science of design." He considered the design process as a rational problem solving process that was amenable to scientific formalism and eventual embodiment into a computer program. This was concurrent with the development of first generation design methods that employed a rational approach to designing [2]. However in the 1970s and early 1980s, the real-world applicability of these methods became suspect with the acknowledgment of design problems as 'wicked' or ill structured [5, 24]. This

N. Sonalkar (✉) · M. Jung · A. Mabogunje · L. Leifer
Center for Design Research, Stanford University, Stanford, USA
e-mail: sonalkar@stanford.edu

A. Chakrabarti and L. T. M. Blessing (eds.), *An Anthology of Theories and Models of Design*, DOI: 10.1007/978-1-4471-6338-1_3,

led to the development of science of design, where researchers considered the inquiry into design to be scientific [7]. This is evident in calls by researchers to move the field toward development of scientific theories [8]. A number of design theories employing formal logic such as the General Design Theory [37], Axiomatic Theory [29], and C-K Theory [14] were developed. These theories have led to prescriptive methods for professional practice, but they have not yet been able to become relevant in the context of the artistry of professional practice that [25] mentioned.

The pull toward greater relevance for professional practice began when first generation design methods were found to be inadequate in real world situations. Recognizing that professional situations are often messy, [25] called for an epistemology of design practice that overcame the limitations of the model of technical rationality derived from the sciences and accounted for the skillful artistry of the practitioner. He proposed a framework of reflection-in-action that describes how designers create theories-in-action, perform on-the-spot experiments to test those theories and reframe to create new theories as "the situation talks back to them." Around the same time, researchers started conducting ethnographic studies of design practice to better understand its messy realities [4, 13]. The 1990s and 2000s witnessed a rise in descriptive studies of design practice using methods such as ethnography, video interaction analysis and conversation analysis derived from the social sciences [18, 20, 33–35]. While these studies advanced the analysis of design practice, the flow of knowledge remained in one direction—from practice to research. Descriptive studies rarely influenced design practice. Blessing et. al [3] pointed out that reasoning based on experience and logical argumentation were more common than descriptive studies as starting points for development of methods and tools.

Thus, the science-practice dichotomy has been playing out in the field of design research with the increasing appearance of scientific theories that miss the realities of professional practice, and of descriptive studies that lack relevance for practitioners. It needs to be noted that theories such as Hatchuel and Weil's C-K theory and Suh's Axiomatic theory have had an influence on practice (see Agogue et al. in this book). But this influence often takes the form of rule-based methods that constrain the moment-to-moment artistry of professional practice. If design research were concerned only with study of design activity, it would have been acceptable to develop scientific theories, or conduct descriptive studies. However, since the improvement of practice is the eventual objective of design research, the theories that are developed need to be rooted in the pragmatics of professional practice that are dealt with by descriptive studies. In this chapter, we propose a structure for design theory that retains scientific formalism while enabling design practitioners to use the theory in their ongoing "conversation" with the evolving design situation as described by [25].

3.2 A Structure for Design Theory

3.2.1 Why Focus on the Structure of Theory

Following Dörner [9], we consider a theory to be a formulation that explains a phenomenon, and a model as an abstraction that simulates a phenomenon. Simply put, models do things while theories explain things. Given this purpose, what could a theory look like? What is its structure?

These questions are relevant in design research, because theory is a term that is often misunderstood. For example, is Pahl and Beitz's [22] prescription of design method, a theory or a model? Given Dörner's definition, it fits the description of a model in that it prescribes the design process but does not give an explanation of why the prescription is the way it is. However, researchers [14] include it in their review of design theories.

The discourse on theory in the field of organizational behavior provides a clue toward why this misunderstanding may exist. Sutton and Staw [32] in their paper 'What theory is not' pointed out the following five ways in which authors tended to confuse the term theory.

1. References as theory
2. Data as theory
3. Variables or constructs as theory
4. Diagrams as theory
5. Hypotheses or predictions as theory.

They acknowledged that theory needs to answer the question why and none of the above five elements answer that question. However, Weick [36] in response to this article pointed out that we need to consider not just theory, but the process of theorizing. Perhaps references, data, constructs, diagrams, and hypotheses could be part of the theorizing process—developments on the way to a complete theory. Similarly, the models, frameworks, and principles that abound in design research could be considered as elements on the way to a theory rather than theory themselves. So what is the end form of the process of theorizing in design?

In order to answer this question, we first examine the structure of scientific theory and the perception–action perspective from ethnographic research.

3.2.2 Structure of Scientific Theory

Our examination of scientific theory derives from the field of philosophy of science in which the structure of scientific theory has been a topic of continuing discourse. The structure of scientific theory has been described in terms of two different views, viz.

1. The received view of scientific theory
2. The semantic view of scientific theory.

The received view of scientific theory

This perspective defines a three-part structure for scientific theory. The first part deals with logical formalism, the second part describes observable constructs and the third part describes theoretical constructs. The three parts are connected by rules of correspondence that hold the mathematical, observable, and theoretical constructs together. See Frederick Suppe's "The structure of scientific theory" [30] for a detailed account of the received view as well as its historical development. The received view of scientific theory has been heavily critiqued and rejected by most philosophers of science [31]. The main reasons for its rejection were the difficulty in conceptually separating physical constructs into observable and theoretical dimensions, the difficulty in creating correspondence rules between formal logic descriptions and observable constructs, and the rigid structure that admitted only logical calculus of the first order to accept as formal descriptions of theory (ibid). The received view of scientific theory gave rise to the semantic view that adopted a less rigid approach to theory description.

The semantic view of scientific theory

Craver [6] specifies the semantic view or the model view of theories as the idea that theories are abstract extra-linguistic structures quite removed from the phenomena in their domains. In this view, theories are not associated with any particular representation. Researchers have a much greater freedom than in the received view to describe their theory in terms of a series of models that explain a set of phenomenon through abstraction constructs that constitute the theory.

Given these two views, we believe that the received view could be adapted to create a structure for design theory. We choose the received view over the semantic view as a point of departure even though the received view has been heavily criticized because of the following reasons.

1. The multiple dimensional structure of the received view is well suited for accommodating both the theoretical constructs and the relationships between them, and the perception–action component that makes the theory relevant to professional practice. The perception–action component is described in greater detail in the following sections.
2. The factors for which the received view was criticized can be overcome in a formulation that avoids rigid definition of logical formulation, and ambiguous separation between theoretical constructs and observable constructs.
3. The existing theories in design research such as C-K Theory, Axiomatic Theory etc., are more suited to formulation in the received view than in the semantic view.
4. The theories of classical physics that underlie engineering analyses as expressed in the received view are more intuitive to engineering design researchers and practitioners.

3.2.3 The Perception–Action Perspective

From an examination of the structure of scientific theory, we move on to examine the elements of professional practice. Our examination is based on Goodwin's description of professional vision [11, 12] and Ingold's [16] study of skill in the development of technology.

Goodwin [11] takes the example of field archeology to describe how perception and understanding of events is socially organized by professional groups into what he calls professional vision. Professional practice is considered to be "a temporally unfolding process that encompasses both human interaction and situated tool use." It is this tool use and the associated human interaction that helps professionals develop a perceptual field that highlights information relevant to a particular professional practice. In the field archeology example, Goodwin points to the Munsell color chart that is a tool used by professional archeologists to detect the color of dirt during digs. The chart is a tool that encodes the theory and practice of previous workers and provides a perceptual aid to highlight information that might otherwise be hidden. The artistry of professionals then lies in the ability to use sophisticated perceptual fields, decode the information that is highlighted and respond to the situation unfolding in front of them. Professionals develop their own action repertoire corresponding to the perceptual fields they encounter, their previous experience, and the theories that underlie professional education.

Ingold [16] too spoke about the role of perception and action in the development of technical skill. He regarded technical processes such as engineering design not as products of intelligence, but as practices of skill, where skill was defined as the coordination of perception and action. Ingold argued that if we want to understand a skill we need to "shift our analytic focus from problem-solving, conceived as a purely cognitive operation distinct from the practical implementation of the solutions reached, to the dynamics of practitioner's engagement, in perception and action, with their environments."

Given the recognition of perception and action in the development of skill relevant to domains such as engineering design, we believe that design theory needs to include a perception–action component that would enable it to be directly relevant to the artistry of professional practice. Combining this perspective with the examination of the structure of scientific theory, we propose the following structure for design theory.

3.2.4 A Structure for Design Theory

We propose a structure with two dimensions.

1. The event-relationship dimension
   The event-relationship dimension mentions the event or sequence of events that the theory attempts to explain by providing the definition of theoretical

constructs—i.e., variables and operators, and the relationships between them. These relationships can be expressed either in natural language such as English or in formal logic.

2. The perception–action dimension
   The perception–action dimension describes the perceptual field and the action repertoire associated with the theoretical constructs in situations of professional practice. Perceptual field and action repertoire were defined in our earlier work [17] as follows.
   We define a perceptual field as sensing organized around a purposeful activity. With the notion of a perceptual field we want to refer to what one notices when one is engaged in the activity of designing. This noticing can refer to things in the environment or to internal states and feelings. A perceptual field can, for example, be set up through disciplinary training, or through certain media that make specific characteristics salient. Re-framing in terms of a perceptual field means shifting from one perceptual field to another perceptual field. This idea of perceptual field is captured in the idea of a "model" or "point-of-view" of the world.
   Analogous to a perceptual field defined with respect to sensing, we define an action repertoire as organized movement within a purposeful activity. With an action repertoire we want to refer to the choices from a corpus of behaviors a designer has when engaged in the practice of designing. An action repertoire can be seen as the corpus of behaviors a designer has at his or her disposal when engaged in a conversation with the situation.

The two dimensions—event-relationship and perception–action characterize the phenomenon the theory explains but in different ways. The event-relationship dimension gives a logical formulation of the involved parameters and their interrelationship, while the perception–action dimension provides the elements necessary for practitioners to actually perceive and act on the parameters involved. It needs to be noted that merely formulating a design theory in this structure does not make it scientific. Following Popper's [23] falsification principle for a theory to be scientific, researchers need to infer hypotheses that test the theory by being amenable to falsification. The two-dimensional structure for design theory provides an opportunity for the necessary scientific testing while being rooted in the perception–action of professional practice. The following section gives an illustration of the proposed structure for design theory as applied to C-K theory.

3.3 Example: C-K Theory Formulated in the Two-Dimensional Structure

The following formulation is derived from the exposition of C-K theory as given in Hatchuel and Weil [14, 15]. The formulation is presented in normal font while our comments are given in italics.

3.3.1 The Event-Relationship Dimension

Event explained: The generation and development of concepts in design practice.

C-K theory as postulated by Hatchuel and Weil [14] *claims to be a unified design theory. However, using a two-dimensional design structure with a perception–action component necessitates a more specific formulation. Hence, we formulate C-K theory not as a unified theory of the entire design activity, but a theory of generation and development of concepts in design practice. This implication of the two-dimensional structure is discussed further in* Sect. 3.4.

Construct definitions:

- Knowledge Space (K)—The space of propositions that have a logical status for a designer D.
- Logical status—An attribute that defines the degree of confidence that D assigns to a proposition. Logical status in standard logic can be true or false. K space propositions are assigned a logical value true/false by the designer D.
- Concept Space (C)—The space of propositions that have no logical status in K.
- Design—The process by which concepts generate other concepts and are ultimately transformed into K space, i.e., propositions that have a logical true/false status.

Further characterizations of C and K spaces as mentioned in Hatchuel and Weil [14] *are not included here, as we do not intend to give a detailed account of C-K theory but use it to illustrate theory structure.*

Relationships:

C and K are mutually exclusive sets that taken together describe the entire universe of propositions expressed when designing. Design progresses because C and K exist. If either ceases to exist, design would itself cease.

Propositions expressed during designing undergo transformations from C to K, C to C, K to K, and K to C.

External operators—Operators transforming propositions between C and K.

- C –> K: Propositions in C are transformed into K when they acquire a logical status of true/false. Hatchuel and Weil describe this operator as corresponding to validation tools or methods such as consulting an expert, doing a test, conducting an experiment, building a physical mock-up or prototype and testing it.
- K –> C: Propositions in K can be extended into C space when they pick up attributes that do not have logical status through the process of concept generation. For example a car has logical status in K space, but a car powered by cold fusion is a proposition in the C space because it has picked up the attribute of cold fusion that itself does not have a true/false status.

Internal operators—Operators transform propositions within C and within K.

- C –> C: Propositions in C can be partitioned further or combined together to form new propositions.
- K –> K: Propositions in K can lead to new propositions in K through knowledge derivation processes such as induction and deduction.

3.3.2 Perception–Action Dimension

This dimension describes the pragmatic situation, i.e., sets of perceptual field and action repertoire that correspond to the theoretical constructs (relevant events and relationships) explained above. For a given set of theoretical constructs there could exist multiple sets of situation—perceptual field—action repertoires depending on the interactions that typically occur in professional practice. We next describe the perceptual field and action repertoire related to the situation of group concept generation.

Situation: A team of engineering designers generating concepts through conversation. The team is situated in a room equipped with whiteboard, markers, table and chairs, and paper for sketching.

Perceptual field of individual designers: Both C and K are articulated by individuals in the group in the form of verbal and non-verbal elements of the conversation. The perceptual field enables designers to distinguish C from K. In our prior work [27], we have identified the certainty expressed through language as an indication of K space and use of conditional language as an indication of C space. Hence, perceptual field for C consists of the following.

1. Expressions that are explicitly called out as concepts either in verbal form or are written down in an explicit list on whiteboard or paper.
2. Expressions indicated by use of conditional terms—could, might, maybe, if. In some cases "would" might also imply conditionality.
3. Expressions indicated by use of 'should' coupled with a conditional term—"maybe we should" or used as a question—"should we do it that way?"
4. Expressions indicated by generative design questions [10]—e.g., scenario creation or proposal creation questions like—"how about…?", "what if…?", "what about…?"
5. Expressions indicated by use of analogies to indicate possible alternatives…"something like Mr. potato head".

The perceptual field for K consists of the following expressions that indicate certainty on part of the individual making that expression.

1. Personal or team narrative—something that happened in the past to an individual or team.
2. General knowledge—something that is accepted to be true, e.g., principles of how a mouthwash works.
3. Personal opinion—opinion including likes and dislikes that were expressed by an individual.
4. Project requirements—expressions related to project requirements.

5. Process—comments on the ideation session going on at the moment.
6. Future certainty—expression of what should happen in the future, rather than what could possibly happen.

The perceptual field enables C and K to be experienced by individual designers in the given situation. Concepts and Knowledge are no longer just abstract constructs but real entities that can be perceived and responded to.

Action-repertoire: Action-repertoire for C-K theory lists the actions designers can take to progress in generation of concepts and their conversion to knowledge. The following list of actions is derived from our previous work [27] on concept generation in engineering design teams.

1. Introduce C or K expressions—Introducing new C or K expressions in a conversation can occur in response to questions by others in a team or as a result of new concepts generated at an individual level.
2. Support C or K expression—Expressing support to C or K expressions given by team members can take the form of nodding and expressing approval, giving a compliment to a C expression, or articulating a narrative that lends further support to a K expression.
3. Build on C expressions—Building on C expressions occurs when an individual listens to a C expression by a team member and adds new attributes to it or develops an analogous C expression. This is the C –> C operator discussed in the theoretical constructs dimension.
4. Blocking C or K expressions—Blocking C or K expression can occur with an individual expressing doubt, disapproval of a C expression, or an opposing K expression.
5. Negotiate blocks to generate new C or K—When a block is given, it needs to be resolved for the conversation to proceed further. This resolution occurs when teams accept the block and reject the C or K that is creating conflict, or generate new C that overcomes the opposition or disapproval, or bring in new K that dissolves the opposition into accepting the C or K in conflict.
6. Introduce questions to generate C or K—Generative design questions direct the conversation to generate new Cs and deep reasoning questions direct the conversation to generate new Ks [10].
7. Generate commitment to evaluate C into K—Individuals can generate consensus on which concepts to prototype and develop further through planning, question asking or encouraging action, e.g., "Let's meet today evening to make a prototype."

The inclusion of an action-repertoire enables designers to not just perceive C and K but also develop the skill of responding to them in situations of professional practice. The following example shows the unfolding of perception–action in a concept generation conversation with the help of the *Interaction Dynamics Notation. Interaction Dynamics Notation* [27] *is a visual notation that captures the C-K dynamics of group conversation. C expressions are captured in red color, K expressions in black color* (Table 3.1).

Table 3.1 An excerpt of conversation of four engineers engaged in concept generation taken from Sonalkar [27]

Transcript
Engineers A, B, C, and D are engaged in generating concepts for a new dental hygiene product
B: So how about sandpaper, we haven't talked about sandpaper much. Like, I guess that's-
C: I feel like whenever I listen to sandpaper I have this really negative mental image of like grinding (B says "Oh yeah" and starting making a grinding sound) away the enamel of your teeth
D: Yeah
B: So what about-
C: So-
B: But-
D: Think of it more as like a loofah
C: Yeah (laughs)
A: But I mean, that doesn't mean we don't have to use real sandpaper obviously something some kind of like (C says "Yeah I know") an abrasive on a sheet or something
B: Let's improve on floss, like the ribbon that you slide (gesturing flossing)
C: It could be floss, sandpaper floss that's a good thing
A: (says something simultaneous with C) slightly abrasive ribbon, yeah
D: Yeah
Interaction dynamics notation of the above conversation (Red indicates expressions in C)
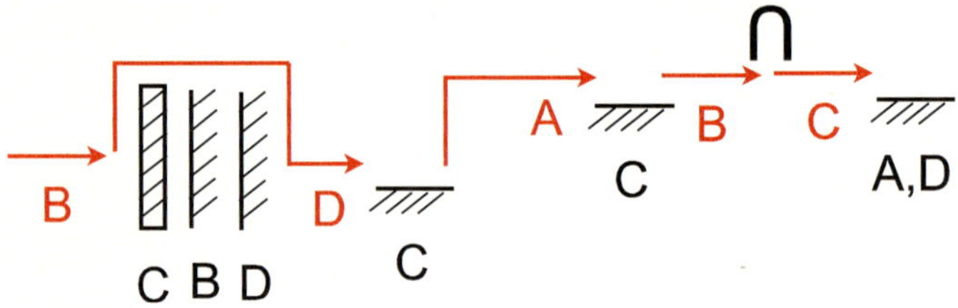
Perception–Action comments
B introduces a new concept "how about sandpaper" into the conversation. However person C expresses dislike and gives a block (indicated by the barrier in the notation). B and D express support to the block. However, D perceives the block to the concept and tries to negotiate the block by person C by proposing an alternative perspective "think of it as a loofah." This is indicated in the notation above by the symbol that goes over and around the barrier. Person A accepting C's dislike tries to introduce another concept, something that is not sandpaper, "an abrasive on a sheet or something." B introduces the concept of something like a floss, person C perceives the new concept expression and builds on it to suggest a sandpaper floss as indicated by the inverted 'U' symbol. The notation is given here as an illustration of a visual C-K representation that captures moment-to-moment dynamics of the situation. The notation and its development are described in detail in Sonalkar et al. [28]

3.4 Implications of the Structure for Design Theory

Proposing an explicit structure for theory formulation with a perception-action dimension in design research has the following implications.

1. From grand theories to specialized theories—The inclusion of perception–action component, which is rooted in situations of professional practice, precludes the formation of grand unified theories of design. If we were to develop a

unified theory of the entire design process, then the perception–action dimension would include all activities that engineering designers perform and it would be prohibitively complex. Instead of grand unified theories, the proposed two-dimensional structure for design theory encourages researchers to develop theories pertaining to specific situations encountered in professional practice. Thus, we could develop a theory of prototyping, a theory of concept generation, a theory of concept evaluation, a theory of requirements analysis etc. We believe that this would enable design researchers to address the complexities of specific situations to develop logically sound theories grounded in professional practice that are amenable to scientific testing.

2. No more pseudo theories—If we know the destination, we would know when we have reached it. If we don't know what the end goal of theory building is, there is a greater possibility of creating pseudo theories by confusing one of the interim stages with the theory itself. Accepting a common structure for design theory has the implication of precluding the development of pseudo theories. Researchers would now have a benchmark as to what is a theory and what is not. Thus, when they are publishing the artifacts of their theorizing that are not yet theory, they could be more open and inviting to receive constructive feedback that could accelerate the development of a specific design theory that is appropriate to their finding.
3. Collaboration between formal theory researchers and observational studies researchers—Including both a perception–action dimension and an event-relationship dimension in the formulation of a design theory implies that researchers conducting ethnographic studies and researchers who traditionally focused on formal theories need to work together. Just developing theoretical constructs or just describing the perception–action in professional practice is not enough. The interplay between formal theoretical constructs and actual empirical events holds the promise to resolve the science-practice dichotomy and create a more cohesive, focused research community.
4. More resources needed to build theories—With the need for greater collaboration between researchers to build theory comes the need for greater resources. Theory building for the two-dimensional design theory would be more expensive in terms of both money and time requirements than building only a formal design theory. However in long term, with greater collaboration and more focused research efforts in the field, the resources required for the field of design research to progress would perhaps be less than those required if the current situation of division and isolation in the field continues.
5. Practitioners can contribute to theory building—With the inclusion of perception–action dimension in design theories, reflective practitioners can make significant contributions to design research. Theory need no longer be a pejorative term of abstract meaningless thinking for practitioners. It can become an indication of intellectual rigor in one's practice that goes hand-in-hand with increased skill and we hope better design outcomes.
6. Encouragement of adaptive design expertise—Most prescriptive models such as Pahl and Beitz [22] that scaffold design practice mention a procedure that

designers need to follow. They do not encourage or even acknowledge the skillful artistry of professional practice that involves perceiving and responding to the situation at hand. Including a perception–action dimension in design theories could promote the acknowledgment and training of the skill needed for professional practice. Neeley [21] defined adaptive design expertise as the ability to combine active engagement with reflective thinking in order to mindfully adapt to a design situation. The two-dimensional structure of design theory that includes both the reflective and the active part could encourage the development of such adaptive design expertise.

3.5 Critique of the Design Theory Structure

The proposed two-dimensional structure is an outcome of our efforts to resolve a perceived dichotomy between the drive toward formal theories with greater scientific rigor, and our experience of design practice that rarely relies on the tools and methods of design research. In 2010, we had proposed the development of a separate category of theories called the perception–action theories [17]. However, we realized that it does not help resolve the science-practice dichotomy. It helps practice oriented researchers create their own set of theories. But this could propagate further divisions within design research without leading to collaborative cohesive efforts to bring in relevance to design practice. The two-dimensional structure for design theory is a step in that direction. However, it raises several concerns, three of which will be discussed here for illustrative purposes.

3.5.1 Integrity of Perception–Action Dimension

If theory is a formulation that explains, then should perception–action be part of it? The perception–action dimension does not explain, but rather gives reflection of the theoretical constructs in situations relevant to practice. So should it not be considered an application of theory?

We argue that perception–action dimension needs to be an integral part of design theory based on (1) the nature of design activity and (2) the purpose of design theory.

(1) Design theories where human agency is a factor are not the same as natural science theories, e.g., the big bang theory in physics, where there is no human agency. Humans are subject to biological and other physical constraints but are not determined by these forces. The two-part theory is consistent with theories in the field of feedback control—where we include humans in the control loop (see the cybernetics systems perspective described in Maier et al. in this book). The perception–action dimension accounts for the human agency in design and hence is an integral part of design theory.

(2) The purpose of theory is to explain a phenomenon. However, in designing its theories a field of study needs to also consider whom this explanation needs to be given to. We believe that the field of engineering design research needs to develop theories that are accessible by and meaningful to both researchers and practitioners. The perception–action dimension enables this by letting the theory be rooted in situations relevant for professional practice. The perception–action dimension also sharpens the theory because it now proposes perceptual fields that theoretical constructs need to be compatible with. There is much tighter coupling between logical relationships and the situational relationships of constructs that design theory uses to explain phenomenon.

3.5.2 Defining a Boundary for the Perception–Action Dimension

Does the perception–action dimension have a boundary in terms of the kinds of situations that need to be included? Without a defined boundary, this dimension could remain open-ended and researchers could simply keep adding perceptual field and action repertoires to a design theory.

The boundary of the perception–action dimension depends on how the constructs of the theory are operationalized in professional practice. The boundary would be better defined in theories that are context specific, and ill-defined in theories that attempt to explain multiple disparate situations in the design process.

The perception–action dimension is grounded in an interaction perspective of the professional world with the underlying assumption that professional practice is socially and materially constructed. It is the interactions that we have with the people around us and with the tools in our environment that define our practice. The kind of interactions we have can themselves be sufficiently abstracted to prevent an overloading of perceptual fields and action repertoires on a design theory. The development of the Interaction Dynamics Notation [27] that abstracts concept generation conversations into a visual notation gives an indication of the methodology through which this abstraction could be achieved. Thus in the example of C-K theory mentioned in this article, we need not mention perceptual fields and action repertoires for all different concept generation conversation occurring in various parts of the world. Instead one set of perceptual field and action repertoire is sufficient to cover such conversational situations.

3.5.3 The Tension Between Generality and Specificity

As mentioned in the implications section, the proposed structure for design theory is biased against the formulations of grand unified theories of design. However, a theory by convention needs to be sufficiently general enough in its explanatory power to cover a number of unique cases of practice. Does this not create a tension within a theory to be both general enough in its theoretical constructs dimension and specific enough in its perception–action dimension?

We believe it is important to distinguish between bounding the phenomenon that a theory attempts to explain and the generality of that explanation. An example of bounding the phenomenon is that we build a theory of prototyping that is a component of the design process, rather than a theory of the entire design process. However, this does not restrict the generality of the explanation that the theory provides for the prototyping phenomenon. Different prototyping situations could be explained by such theories. The development of the perception–action dimension could encompass such different situations. As mentioned above, the interactions that occur during prototyping could be sufficiently abstracted to develop relevant perceptual fields and action repertoires.

3.6 Summary

In this chapter, we have described a science-practice dichotomy in design research that is manifest in a drive for scientific rigor and formal theories on the one hand and a greater relevance to professional practice on the other. We pointed out the need to not only create prescriptive models for practitioners, but also recognize and enable their artistry in dealing with the messy realities of practice. To enable this, we proposed a two-dimensional structure for design theory that includes an event-relationship dimension and a perception–action dimension. We explained the origin of this structure, its description, its implications, and commented on some of the concerns it could raise for design researchers. We hope that the proposed structure for design theory will start a discussion among design researchers that will eventually lead to better design theories with greater relevance to design practice.

References

1. Agogue M, Kazakci A (2013) 10 years of C-K theory: a survey on the academic and industrial impacts of a design theory. In: Chakrabarti A, Blessing L (eds) An anthology of theories and models of design. Springer, London (in press)
2. Bayazit N (2004) Investigating design: a review of forty years of design research. Design Issues 20(1):16–29
3. Blessing LTM, Chakrabarti A, Wallace KM (1998) An overview of descriptive studies in relation to a general design research methodology. The Key to Successful Product Development, Designers, pp 42–56
4. Bucciarelli LL (1984) Reflective practice in engineering design. Des Stud 5(3):185–190
5. Buchanan R (1992) Wicked problems in design thinking. Des issues, pp 5–21
6. Craver CF (2002) Structures of scientific theories. The Blackwell Guide, p 55
7. Cross N (1993) Science and design methodology: a review. Res Eng Design 5(2):63–69
8. Dixon JR (1987) On research methodology towards a scientific theory of engineering design. Artif Intell Eng Des Anal Manuf 1(3):145–157

9. Dörner D (1994) Heuristik der Theorienbildung. Enzyklopädie der Psychologie 1:343–388
10. Eris O (2002) Perceiving, Comprehending, and Measuring Design Activity through the Questions Asked while Designing. Ph. D. Thesis, Stanford University
11. Goodwin C (1994) Professional vision. Am Anthropologist 96(3):606–633
12. Goodwin C (2000) Action and embodiment within situated human interaction. J pragmatics 32(10):1489–1522
13. Hales C (1987) Analysis of an engineering design process in an industrial context. Department of Engineering, University of Cambridge
14. Hatchuel A, Weil B (2003) A new approach of innovative design: an introduction to CK theory. Paper presented at the international conference on engineering design, Stockholm, Sweden
15. Hatchuel A, Weil B (2009) CK design theory: an advanced formulation. Res Eng Des 19(4):181–192
16. Ingold T (2001) Beyond art and technology: the anthropology of skill. Anthropological perspectives on technology, pp 17–31
17. Jung M, Sonalkar N, Mabogunje A, Banerjee B, Lande M, Han C, Leifer L (2010) Designing perception action theories: theory building for design practice. Paper submitted to the 8th design thinking research symposium (DTRS8), Sydney
18. Mabogunje A (1997) Measuring conceptual design process performance in mechanical engineering: a question based approach. Ph. D. Thesis, Stanford University
19. Maier A, Wynn D, Howard T, Andreasen M (2013) Perceiving design as modelling: a cybernetic systems perspective. In: Chakrabarti A, Blessing L (eds) An anthology of theories and models of design. Springer, London (in press)
20. Minneman S (1991) The social construction of a technical reality: empirical studies of group engineering design practice. Ph.D. Thesis, Stanford University
21. Neeley Jr W (2007) Adaptive design expertise: a theory of design thinking and innovation. PhD. Thesis, Stanford University
22. Pahl G, Beitz W (1986) Engineering design. The Design Council, London
23. Popper KR (1963) Science as falsification. Conjectures and refutations, pp 33–39
24. Rittel HW, Webber MM (1973) Dilemmas in a general theory of planning. Policy Sci 4(2):155–169
25. Schön DA (1983) The reflective practitioner: How professionals think in action. Basic books, New York
26. Simon HA (1969) The Sciences of the Artificial. MIT Press, Cambridge, MA
27. Sonalkar N (2012) A visual representation to characterize moment-to-moment concept generation through interpersonal interactions in engineering design teams. PhD Thesis, Stanford University
28. Sonalkar N, Mabogunje A, Leifer L (2013) Developing a visual representation to characterize moment-to-moment concept generation in design teams. Int J Des Creativity Innov 1(2):93–108
29. Suh NP (1990) The principles of design, vol 990. Oxford University Press, New York
30. Suppe F (ed) (1977). The structure of scientific theories. University of Illinois Press
31. Suppe F (1972) What's wrong with the received view on the structure of scientific theories?. Philosophy of Science, pp 1–19
32. Sutton RI, Staw BM (1995) What theory is not. Adm Sci Q, PP 371–384
33. Tang J, Leifer L (1991) An observational methodology for studying group design activity. Res Eng Design 2(4):209–219
34. Ullman D, Herling D, Sinton A (1996) Analysis of protocol data to identify product information evolution and decision making process. In: Cross N, Christiaans H, Dorst K (eds) Analysing design activity, Wiley and Sons, pp 169–185
35. Valkenburg R (2000) The reflective practice in product design teams. Delft University of Technology, Delft
36. Weick KE (1995) What theory is not, theorizing is. Adm Sci Q 40(3):385–390
37. Yoshikawa H (1981) General design theory and a CAD system. Man-Machine Communications in CAD/CAM

Chapter 4
Motive of Design: Roles of Pre- and Post-design in Highly Advanced Products

Toshiharu Taura

4.1 Introduction

As we have seen, technology has become highly advanced. Our living environment is filled with products. Thus, we face the following issues: How can we create excellent products that do not yet exist? How can we accept high risks that may cause significant damage to human beings? To address these issues, the author focuses on the conception of *motive of design*. We must reexamine the origins of design to gain a critical understanding of these fundamental issues. Here, the author uses the term, ***motive,*** in the context of design, to imply "an underlying reason for the design of a product." This statement refers to the definition, "a reason for doing something" provided by the online Oxford Dictionaries [1]. We must distinguish the term ***motive*** from ***motivation,*** which is defined as "the desire or willingness to do something" as per the online Oxford Dictionaries.

In general, the term "motive" is considered a conception that can be applied to one individual. However, in addition to the *personal motive* of the designer, individuals' feelings and criteria for products, or awareness of a *problem* that is overtly or covertly created and contained in society, may also be a reason (*motive*) for the design of new products. Hence, the author proposes the conception of *social motive*. In this chapter, the ***social motive of design*** is defined as an individual's feelings and criteria for products, or awareness of a *problem* that can be shared in society; the ***personal motive of design*** is defined as an individual's feelings and criteria for products, or awareness of a *problem* that exists deep in a designer's mind. Further, the ***motive of design*** is defined as a conception that includes both the *social motive of design* and the *personal motive of design*. Hereafter, the *social motive of design*, the *personal motive of design*, and the *motive of design* are described more simply as ***social motive***, ***personal motive***, and ***motive***, respectively.

T. Taura (✉)
Kobe University, Kobe, Japan
e-mail: taura@kobe-u.ac.jp

A. Chakrabarti and L. T. M. Blessing (eds.), *An Anthology of Theories and Models of Design*, DOI: 10.1007/978-1-4471-6338-1_4,

The ***motive of design*** differs from so-called "needs." Needs are a part of the ***motive of design***. However, "design" occurs in situations in which needs do not yet exist. The *motive of design* exists underneath the "needs" and drives them. Further, in this chapter, we do not address the guidelines provided by an organization that govern designers' behaviors within that organization (e.g., a company). Rather, we discuss factors that operate behind these guidelines that implicitly encourage the product design process.

In March 2011, Japan experienced a nuclear power plant disaster. The nuclear reactor or power supply equipment was designed by each nuclear power plant design company according to given specifications. However, positive or negative feelings (*social motive*) about nuclear power generation might have been premature and were not made explicitly clear to society. The nuclear power plant had already been "designed." However, the fundamental reason why we create and use a product that presents extremely high risk had not been clarified. Should this discussion be included in the "design" process? If so, what types of design issues must be examined in the creation of a nuclear power plant?

On the other hand, some advanced technologies (e.g., some components of the Hard Disk Drive) can be considered "products that can be produced by only a limited number of companies." In many cases, these products are not developed based on concepts that users can deduce or induce (*social motive*). Thus, we must ask, "Why was this product designed?" It is assumed that these advanced products are developed on the basis of a designer's highly creative and mysterious thoughts.

The two cases mentioned above appear to have completely different issues. However, the author believes they both can benefit from a discussion on the same theme: *motive of design*; although the *motive of design* has not yet been explicitly discussed.

In the following sections, the author first outlines the framework of the design process used in capturing *motive*. Second, the relationship between *personal motive* and *social motive* is illustrated from the viewpoints of *inner sense* and *problem*. Third, both the designer's and the consumer's *inner sense* is discussed as an aspect of *motive of design*. Fourth, the *latent functions* of products are discussed as an additional aspect of *motive of design*. Finally, this chapter describes how the findings and considerations related to the personal *inner sense* and *latent functions* can serve as approaches that might enhance the *motive of design*; although the *motive of design*

4.2 Pre-design, Design, and Post-design

The conventional design process consists of the following phases: conceptual design, embodiment design, and detailed design [2]. However, the author believes a simple examination of these three design phases cannot capture the essential nature of *motive*. We must also observe the design stages that occur before and after these three phases and examine the relationships among the stages. We call the former the Pre-design stage and the latter, the Post-design stage.

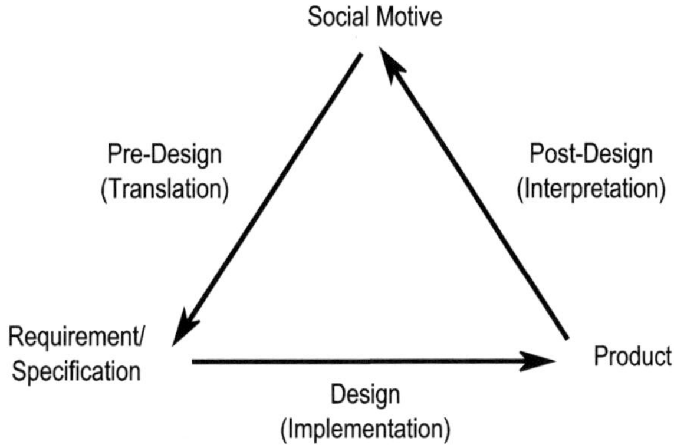

Fig. 4.1 The Pre-design stage, the Design stage, and the Post-design stage

The Design stage is a so-called conventional design stage. It consists of the phases of conceptual design, embodiment design, and detailed design. During this stage, a new product is developed to satisfy the requirements or specifications explicitly proposed for new products. This stage is described as a process of "implementation" during which the requirements and specifications are implemented into "products."

The **Post-design stage** is a process in which products are used by consumers. On the basis of consumer-product interactions (experiences of utility or accident), or, in some cases, without reliance on consumer-product interactions, consumers' feelings and criteria for products, or awareness of *problems* with existing and/or new products are overtly or covertly created and contained in society. These newly formed feelings, criteria, or awareness of *problems* create the *social motive*. This stage is described as a process of "interpretation" during which the meaning of "products" are interpreted into *social motive*.

The Pre-design stage is a process in which the concrete requirements or specifications for new products that we might expect society to accept are created on the basis of the designer's *personal motives* or the *social motives*. This stage is described as a "translation" process during which *motive* is translated into the concrete requirements or specifications for new products.

Figure 4.1 illustrates the relationship between the Pre-design stage, the Design stage, the Post-design stage, and *social motive*.

4.2.1 Categories of the Pre-design Stage

Applying the conception of inference types, the Pre-design stage can be categorized into three types from the deterministic level to the non-deterministic level: deduction that means to infer (deduce) an individual instance from a general principle or law (deterministic); induction that means to generalize (induce) a set of instances or observations (semi-nondeterministic); and abduction that means to create a possible hypothesis that explains a set of observations (non-deterministic).

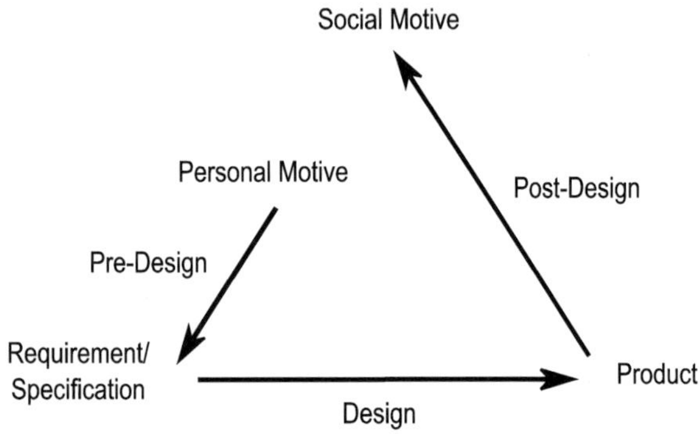

Fig. 4.2 Abductive pre-design process

The **Deductive Pre-design** is a process in which the *social motive* is created and contained "overtly" in society. The requirements and specifications for new products are deduced (translated) from the explicitly expressed *social motive*. An example of the Deductive Pre-design process would be the improvement of utility or efficiency on the basis of data obtained through the usage of new products.

The **Inductive Pre-design** is a process in which the *social motive* is created and contained "covertly" in society. The requirements and specifications for new products are induced or extracted (translated) from the implicit *social motive*. Market surveys, antenna shops, or observations of users' behaviors are the general methods employed during the Inductive Pre-design process.

The **Abductive Pre-design** is a process in which the requirements or specifications for new products can neither be deduced nor induced from the present *social motive*. Because the *social motive* that corresponds to the requirements or specifications of the products to be obtained during the Pre-design stage is "premature or non-existing," the designer's *personal motive* may play an important role in the Abductive Pre-design process (see, Fig. 4.2).

In other words, the requirements and specifications of the new products are translated from the designer's *personal motive*. There are two ways to perform the Abductive Pre-design process. One way is to pre-create *social motive* that is expected to be accepted by society. The other way is to state the requirements or specifications of the new products directly, that is, without giving consideration to the *social motive*. Products that result from extreme innovation that fall beyond the existing categories, and high-risk products that society is unfamiliar with in terms of utility or accident, are believed to be designed by the Abductive Pre-design process.

4.2.2 Categories of the Post-design Stage

Similar to the Pre-design stage, the Post-design stage can also be categorized into deductive, inductive, and abductive processes.

The **Deductive Post-design** is a process in which the *social motive* is created through "direct" consumer-product interactions (experience of utility or accident) under the condition that the newly created *social motive* remains "in accordance with" the designer's intention. Although new products are used (interpreted) without uncertainty during this process, society may detect and recognize improvable features that will become the *social motives* of the products. An example of the Deductive Post-design process is a recongnition of improved utility during use of the product by referring to the product manual.

The **Inductive Post-design** is a process in which the *social motive* is created through "indirect" consumer-product interactions (experience of utility or accident) under the condition that the newly created *social motive* remains "in accordance with" the designer's intention. An example of the Inductive Post-design process is trends in fashion, food, and IT products that are spread by word-of-mouth.

The **Abductive Post-design** is a process in which the *social motive* is created "without reliance on" consumer–product interactions (experience of utility or accident). Alternatively, a process wherein the newly discovered use of an existing product falls "beyond" the designer's intention. Usually, very highly advanced, high-risk products are expected to appear without accident. Hence, the *social motive* must be created without experience of the products' risks. In addition, unexpected usage may yield a new *social motive* that is spread by word-of mouth and can be used in the design of future products. However, in some cases, these types of usage can be dangerous. In these cases, they should be prohibited.

4.2.3 Categories of the Design Stage

The Design stage consists of a conventional design process. It can also be categorized into deductive, inductive, and abductive processes.

Incidentally, Pahl and Beitz [2] have classified "design" into three types:

Original design: This involves the elaboration of an original solution principle for a system (e.g., plant, machine, or assembly) that involves the same, a similar, or a new task.

Adaptive design: This involves the adaptation of a known system (the solution principle remains the same) to a changed task. In this case, the original design of parts or assemblies may be necessary.

Variant design: This involves variation of the size and/or arrangement of certain aspects of the chosen system. However, the system's function and solution principle remain unchanged. No new problems arise because of changes in materials, constraints, or technological factors.

When we consider the above-mentioned classifications, we see that the Deductive, Inductive, and Abductive Design processes correspond to the variant design, adaptive design, and original design types, respectively.

Table 4.1 Categories of *the Pre-design*, *Design*, and *Post-design stages*

	Pre-design	Design	Post-design
Deductive	Explicit social motive is translated into the requirements or specifications for the new products	The size and/or arrangement of certain aspects are varied. The solution principle remains unchanged	Social motive is created by direct consumer-product interactions (experience of utility or accident) within the designer's intention
Inductive	Implicit social motive is translated into the requirements or specifications for the new products	A known system is adapted. The solution principle remains unchanged	Social motive is created by indirect consumer-product interactions (experience of utility or accident) within the designer's intention
Abductive	Personal motive is translated into requirements or specifications	An original solution principle is elaborated	Social motive is created without consumer-product interactions (experience of utility or accident) or the newly discovered use falls beyond the designer's intention

Table 4.2 Methods employed during *the Pre-design, Design,* and *Post-design stages*

	Pre-design	Design	Post-design
Deductive	Analysis of utility or efficiency	Variant design	Usage of guidebook
Inductive	Market survey, Antenna shop, observation of user's behavior	Adaptive design	Methods employed based on customer's indirect experience Word-of-mouth
Abductive	Method employed based on designer's personal ability	Original design	Methods employed based on customer's ability Word-of-mouth

In sum, as mentioned above, the Pre-design, Design, and Post-design stages can be classified into deductive, inductive, and abductive processes, respectively (see, Table 4.1). Further, methods employed during the Pre-design, Design, and Post-design processes are summarized in Table 4.2.

We can see that abductive processes during the Pre-design, Design, and Post-design stages involve the essential nature of design mentioned in Sect. 4.1. Further, the missing link between the Post-design stage and the Pre-design stage appears to exist for highly advanced products.

In the following sections, the Abductive Pre-design and the Abductive Post-design will be discussed in more depth.

Up to this point, the Pre-design stage has been examined within the framework of idea generation, concept generation, market survey, risk-management, and so on. The Post-design stage has been studied within the framework of product usability, emotional design, user-centered design, and so on. However, these areas have been approached independently; they have not yet been systematized on deeper level. In particular, little attention has been paid to the often unrecognized link between the Pre-design stage and the Post-design stage, during which the conception of *social motive* can be expected to play an important role.

4.3 Outer Motive and Inner Motive

We can assume that two types of *motives of design* exist: one in terms of the product and the other in the designer's mind. In this chapter, the former is considered the ***outer motive;*** the latter is considered the ***inner motive***. This discrimination relates to the discussion of concept generation presented in our previous publication [3]. In that book, we proposed that two "bases" are sources for concept generation: *problem* and *inner sense*. Here, a ***problem*** is described as the gap that exists between an object's goal and its existing situation. ***Inner sense*** is that which involves *inner criteria* and *intrinsic motivation* that can form the basis for the generation of a new concept through reference to existing concepts. ***Inner criteria*** are that which are explicitly or implicitly underlying in the designer's mind and that guide the process of concept generation. The details of *inner criteria* will be discussed in the next section. ***Intrinsic motivation*** can be explained in the following manner [4]. When intrinsically motivated, people perform activities for the sake of performing, usually because they derive pleasure from being engaged in it (for example, a person may enjoy playing music) and perform these activities with no expectation of a tangible external reward. People typically report greater enjoyment and satisfaction when performing activities for which they are intrinsically, rather than extrinsically, motivated. Intrinsic motivation is typically associated with greater commitment to the activity (e.g., greater chances of spontaneous resumption after an interruption). In addition, it has been suggested that intrinsic motivation enhances creativity. Intrinsic motivation refers to performing an activity for the inherent satisfaction derived from the activity itself [5].

The author believes the two above-mentioned bases, *problem* and *inner sense*, correspond to *outer motive* and *inner motive*, respectively. "Feeling" and "criteria" found in the definition of *motive* may correspond to *inner sense*, and "awareness of *problem*" may correspond to *problem*. If *problem* or *inner sense* are shared in society, then they can become *social motives*. On the basis of these discussions, a summary of the characteristics of *personal motive* and *social motive* is provided in Table 4.3.

Here, the author notes that these two types of *motives* (*outer motive* and *inner motive*) are not exclusive. Every *motive* can be recognized as both an *outer motive* and an *inner motive*. Every product can affect (change) an individual's and society's behavior in a variety of ways. This change can appear to be a solution for a *problem* (*outer motive*). On the other hand, every product is affected by the designer's mind. This appears to be a kind of *inner motive*. We should distinguish

Table 4.3 *Outer motive* and *inner motive*

	Outer motive	Inner motive
Social motive	Social problem	Social inner sense
Personal motive	Personal problem	Personal inner sense

between the proposition that examines a type of *motive* from that which describes a *motive's* appearance.

Among the four types of *motives* shown in Table 4.3, the author will now focus on *social problem* for *social motive* and *personal inner sense* for *personal motive*, because *inner sense* is difficult, and sometimes impossible, to share with others. A *problem* can easily be shared by many.

We will focus on the conception of the ***force*** with which a product can autonomously affect its circumstances in our discussion of the essence of *motive*. The *force* of the product provides society with a kind of ***standard***. The ***field*** is the area within the *force* that can be affected. For example, the product "car" provides a society with a *standard* that requires roads to be relatively flat (flatter than the roads found in a society that has no cars). In addition, the same physical distance must be perceived to be shorter (of lesser perceived distance than that found in a society that has no cars). This *standard* and its *field* are always changing. Products can be accepted or rejected by society on the basis of whether the *standard* and *field* are acceptable to that society. Hence, the conceptions of *force*, *standard*, and *field* are strongly related to *motive of design*.

The *field* can be classified into the following three types:

The first type is the ***physical field***. In general, products are employed under determined physical conditions that include voltage, temperature, humidity, and so on.

The second type is the ***scenic field***. Products are employed to carry out actions of individuals in certain scenarios. For example, a mobile phone can be used to make calls in one scenario. It can also take pictures in another scenario.

The third type is the ***semantic field***. For example, a hobby may be meaningful to hobbyists or enthusiasts, but it may be meaningless to others.

In this chapter, the author focuses on the ***function*** of a product. This particular role assigned to a product serves as a *force* of that product. This is the most general and universal *force* of products. In addition, in many cases, the requirements and specifications of products is described by the term *function*.

4.4 Personal Inner Sense

In our previous study [6], to capture the nature of the *inner criteria* in terms of how people receive explicit or implicit impressions from products, we developed a method to construct "virtual impression networks" by using a semantic network. To construct a virtual impression network, we used words to express the

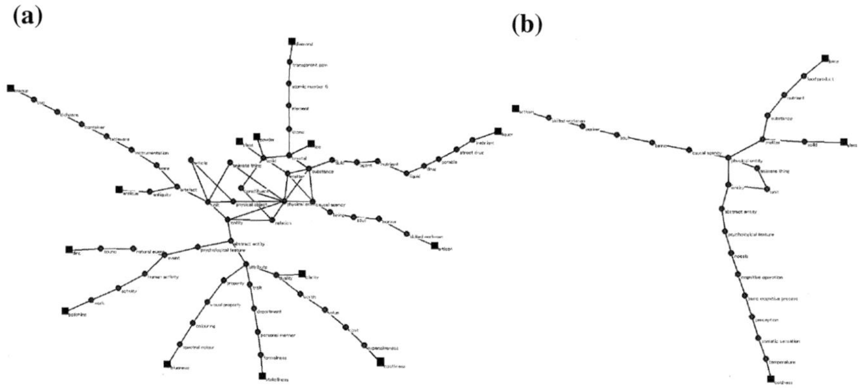

Fig. 4.3 Examples of virtual impression networks for “like” (**a**) and “dislike” (**b**). ■ is the explicit impression word; ● is the node that appears in the path between the explicit impression words

impressions held by a user who was asked to describe his/her impressions of a product using certain words. We referred to these words as “explicit impression words.” A semantic network (in this case, WordNet) is used to trace a virtual chain of nodes that represent the meaning of words inserted between every pair of explicit impression words. This virtual chain is a path that connects one explicit impression word to another. We assumed that the nodes that appear in the paths were inexplicit impressions. We constructed a network in which the traced paths could serve as a representation of a virtual impression network. Thus, a virtual impression network consists of two types of nodes: explicit impressions and inexplicit impressions. In particular, we attempted to understand the *inner criteria* individuals rely upon to form impressions (“like” or “dislike”). Our results show that the difference between the *inner criteria* of “like” and “dislike” can be explained by using several structural criteria found in impression networks. The network of “like” is more intricately intertwined than that of “dislike.” These differences are shown in Fig. 4.3. The results reveal that the difference between “like” and “dislike” originates from the nature of the impression process rather than a specific image or shape because the difference between “like” and “dislike” is related to the differences found in the structure of the impression network itself. In other words, the *inner criteria* for the way in which impressions of products are received are not portrayed by certain specific images or shapes. Rather, they are contained in the deep underlying nature of way in which the impressions are received. When an impression expands in an intricately intertwined way, we receive a good impression. On the other hand, when an impression fails to expand, we receive a bad impression.

Further, we simulated the concept generation process by using the same method we used to simulate the “virtual impression network” [7]. In this study, we simulated the generation process for a new idea by combining two initial concepts. Specifically, we connected the paths between the two initial concepts that were

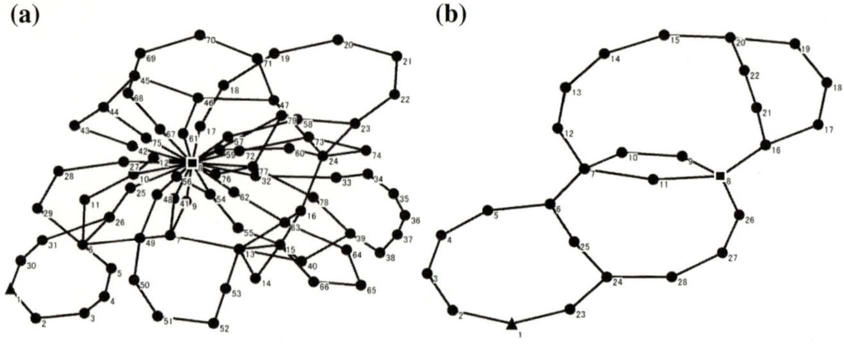

Fig. 4.4 Examples of virtual concept generation process networks with high originality (**a**) and low originality (**b**). ▲ is the node of the two initial concepts; ■ is the node of a set of words subjects used to describe design outcomes; ● is the node that appears in the path between the node of the two initial concepts and the node of a set of words subjects used to describe design outcomes

expressed with a single word and a set of words that subjects used to describe design outcomes. Our results show that it is possible to explain the difference between a product that raters evaluate as highly original and a one that raters evaluate as less original by using several structural criteria found in concept generation networks. The network for the "highly creative thinking process" is more intricately intertwined than the network for the "less creative thinking process." These networks are shown in Fig. 4.4. This result indicates that the *inner criteria* of creative thought do not exist in a certain specific seed of an idea. Rather, they exist in the deep underlying nature of the expanding thought space. When the thought space expands in an intricately intertwined way, we can generate a creative idea. On the other hand, when the thought space fails to expand, we cannot generate a creative idea.

These findings and considerations indicate that the same type of *inner criteria* is active during the process of the receipt of impressions that initiates the Post-design stage and during the process of generation of a new idea that occurs during the Pre-design stage. That the *inner criteria* for the Post-design and Pre-design stages involve the same mechanism implies that the possibility to bridge the two stages exists. That is, in the Abductive Pre-design process, an individual can generate an idea that will be accepted by society (*social motive*) if he or she follows his/her deep underlying *inner criteria*. Furthermore, the generation of a new idea or the receipt of impressions in an intricately intertwined way can be assumed to form a *motive* for design because both can be expected to lead to the development of highly original products that make good impressions. Alternatively, the process of feeling the intricately intertwined way itself is assumed to form the *motive*.

Based on the above discussions, we can infer that the *personal inner motive* for design is the explicit or implicit consciousness that is activated by *intrinsic motivation* that can inspire an individual to desire a product that stimulates the *inner criteria* of both the user and designer in an intricately intertwined way.

4.5 Latent Function

Natural disasters, such as typhoons, hurricanes, earthquakes, droughts, or floods, are conventional *social problems* that generate product designs. The development of methods to cope with these disasters is considered a *sharable motive* in society. These motives may result in the construction of many structures such as bridges and dams. In addition, the conception of labor saving, which involves higher performance of labor and a fewer working hours, is considered a *sharable motive* in society. Amenities such as water and gas supply systems, railroads, and roads have been developed, and products such as computers, cars, washing machines, vacuum cleaners, and microwave ovens have been invented to save time and effort. Further, in recent times, many entertainment products, such as radios, televisions, audio devices, and video games, have been created. We can say that "safety," "efficiency," and "comfort" were the *motives* for the design of these products. These are considered *social motives* because they can be shared in society.

On the other hand, we may possess different kinds of feelings, criteria, or awareness of a *problem* during our use of objects developed to cope with the above-mentioned *problems* because those objects may manifest other *functions* in addition to their originally intended *functions*. For example, cars are beneficial. However, they cause air pollution; drivers cause traffic accidents, and many persons can be killed. In addition, the use of cars can change our perception of distance and may weaken our impetus to walk. Technological products may have another *function* in terms of both their tangible and intangible aspects. These products are assumed to be accepted by society through consumer–product interactions. Consumers' feelings and criteria for the products are generated by their experiences of negative *functions* that were not originally intended, as well as by their experiences of positive *functions* that were originally intended (utility and accident) during the Post-design stage. On the other hand, during this process, consumers' awareness of *functions* (both positive and negative), not intended originally may be considered the *motive* to develop a new product. For example, in addition to obvious *motives* for product development such as safety improvement and the reduction of energy consumption, a new *function* discovered by consumers may serve as a *motive* for the development of a new product. Additional perceptions of the *function* that were or were not originally intended may also serve as *motives* for the development of a new product.

In this section, the author will focus on the *function* of products and on the *field* in which the *function* is manifested.

Similar to *force,* mentioned in Sect. 4.3, *function* manifests in a limited *field.* For example, a car's *function* is to "run" within these *fields*: on a flat road, operated by a driver, and supplied with gasoline. On the other hand, in the scene in which another car breaks down on the road, the first car can manifest another *function:* "pulling a car." Thus, products manifest other, different *functions* within different fields. On the basis of the General Design Theory [8], these phenomena can be addressed as follows: When an entity is exposed to a circumstance, a

Fig. 4.5 The relationship among visible function, latent function, visible field, and latent field

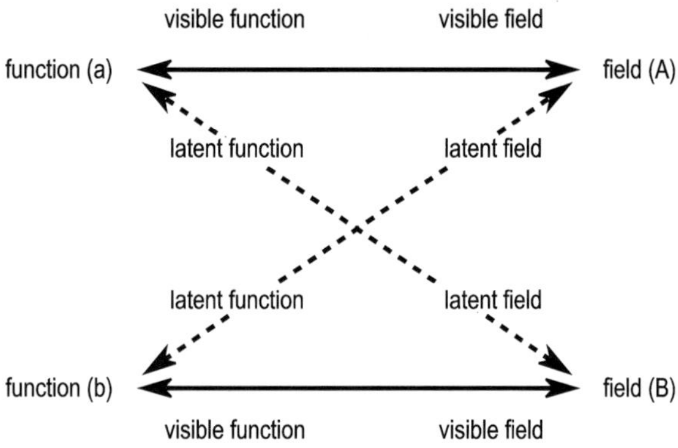

peculiar behavior manifests that corresponds to the circumstance. This behavior is referred to as a *visible function*. Different behaviors are observed in different circumstances. The sum of these behaviors is referred to as a *latent function*. Both functions can be referred to as *functions*. At times, we use the term, *function,* to mean *visible function* when there is no opportunity for confusion. The term *field* represents the circumstance that allows a *function* to manifest. On the basis of this definition, *visible function*, *latent function*, and *field* are defined as follows: The *function* (a) that is manifested in the *field* (A) is defined as the ***visible function*** in the *field* (A), and the other, different *function* (b) that is manifested by the same product in the *field* (B) is defined as the ***latent function*** of the product for the *field* (A). In this chapter, the author will extend this conception of *visible function* and *latent function* to the conceptions of *visible field* and *latent field* as follows: The *field* (A) in which the *function* (a) is manifested is defined as the ***visible field*** for the *visible function* (a), and the *field* (B) in which the other, different *function* (b) is manifested by the same product is defined as the ***latent field*** for the *visible function* (a) of the product. The relationship among *visible function*, *latent function*, *visible field*, and *latent field* is illustrated in Fig. 4.5.

Some *latent functions* and *latent fields* can be assumed by the designer in advance. However, many *latent functions* and *latent fields* are discovered by customers during consumer-product interactions (by the process of interpretation of the product). For example, a traveler's use of a hair dryer to dry socks he or she washed in a hotel room because he or she had no clean socks would not be a *function* assumed in advance by the designer. However, a cell motor directly powering a car when it stops at a railroad crossing is a *function* assumed by the designer. The *latent function* and *latent field* sometimes cause a less-than-desirable use for a product, although these functions may often cause desirable uses. For example, a knife can be used as a weapon. As mentioned above, *functions* that manifest in different *fields* are essential characteristics inherent in a product. The experience of these different *functions* can generate feelings, criteria, or awareness of *problems*.

Next, the author will suggest a method to infer *latent functions* and *latent fields*. In the Post-design stage, a new use for a product (a new interpretation) is discovered. If we could recognize these *latent functions* and *latent fields* in advance,

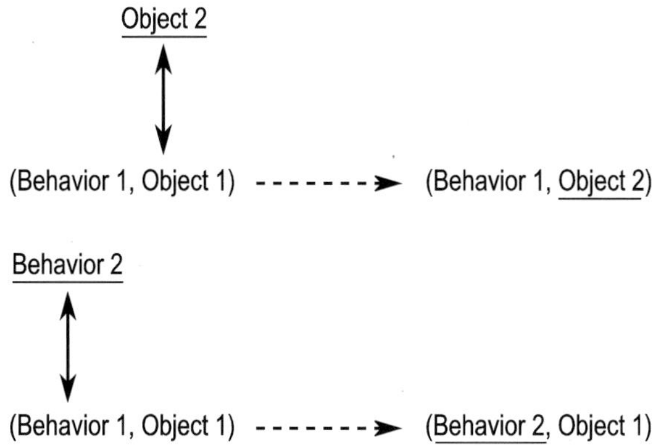

Fig. 4.6 Inference of a latent function by replacing a word with a different word

deeper feelings and criteria may be developed for products without requiring a process that relies on consumer-product interactions. In addition, accidents that might be difficult to predict may be avoided by considering the underlying risks. Accordingly, we might expect that inferring *latent functions* would promote Abductive Post-design. On the other hand, innovative requirements or specifications for products that can neither be deduced nor induced from the present *social motive* can be created during the process of inferring *latent functions.* This creation of innovative requirements or specifications implies that inferring *latent functions* is also effective for Abductive Pre-design.

Two methods may help us better infer *latent functions* and *latent fields*. In the first method, one word is replaced with a different word (see, Fig. 4.6).

In many cases, a *function* is represented within the sets of "behavior" and "object." Here, "behavior" can be represented by the word, "verb," and "object" can be represented by the word "noun." For example, the *function* of "knife" is represented by the words "cut food." Within this framework, a *latent function* is obtained by replacing the word "object" with a new word. Digital cameras were used to take pictures of landscapes when they were first released in the market. Now, however, digital cameras are also used to record memoranda written on whiteboards during meetings. Digital cameras' *latent function* to "record memoranda" is obtained by replacing the word "landscape" with the word "memoranda." In addition, the *latent function* "dry socks" with a hairdryer is obtained by replacing the word "hair" in the *visible function* with the word "socks." Another method involves replacing the word "behavior" with a new word. If we store vegetables in a refrigerator, the vegetables may become dehydrated even though they remain cool. In this case, the *latent function* of the refrigerator is "to dehydrate vegetables." Its *visible function* is "to cool vegetables." This kind of *latent function* is considered a "side effect." Further, a more innovative *function* can be obtained by replacing both "behavior" and "object" with new words. The abovementioned methods can be achieved by the search for newly inserted words.

The second method requires that we apply the conception of analogy (see, Fig. 4.7).

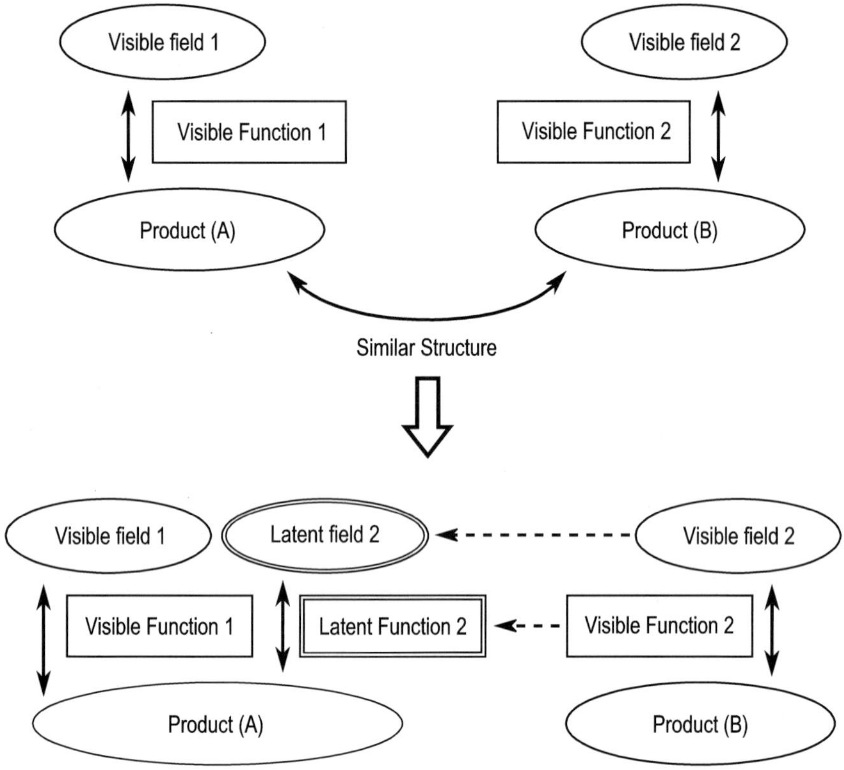

Fig. 4.7 Inference of a latent function by applying the conception of analogy

This method is based on the idea that the *visible function* and *visible field* of a product (B) can be a *latent function* and *latent field* of another product (A) if both products (A) and (B) contain similar structures [9]. For example, a desk and a stool share a similar structure. Hence, an individual can sometimes step on the desk when he or she needs to reach for something on a high shelf. In this situation, the *function* of "support a person" also must be manifested by the desk. On the basis of this method, we can find the *latent function* and *latent field* for an existing product. For example, let us consider a situation where a laptop computer is used in an uncommon circumstance that differs from its normal *physical field*. Let us assume that we are in an abnormal *physical field* (in terms of temperature, humidity, etc.) and an abnormal *scenic field* (e.g., typhoon, hurricane, earthquake, drought, or flood). Then, let us consider the existence of a laptop computer in these *fields*. If we consider that a cell phone, radio, digital camera, or digital watch contain somewhat similar structures to the laptop computer, we can then infer that a laptop computer may manifest the following *latent functions*: the laptop might light the environment during a blackout, receive radio and television signals, supply power at the highest power-saving mode, and remain waterproof. This type of laptop computer might meet new requirements or specifications for products that can be used during a disaster (i.e., the laptop's *latent field).*

As mentioned above, consideration of the *latent functions* and *latent fields* of a product can yield feelings and criteria for products, or awareness of possible *problems* inherent in a product without requiring a process that relies on consumer-product interactions that may promote Abductive Post-design and Abductive Pre-design, respectively. The fact that considerations of *latent functions* and *latent fields* are effective for both Abductive Post-design, as well as Abductive Pre-design implies that it may be possible to bridge both processes and enhance the *motive of design.*

4.6 Discussion

In this chapter, the *motive of design* was classified into *personal motive* and *social motive.* The framework of the Pre-design stage, Design stage, and Post-design stage was outlined. Specifically, the author discussed the *inner senses* and the *latent functions* as aspects of the *motive of design.* The findings revealed that the deep underlying nature of association in consumers' minds as well as *latent functions* and *latent fields* may bridge the Abductive Post-design and Abductive Pre-design processes and enhance the *motive of design.* These results indicate that the missing link that often goes unrecognized between Post-design and Pre-design might be connected by the above-noted two aspects.

However, many issues remain unexplored. First, other aspects of *motives* may exist. For example, a greater physical desire or eagerness can be considered a *personal motive.* Second, the author has not yet discussed the ways in which a *problem* can be shared in society. To engage in fruitful discussion, we should extend these discussions to include the history and culture of society.

Discussion related to the *motive of design* can serve as theories or models of design when the design is captured in a broad view. I believe a theory or model of design is expected to extract the essences of phenomena within the real design process, as well as to predict and lead future new design methods. In this sense, the discussion related to *motive of design* has the potential to extend existing methods and to develop products that will be more readily acceptable to society.

The discussions in this chapter are merely in the beginning stages. We must devote further study to these issues.

Acknowledgment The conceptions of "Pre-design," "Post-design," and "*social motive*" introduced in this chapter are based on collaborative research performed with Prof. Michio Ito (Tokyo University of Agriculture and Technology, Japan). The conception of "*inner sense*" is based on collaborative research performed with Prof. Yukari Nagai (Japan Advanced Institute of Science and Technology). This work was supported by KAKEN (24603012).

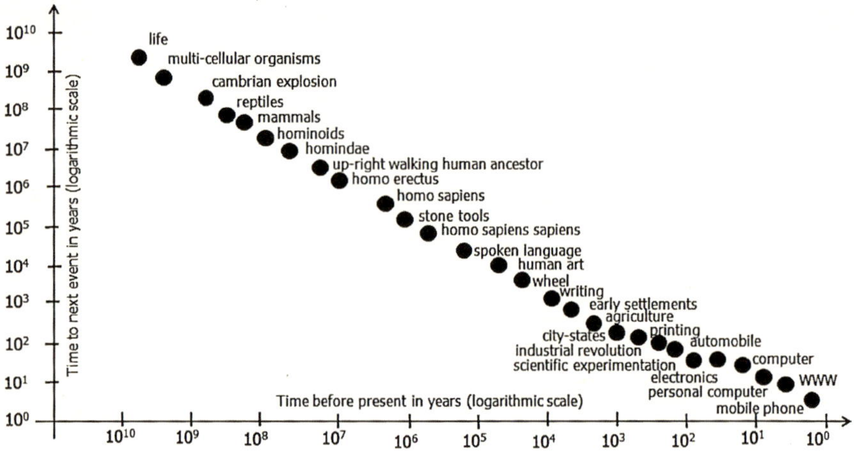

Fig. 5.1 Accelerating physical, biological, social and technological evolution

(i.e. information-based reconstruction of quantum theory) may provide a new theoretical basis for dealing with this dichotomy [3].

After this, let us have a look at the modes of information encapsulation and the changing roles of information in the process of PBST evolution. By doing this, we can create a platform for our system engineering oriented follow-up discussions. As shown in Fig. 5.2, at the beginning of everything, information basically resided in atomic structures [4]. When genetic materials (such as deoxyribonucleic acid) have evolved, information has been coded, among others, in DNA [5]. When the human brain evolved, information has also become embedded in neural patterns [6]. In the process of formation of human intelligence, capabilities have been developed to externalize and disseminate information by various primary and secondary means of human communication [7]. This was a crucial advancement not only from a cultural point of view, but also from the aspect of aggregating technology-related commonsensical and scientific human knowledge. In the age of industrialization, this aggregation, multiplication and conversion of information to technologies has enabled society-level creation and making, and later on, production of artefacts, systems and processes [8].

In our modern time, human engineered systems not only encapsulate information and knowledge, but also acquire the potential and abilities to regenerate information and to convert it into operative intelligence. As technology and intelligence continue to integrate, systems with a high-level working intelligence, even with a self-reproductive intelligence, can be expected [9]. It has been predicted that it becomes possible for intelligence to reside and evolve in multi-scale engineered systems already in the near future, but surely in the further future. By that time, it will facilitate not only the human outreach to the solar system, but also the human presence in the nearby part of the universe.

What is happening in our days is a kind of unrestricted integration of human acquired and artificially generated information with human created artefacts. This

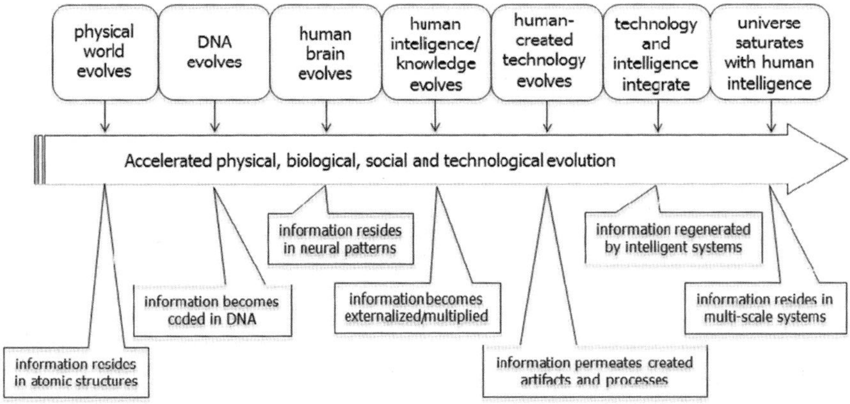

Fig. 5.2 The changing position/role of information in the PBST evolution

is supported by the fast development of digital computing and remote communication technologies which, together, form one of the current major assets of human society. We have been witnessing the emergence and consolidation of four digital computing paradigms in the last 60 years (Fig. 5.3). The history of computing commenced with the paradigm of mainframe computing. Not more than three decades later, this has been made obsolete by that of networked personal computing. Two decades later, the latter has started its growing into the paradigm of embedded and portable ubiquitous computing. Though this paradigm is still far from being fully exploited or exhausted, the new paradigm of cyber-physical computing is already with us and rapidly evolving [10]. Actually, the first results are already out from the research and development laboratories and getting acceptance in the daily practice.

Blending information and knowledge with physical artefacts has reached a very high level with the advent of the concept of cyber-physical systems. This level is referred to as synergetic integration. It has to be noted that the emergence of the next possible information processing paradigms has also started. The paradigms of quantum computing and biological computing are progressing with a large pace and large expectations have been formulated [11]. Though they are still in their infancy, these paradigms already have a large influence on scientific research and technology development. Experts forecast that they will have a never-before-experienced impact on generating and handling information, in particular by artificial systems. Likely, they will permeate and saturate our natural and created environments with qubits-based computing and communication capacity.

As it seems, blending information with physicality in artificial systems continues. What does it mean in the context of a design theory of cyber-physical systems? This question is addressed below. The rest of the section is structured as follows: In order to cast light on complex (application) systems, first a concise survey of the chronological and conceptual developments is provided in Sect. 5.2.

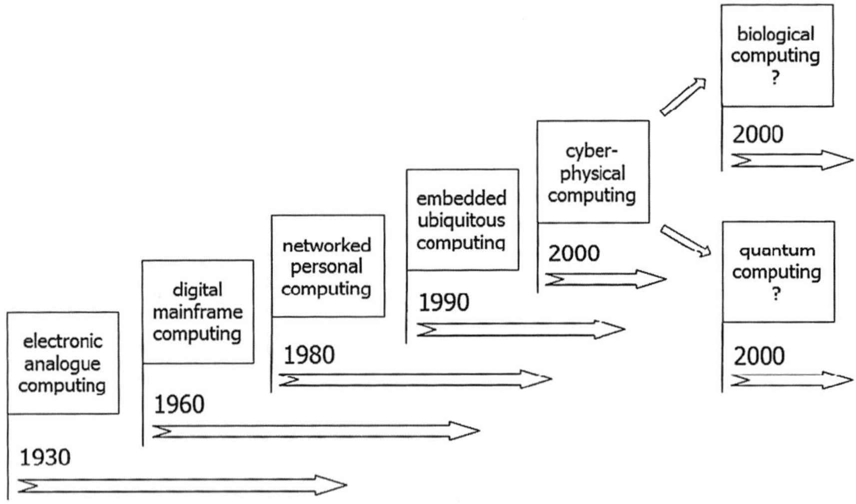

Fig. 5.3 Shifting paradigms of digital computing

Section 5.3 gives an overview of the principal characteristics of cyber-physical systems, focusing on high-end implementations, rather than on low-end ones. Section 5.4 discusses the social and cognitive aspects, and the progression of cyber-physical systems in these directions. Section 5.5 elaborates on the need for a design theory for social-cyber-physical systems. Section 5.6 introduces and briefly looks into some substantial and challenging phenomena that the design theory of social-cyber-physical systems should describe, explain and predict. Finally, some conclusions are offered and future research work is stimulated.

5.2 Illuminating the Road to Cyber-Physical Systems

Below, we give a brief overview of the successive developments that have leaded us to cyber-physical systems. As analysed by Isermann, technical systems were purely mechanical before the second industrial revolution, the major feature of which was exploitation of electromagnetism in various forms [12]. This gave floor to the emergence of mechanical systems with electromechanical drives. The next phase of development, at the beginning of the 1930s, witnessed the appearance of electromechanical systems with analogue control. The third technological revolution, which was driven by the new digital control and computing technologies in the 1950s, made it possible to include digital processors and computers in the control of electromechanical systems [13]. Incorporation of digital computing commenced with electronic control at the beginning of 1970s and was remarkably accelerated by the introduction of the microprocessor in the early 1980s. Actually, this lent itself to the formation of the discipline of mechatronics [14]. It was jointly

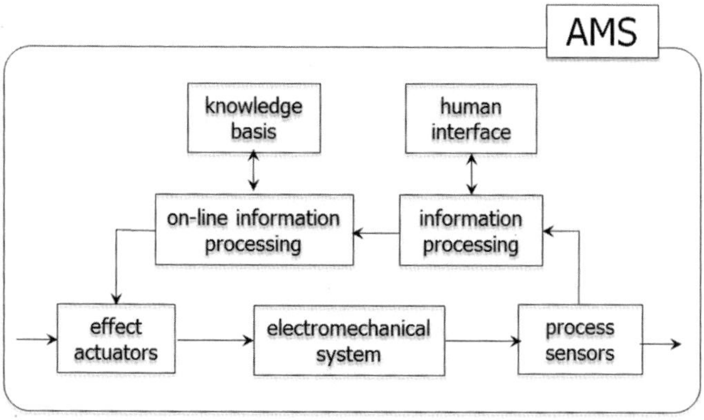

Fig. 5.4 Generic architecture of advanced mechatronics systems

stimulated by the affordances offered by combinations of mechanical, electronic, control and computational technologies, and the growing societal need for more sophisticated industrial systems and infrastructural solutions [15].

Interestingly, in the late 1970s, the Japan Society for the Promotion of Machine Industry (JSPMI) classified mechatronics products into four categories: (i): Class I: primarily mechanical products with electronics incorporated to enhance functionality, (ii) Class II: traditional mechanical systems with significantly updated internal devices incorporating electronics (iii) Class III: systems that retain the functionality of the traditional mechanical system, but the internal mechanisms are replaced by electronics, and (iv) Class IV: products designed with mechanical and electronic technologies through a form of integration that allows enhancing the effectiveness of each other. Over the last three decades, the above classification has become obsolete, in particular, due to the recent developments of digital computing and communication. In this time period, mechatronics systems and products have gone through a kind of metamorphosis.

Advanced mechatronics products such as humanoid robots and service equipment, have been equipped with sophisticated sensors, interfaces, processors, actuators, as well as with complex control algorithms, software agents, and communication means. Exploitation of these resources and knowledge-intensiveness has been the objective of advanced mechatronics since the mid-1980s, with the intent to achieve high level of flexibility and adaptability based on the functionality of the control software [16]. The generic architecture of advanced mechatronics systems is shown in Fig. 5.4. Enablers of the development in this direction were not only advanced software design and programming technologies and tools, but also new software architecting concepts such as agent- and component-based implementation. As a result, products are now showing a much higher level of functional integration and implementation complexity.

The affordances of the above technologies and the increased expectations for complex functions and sophisticated structures gave birth to embedded systems

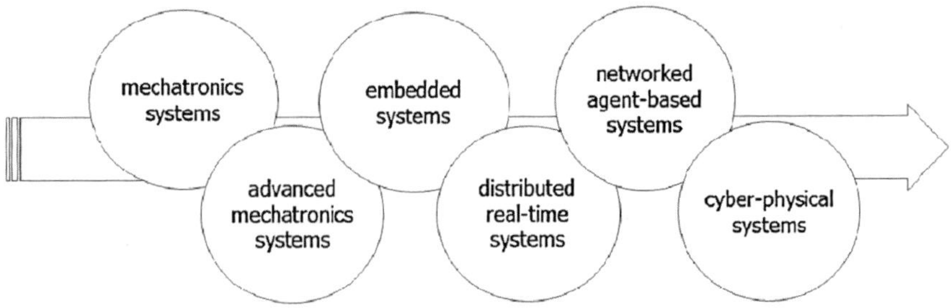

Fig. 5.5 Development path from mechatronics systems to cyber-physical systems

(ESs) (Fig. 5.5). The main objectives of ESs research and development have been to develop functionally smart, structurally adaptive, partially autonomous, and reprogrammable systems [17]. In ESs, computers (more precisely, embedded microprocessors and software means) are used as components to implement the above attributes and functions. While physical processes were controlled by the computational elements in the case of traditional (totally hardwired) electronic feedback systems based upon local and remote computational models and algorithms, in the case of ESs, physical processes are monitored and optimized by the computational elements based on sensor information.

The traditional feedback-based control systems were designed as closed systems, without operational interfaces. The research in ESs largely contributed to moving from closed boundary systems with limited scalability, through cross-boundary systems, to open systems expected to be fully-scalable in the near future [18]. ESs are pre-programmed to do specific functions, required to show real-time behaviour, but also constrained in terms of certain resources (e.g. battery-operated). Incorporation of programmable processors in circuits makes the design more robust and thus reduces the design time cycle.

Enabled by digital computing and control, another branch of system development has been real-time systems (RTSs) [19]. This family of systems has its legacy for the reason that in certain information-intensive engineering systems, such as robots, vehicles and medical equipment, it is important not only to provide right output, but also to compute it fast at the right time. Actually, correctness of the control data is a function of the time of delivery (though consistency of the results may be more important than the raw computing speed) [20]. Centralized RTSs require real time operating systems. One of the most popular one in use today is QNX, which uses a micro kernel for implementing basic system calls, but system level functions such as device drivers, are not part of it. RTSs are either (i) transformational systems (T-RTSs), which take input from the environment at a given time, transform these inputs, and terminate giving the outputs, or (ii) reactive systems (R-RTSs) that have continuous interaction with their environment [21]. While the reaction of R-RTSs on regular (periodic) events can be statically scheduled, random (aperiodic) events must be dynamically recognized, or statistically predicted, when possible [22].

It has been realized that centralized systems are not optimal in many applications because of their: (i) high reliance on centralized communication, (ii) the technical challenges of large-scale integration, (iii) lack of scalability, and (iv) the high cost of integration. The use of distributed intelligence technologies avoids these weaknesses. Distributed intelligence systems (DISs) are usually based on physical and software agents that: (i) operate autonomously, (ii) handle specialized tasks independently, (iii) cooperate to satisfy system-level goals, and (iv) achieve a high degree of flexibility. One sub-family of DISs is sensor network systems (SNSs), which implement collaborative signal (information) processing on the basis of large-scale, distributed macro- and micro-sensor technologies and connectivity (transmission and networking) technologies [23]. Other sub-family is intelligent agents systems (IASs), which manifests in dynamically changing, functionally decentralized, and networked multi-agents enabled environments of high robustness and scalability, such as distributed energy systems. In some publications, systems with these characteristics are also referred to as distributed autonomous decision making systems.

All of the above mentioned disciplines (and system concepts) are moving towards a higher level of integration of the material world and the cyber world. The paradigm of cyber-physical systems (CPSs), which is sweeping the society since 2005, intends to achieve the currently known highest possible level [24]. As discussed later, CPSs feature extensive functional integration, increasing complexity, emergent intelligence, adaptive structure and behaviour, and make a huge impact on humans and the environments [25]. The notional constituents of CPSs are shown in Fig. 5.6. In CPSs, human users can be both in- and out-of-the-loop [26].

The phrase 'cyber-physical systems' has been introduced in the USA by the NSF [27]. As a counterpart of this, systems with practically congruent characteristics have been called collaborative adaptive systems (CASs) in Europe. CASs differ from the current generation of open control systems in two important aspects, namely in terms of collectiveness and multi-scaling [28]. They typically comprise very large number of multi-objective units, which have autonomy in their own individual properties, objectives and actions. Decision-making is highly dispersed and the variety of interactions amongst the units may lead to the emergence of new and/or unexpected phenomena and behaviours. The concept of CPSs should be demarcated from that of the Internet of Things (IoT), which assumes that things interact and exchange information, and that this gives a basis for future pervasive computing environments [29]. However, its objectives are more infrastructural, than problem solving and application context orientated.

The paradigm of CPSs is still in evolution. Therefore, we may come across with rather different interpretations and forms of implementations. According to the classical NSF definition, CPSs are 'physical and engineered systems whose operations are monitored, coordinated, controlled and tightly integrated by a computing and communication core at all scales and levels'. The cyber sub-system is responsible for computation, communication and control, and is discrete, logic-based and event-oriented, while the physical sub-system incorporates natural and human-made components that are governed by the laws of physics, and that

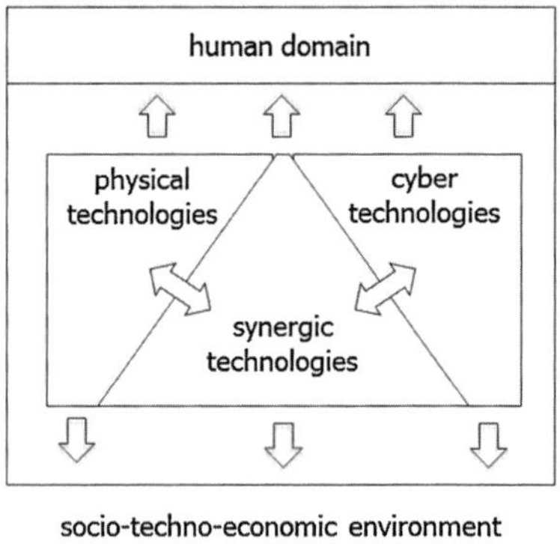

Fig. 5.6 Notional components of cyber-physical systems

operate in continuous time [30]. We should mention that this definition is already deemed as a somewhat conservative one nowadays. One of our previous papers proposed a definition that gives more consideration to decentralization, dynamism and evolutionary nature of CPSs. This definition circumscribes CPSs as structurally and functionally open, context-sensitive, intelligent and self-managing engineered systems in which the physical and the cyber constituents evolve cooperatively, and which gradually penetrate into the social world, as well as into the mental world of humans. Structural openness means that they may include collaborative sub-systems of varying spatial scales and complexity scales—both in time and in space. Functional openness implies that they may consist of units that happen to enter or leave the collective at any time. The units (i) can be highly heterogeneous (computers, agents, devices, humans, networks, etc.), (ii) may operate at different temporal and spatial scales, and (iii) may have different (potentially conflicting) objectives and goals.

A generic architecture of cyber-physical systems is shown in Fig. 5.7. The families of the current enabling technologies are shown in Fig. 5.8. There have to be a technological synergy among the enabling technologies as there is a functional synergy among the physical and the cyber components. A rapid and wide proliferation of synergic technologies can be expected already in the near future. The interactions among the geographically remote components of CPSs happen in real time, under emergent constraints, and often towards non-predefined objectives. Combined with structural variability, these characteristics introduce uncertainty that is difficult to handle by traditional design methods and implementation technologies. The main source of uncertainty originates in the capability of CPSs to change their structure and behaviour by learning and adaptation in operation.

Many authors differentiate low-end (i.e. ordinary or complex) implementations and high-end (i.e. complicated, adaptive, evolving and reproductive) implementations of cyber physical systems. The former ones are distributed and networked systems, equipped with sensors, enabled by embedded and smart computing, and controlled by situation dependent feedback. The 'follow-me' printing environment shown in Fig. 5.9 is a practical implementation of cyber-physical systems, but

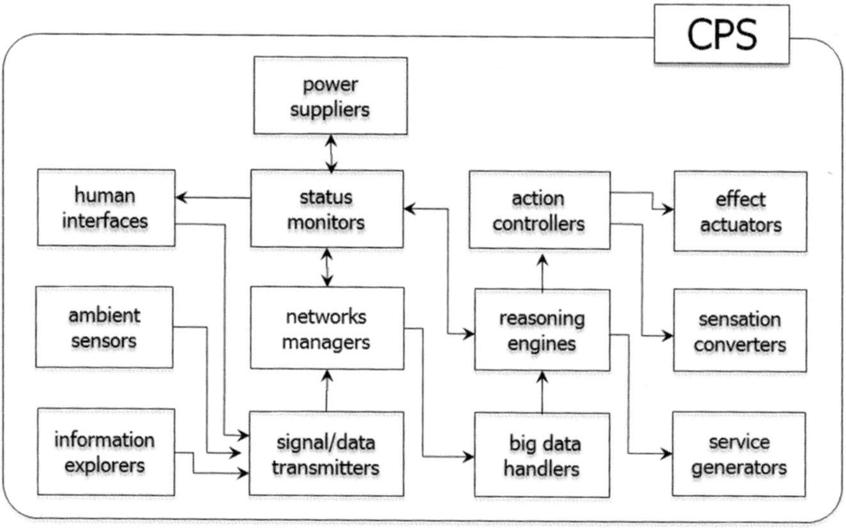

Fig. 5.7 Generic architecture of cyber-physical systems

Fig. 5.8 Families of technologies enabling the physical and cyber components of CPSs

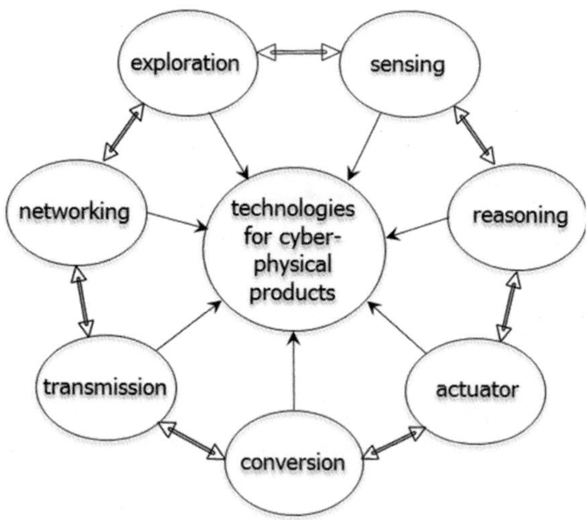

there are many similar examples reported in the literature [31, 32, 33]. The high-end implementations are seen as largely complex, open, multi-scale, heterogeneous, intelligent, self-managing, and partly autonomous (even reproductive) systems [34, 35, 36]. The specific characteristics of both categories of CPSs are discussed below.

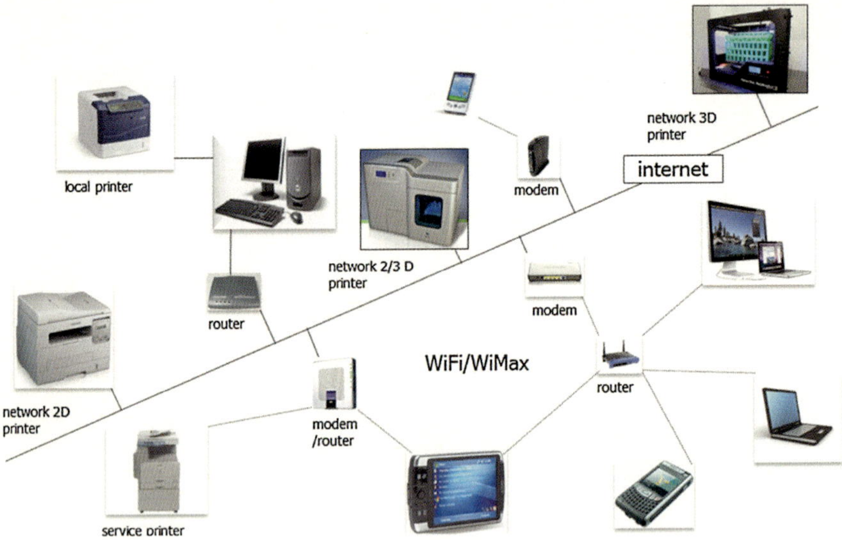

Fig. 5.9 A 'follow-me' printing environment—a low-end cyber-physical system

5.3 Main Features of Cyber-Physical Systems

As indicated above, cyber-physical systems have a large number of specific characteristics. Some of them are also characteristics of other traditional systems, but the whole set of these characteristics can only be recognized on systems that belong to the family of CPSs. Below this distinguishing set of characteristics is discussed:

C1 CPSs are designed and implemented in order to support human activities and well-being by decentralized cooperative problem solving, in harmony with the techno-econo-social environment,

C2 CPSs consist of a digital cyber-part and an analogue physical-part, which are supposed to work together towards the highest possible level of functional and structural synergy,

C3 CPSs are functionally decentralized and geographically distributed open systems with blurred overall system boundaries,

C4 CPSs are capable not only to dynamically reconfigure their internal structure and reorganize their functionality/behaviour, but also to change their boundaries,

C5 CPSs are constructed of very heterogeneous sets of active components, which can enter and leave the collective at any time, and may encounter other systems with similar or conflicting objectives,

C6 CPSs, as well as their components, may work in extreme temporal ranges (from instantaneous to quasi-infinite, and beyond), and manifest on various spatial scales (from intercontinental to nano-scales),

C7 components are typically hybrid structures, encapsulating various compositions of hardware (e.g. transformer and actuator) entities and embedded cyber (e.g. software and knowledge) entities,
C8 components may have predefined, emergent or ad-hoc functional connections, or all, with other interoperable components at multiple levels,
C9 components may operate according to different problem solving strategies (plans) towards achieving the overall objective of the system,
C10 components are knowledge-intensive and able to handle built-in formal knowledge, knowledge obtained by sensors, and knowledge generated by reasoning and learning mechanisms,
C11 components are able to make situated decisions and strive for automated problem solving by gathering descriptive information and applying context-dependent causal and procedural reasoning,
C12 components are able to memorize and learn from history and situations in an unsupervised manner and to specialize themselves based on smart software agents and emergent intelligence,
C13 components are able to adapt to unpredictable system states or emergent environmental circumstances, as well as to execute non-planned functional interactions and to act proactively,
C14 overall decision-making is distributed over a large number of components (agents), and is based on the reflexive interactions among the components and multi-criteria analysis (optimization),
C15 contrary to their distributed and decentralized nature, CPSs need to operate and communicate in real-time and in a synchronized manner,
C16 system resources are managed by different sophisticated strategies and maintain security, integrity and reliability of the components and the CPSs as a whole

It has to be mentioned that, in addition to the above characteristics, many researchers have already argued that:
C17 next-generation (molecular and bio-computing-based) CPSs can be supposed to have some level of reproductive intelligence

There have been many possible application domains circumscribed for CPSs such as (i) situated intervention (e.g., collision avoidance), (ii) operation in dangerous environments (e.g., fire fighting), (iii) exploration in inaccessible environments (e.g. deep-sea), (iv) precision operation (e.g., robotic surgery and nano-manufacturing), (v) flow coordination (e.g., traffic control, goods manipulation), (vi) efficiency enhancement (e.g., zero-net energy buildings), (vii) augmentation of capabilities (e.g., healthcare monitoring), etc. Actually, only human imagination can be a limit of exploring high-potential applications and innovative solutions.

Some implementations of CPSs may not show the entire set of the above characteristics, or may just incompletely realize them. In these cases, we speak about partial compliance with the paradigm of cyber-physical systems. For

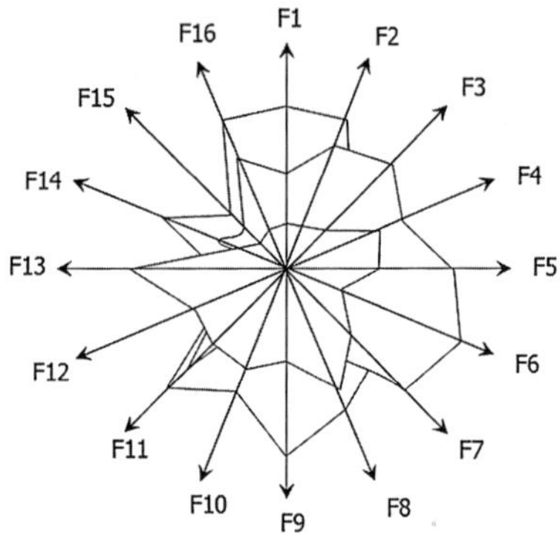

Fig. 5.10 Characteristics profiles of various CPS instances

instance, though it stands in general, CPSs should not always be structurally open, fully autonomic, multi-scaled, or functionally decentralized. The distribution and the measure of partial compliance can be graphically represented and analysed based on a so-called characteristics profile diagram. A case-independent one is shown in Fig. 5.10. By defining qualitative requirements or quantitative criteria for each of the system characteristics, it can be decided if a particular system concept or implementations can (or should) be considered a CPS, or to what extent it complies with the paradigm of CPSs.

5.4 Towards Social-Cyber-Physical Systems

We are approaching a point where CPSs cease to be just technical systems. They are progressively becoming part of the socio-technical fabric of society. CPSs strongly interact with the human domain and the embedding environment, even if it not always happens in an explicit form. These two domains of interaction form two interrelated dimensions of socialization. Therefore, they should be seen as complex socio-technical systems, in which human and technical aspects are massively intertwined. Social-cyber-physical systems (SCPS) should work, on the one hand, according to the expectations of humans, communities and society, and on the other hand, under the constraints and conditions imposed by the embedding environment. However, no matter how good the original design specification was, systems become less well adapted to users and environment over time due to changing requirements of the changing users or environment, or to the evolution of the system itself. Therefore, SCPSs are supposed to flexibly adapt to the environment, and to the (communities of) users. These can be achieved based on situation cognizance and context awareness (Fig. 5.11). In this context, four additional system characteristics can be stated:

C18 Overall, SCPSs are able to become aware of the users and their personal and social contexts, and to adapt themselves towards and optimal symbiosis

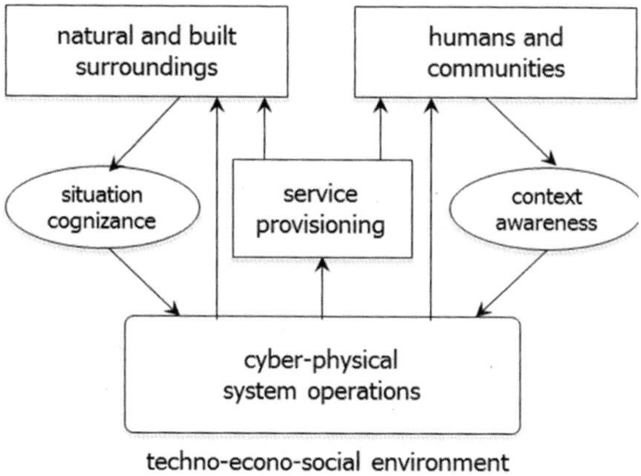

Fig. 5.11 CPS operations in social contexts

C19 SCPSs are able to achieve the highest possible level of dependability (trustworthiness and confidence), accountability, security, accessibility, and maintainability

C20 SCPSs strive for operating as a self-organizing holarchic open systems, with a minimal environmental impact and sustainability from ecological, economic and social viewpoints

C21 SCPSs are able to achieve a balance between overheads and outputs, demand and usage of resources, and wastes and gains

Nevertheless, current technological limitations make CPSs intrusive. They are more syntactic, than semantic - therefore they create a mismatch with regard to the human way of thinking and doing. SCPSs should have some basic social abilities such as: (i) detecting users and the social connections between them, (ii) accessing users' data, (iii) inferring the social context according to users' networks topology, preferences and features, (iv) inferring social goals according to the social context and the user model, (v) coordinating their behaviour, and (vi) providing a context-driven output [37]. The awareness of SCPSs should extend to the intangibles of social context, which includes social culture and norms, personal believes and attitudes, and informal institutions of social interactions.

5.5 The Need for a Design Theory for Social-Cyber-Physical Systems

There are two fundamental categories of systems: natural and artificial systems. As shown in Fig. 5.12, the range of natural systems extends from physical systems through biological systems to a part of social systems. The other part of social systems and all technical systems are human engineered artificial systems. They serve very different purposes and show rather different behaviours. As introduced

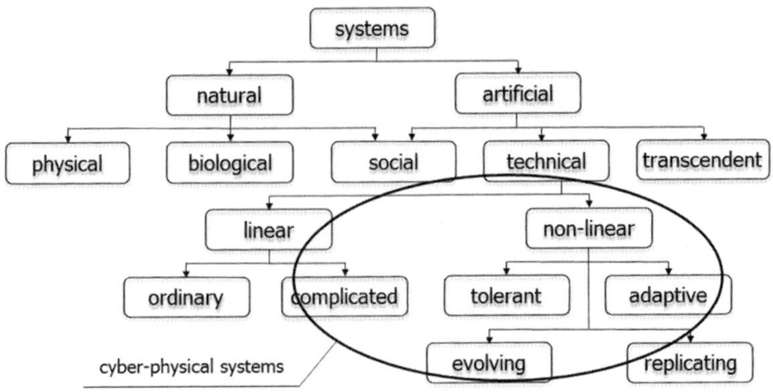

Fig. 5.12 Categorization of systems

above, in case of low-end systems: (i) all operations are determined by physical laws, (ii) the number of components is typically small, (iii) the attributes of the components are predetermined, (iv) the number of interactions among the components is few, (v) the functional interoperation among the components is highly organized, (vi) the components do not pursue their own goals, (vii), the overall behaviour of these systems is linear and predictable, (viii) interaction of the system with the environment is limited and reactive (definitive). That is, low-end implementations do not have the capability to adapt or evolve - they behave as linear systems. System science considers these technical systems as aggregative and reducible systems.

On the other hand, in case of high-end implementations: (i) the number of components is large, (ii) the attributes of the components are not predetermined, (iii) the number of interactions among the components can be extremely large, (iv) the functional interoperation among the components is loosely organized and can be intricate, (v) the components may work autonomously and generate their own goals, (vi) the overall system behaviour is probabilistic and can be highly non-linear, (vii) the interaction of the system with their environments is extensive, proactive and may be continuous, (viii) understanding of the contexts results in a context-sensitive operation. Since these systems have the capability to dynamically adapt or evolve over time, their behaviour is non-linear. System science regards them as holistic and non-reducible systems.

Traditional system design theories have been developed to provide knowledge and understanding of linear systems that do, in simple words, what they have been designed for. Due to the non-trivial relationships between the system components and the system's macroscopic properties, non-linear cyber physical systems display emergent properties [38]. The phenomenon of emergence is not addressed by traditional system design theories [39]. This is the main reason why we do need a dedicated (proper) design theory for social-cyber physical systems. This is an urgent need which is stimulated by the rapid development and widespread proliferation of these systems. The new theory is supposed to address all design aspects in a non-specialized and comprehensive manner. It is obvious that the increase of complexity of systems goes together not only with an increase of intelligence, adaptability and autonomy, but also of with an increase of uncertainty and unpredictability.

The addressed family of engineered systems lends itself to emergent behaviours well beyond that have been conceived or intended by their designers. Managing emergent attributes and behaviour is a new challenge for designers and engineers of high-end cyber-physical systems. The challenge manifests in multiple forms. First, the generic phenomenon of emergence of attributes in engineered systems is not sufficiently understood. Second, the specific mechanisms of structure, functionality, interaction, and behaviour emergence are not sufficiently known. Third, the strategies and principles of designing complex systems for emerging attributes and behaviour have not been elaborated yet. Fourth, no systematic methodologies and technological solutions have been developed for anticipating, influencing and controlling the self-learning and self-adaptation in various contexts.

5.6 What the Design Theory of Social-Cyber-Physical Systems Should Describe, Explain and Predict?

The simple answer is: a lot. For the reason that many aspects are not yet disclosed by scientific research, or need further articulation or contextualization, it is not easy to consider all aspects that a design theory of social-cyber-physical systems has to cover. On the other hand, it may be possible to devise an initial theory that can explain what CPSs are and how they should be designed in various contexts by pulling together the seemingly unlimited accessible information and knowledge. As generic objectives for research: (i) understanding the fundamental features of various complex (complicated, adaptive, evolving, reproducing) systems, (ii) getting deeper insights in the required synergy between the cyber and the physical parts, (iii) investigation of the effects of interactions with human stakeholders and social environments, and (iv) providing a unified design theory and methodology that facilitates addressing the issues of both worlds in concert have been proposed. Below we give an overview of some essential aspects that the sought for design theory is supposed to address as first objectives.

5.6.1 Aggregative Complexity

The term ‘complexity’ is used to characterize something with many parts in an intricate arrangement [40]. Complexity of CPSs increases with: (i) the number of (real or potential) functional components, (ii) the complexities of the distinct functional components, (iii) the heterogeneity of the structural components, (iv) the multiplicity of scales of the structural components, (v) the number of connections between them, and (vi) the complexities of the connections among components [41]. Five types of complexities can be differentiated: (i) static complexity (the number and relationships of components that do not change with time), (ii) dynamic complexity (the number and relationships of components that change with time), (iii) self-organizing complexity (open systems reorganize themselves to different systems), (iv) evolving complexity (open systems evolve through time into different systems) and (v) co-evolving complexity (two-way interplay between the changing system and its environment). When all these types and forms of complexities are present, we talk about aggregative complexity. Current knowledge offered by complexity science is in its infancy and unable to explain how to reduce and manage aggregative complexities [42, 43]. This should also be projected to multiple-scale systems, whose physical scales may range from nano-scale (10^{-9}) to mega-scale (10^{+6}) [44]. These systems are complicated not only due to the interfacing problems caused by the different physical sizes of the sub-systems, but also by the matching problems that are caused by the different information contents to be processed. Complexity also increases with the evolution of CPSs [45].

5.6.2 Emergent Attributes/Behaviour

While in case of a linear system the effects (outputs) are proportional to their causes (inputs) and subject to superposition, the operation of a non-linear system cannot be expressed as a sum of the operations of its parts (or of their multiples). Emergence is typically the major cause of non-linear characteristics of CPSs [46]. It may change the attributes of a system as well as its overall behaviour [47]. Ultimately, emergence may manifest in many different forms and scales in CPSs. Complexity theory states that critically interacting components self-organize to form potentially evolving structures exhibiting a hierarchy of emergent system properties [48]. At the bottom line, the question is how we can architect and engineer CPSs with evolutionary capabilities and under varying operational circumstances to ensure purposeful and secure behaviour? The principles of how to forecast the emergent characteristics and behaviour, and how to integrate, regulate and benefit from them are hardly known now [49]. Formalization and handling emergence is a challenge in itself because of the difficulty of: (i) capturing and modelling all components and relationships (interactions), (ii) managing interactions when everything affects everything else, (iii) considering all potential non-predefined system states, (iv) quantifying risk/uncertainty for very integrated systems, (v) handling disturbances safely and effectively, and (vi) working with rigid design constraints and tight design space [50, 51].

5.6.3 Compositional Synergy

Composition is becoming a generic design and implementation principle in engineering disciplines [52]. Nowadays, many kinds of systems are conceptualized, designed and implemented by exploiting the benefits of component-based approaches [53, 54, 55]. Component-based design (CBD) involves the creation, integration and re-use of hardware, software and knowledgeware components. The feasibility of component-based system design depends on two key conditions: composability and compositionality [56]. Composability expresses that component properties are not changing as a result of their interactions with other components within the system. It is a measure of the degree to which components can be assembled in various combinations to satisfy specific user requirements. Compositionality determines if synergic system-level properties can be established by local properties of components [57]. A CPS is compositional if its emergent behaviour may be derived from the behaviour of its constituent components. Lack of compositionality causes systems that do not behave well outside a small operational envelope. It is known that CBD helps manage complexity, increases dependability, decreases time-to-market, and optimizes costs, but the principles and methodologies for compositionality in heterogeneous adaptive and evolving systems are not explored yet.

5.6.4 Multi-Abstraction-Based Specification

The objective of abstraction is to facilitate coping with the structural and functional complexities and heterogeneity of CPSs [58]. We can identify subjects, aspects and levels of abstraction. Subject of abstraction can be: (i) a system of systems, (ii) a particular system, (iii) a sub-system and (iv) a component. Abstraction can be applied, among others, from the aspect of (i) architecture (platform), (ii) procedure (operation), (iii) hardware, (iv) software, (v) networking, (vi) interfacing, (vii) programming, and (viii) computation. The levels of abstraction can be: (i) entity, (ii) group, (iii) neighbourhood, and (iv) cluster abstraction. From the viewpoint of components, architectural abstractions can be top down (supporting composability), or bottom up (supporting compositionality). Abstraction should be applied on both component and system level. Abstraction of a component results a structural model, a behavioural model and an interaction model that are superimposed. Component abstractions ignore implementation details and describe properties of components relevant to their composition, e.g. transfer functions, user interfaces. As explained by Lee, components at any level of abstraction should be made predictable and reliable, and the system level of abstraction should compensate for the lack of robustness on a lower level of abstraction [55]. These indicate the need for a generic theory of abstractions.

5.6.5 Dynamic Scalability

Scaling is about the specification of the properties, control and behaviour of CPSs as their size is varied. This issue is poorly addressed in the literature. In case of simple linear systems, scaling would mean application of certain scaling laws.

Scalability may be contraction (down-scaling) or expansion (up-scaling). In the context of CPSs we usually face the problem of up-scaling that assumes the system to have the ability to be enlarged or to handle growing amounts of work in a regular manner. Various forms of scalability have been identified and studied in the literature. Functional scalability is about enhancing the system by adding new functionality at minimal effort. Geographic scalability involves maintaining performance, usefulness, or usability regardless of expansion from concentration in a local area to a more distributed geographic pattern. Loading scalability means expanding and contracting the resource pool to accommodate heavier or lighter loads or number of inputs. Administrative scalability concerns increasing the number of users or organizations to easily share a single distributed system. Finally, instrumental scalability is enhancing the ease with which a system or component can be modified, added or removed. The primary design question is how to architect a complex system to be extendable to multiple arbitrary scales in time and space. A system, whose performance increases to the requested level proportionally to the capacity added, is considered to be a scalable system. If the system fails to achieve it, it does not scale. In case of adaptive or evolving systems it may be the case due to the exponentially increasing number of functional relationships among a linearly growing number of distinct components. The design theory should explain the scalability mechanisms, as well as the opportunities in various contexts [59].

5.6.6 Multi-Modal Prototyping

Though model-based testing and virtual engineering are effective approaches of traditional systems engineering, they are not able to support all aspects of realization of CPSs [60]. Due to the lack of dedicated prototyping methodologies and means early prototyping of CPSs is complicated. Many characteristics, e.g. geographical distribution, diversity and number of components, interaction of multi-scale sub-systems, and operation according to numerous possible scenarios cannot be investigated with the virtual simulation and optimization resources of traditional engineering. They also pose many limitations due to the needed long preparatory times and uncertainties of knowledge. Physical (empirical) testing of large-scale CPSs is an unsolved issue. We do not have system prototyping approaches for multiple-scale heterogeneous systems where every scale is different in nature and secondary behavioural effects (e.g. interference) that may influence the integrity of system performance are to be considered [61]. Early and rapid system prototyping methodologies and technologies are needed that: (i) complement the conventional technologies, (ii) enable the investigation of dependability, functional integrity, technical feasibility, accuracy, etc., (iii) reduce development time and costs, and (iv) allow testing many other factors as a function of design variables. The needed methodology is also supposed to apply a kind of a correct-by-construction strategy.

5.6.7 Verification and Validation

Proving dependability, reliability and maintainability needs dedicated verification and validation methodologies, which are currently either in a premature stage or non-existent at all. In principle, system integrity verification and behaviour validation can be conducted directly on the implemented system and indirectly, using various aspect models. As discussed above, empirical (experimental) testing can be based on functional prototypes or mock-ups of the system, or on the fully-featured implementation of a system. This allows testing both the physical part and the cyber part of CPSs on component and system level. Due to the current limitations, the concept of model-based verification and validation (MBVV) has been proposed and applied by many researches. MBVV decomposes to two phases. First, a correct system model has to be developed and tested to see if it provides the necessary and sufficient fidelity for functional, structural, behavioural, and utility verification and validation [62]. Second, the model is used for verification and validation of the system in a scrutinized procedure [63]. It means that the model itself should be logically verified and validated for appropriateness. Besides functional, structural, behavioural, and utility verification and validation of CPPs, an emerging issue is validation of the system performance in social contexts. Cyber-physical systems are rapidly penetrating into human cognitive processes. However, for instance, recognizing human behavioural patterns in real-life and generalizing them into models are not well understood and not implemented in computers. Therefore, there is a need for new insights in the motor, perceptive, cognitive and affective cooperation of humans with CPSs as well as to ever-green interface development issues.

5.7 Conclusions

The new features and design challenges of cyber-physical systems have to be addressed by proper design theories. Several definitions of CPSs have been published and many systems have been realized, but overall understanding of the nature and phenomena of these systems is still in its infancy. In general, we are still a considerable way away from having a transdisciplinary theoretical framework for true CPSs and SCPSs, or even from elucidating the major principles by which they should operate. For this reason, design, implementation and utilization of these systems are still perplexing, not to mention the relatively low awareness of their possible impacts on the environments, society and people. There seems to be a huge knowledge gap concerning the design and engineering principles of realizing non-linear CPSs. We do not know how to design for long-term self-learning, self-adaptation, and self-evolution, not to mention self-reproduction, of CPSs. There are no tested design methodologies that could provide guidance in designing for semi-autonomous or fully-autonomous operation. Next generation CPSs are envisioned to be a horizontally and vertically heterogeneous system of systems, having some level of reproductive intelligence. As part of the design theory of CPSs, new system abstraction, modelling, prototyping, and testing theories are needed, together with genuine system adaptation, evolution, and reproduction theories. In order to advance the state-of-the-art, both transdisciplinary insights and multi-disciplinary operative knowledge synthesis are needed.

References

1. Goyal P (2012) Information physics—towards a new conception of physical reality. Inf 3:567–594
2. Wheeler JA (1989) It from bit. In: Proceedings of the 3rd International Symposium on the Foundations of Quantum Mechanics, Tokyo
3. Deutsch D (2003) It from Qubit. In: Barrow J, Davies P, Harper C (eds) Science and ultimate reality. Cambridge University Press, Cambridge, pp 1–16
4. Vedral V (2012) Information and physics. Inf 3:219–223
5. Auletta G (2010) A paradigm shift in biology? Inf 1:28–59
6. Biro JC (2011) Biological information—definitions from a biological perspective. Inf 2:117–139
7. del Moral R, González M, Navarro J, Marijuán PC (2011) From genomics to scientomics: expanding the bioinformation paradigm. Inf 2:651–671
8. Yan X-S (2011) Information science: its past, present and future. Inf 2:510–527
9. Beckerman LP (2000) Application of complex systems science to systems engineering. Syst Eng 3(2):96–102
10. Wu FJ, Kao YF, Tseng YC (2011) From wireless sensor networks towards cyber physical systems. Pervasive Mob Comput 7(4):397–413
11. Horváth I, Gerritsen BHM (2012) Cyber-physical system: concepts, technologies and manifestation. In: Proceedings of the TMCE, Karlsruhe, pp 1–16, 2012
12. Isermann R (2008) Mechatronic systems—innovative products with embedded control. Control Eng Pract 16:14–29
13. Harashima F, Tomizuka M (1996) Mechatronics—"what it is, why and how?". IEEE/ASME Trans Mechatron 1:1–2
14. Bishop RH, Ramasubramanian MK (2002) What is mechatronics?. In: Overview of mechatronics, CRC Press LLC, Boca Raton, pp 1–11
15. Kyura N, Oho H (1996) Mechatronics—an industrial perspective. IEEE/ASME Trans Mechatron 1:10–15
16. Bhave A, Garlan D, Krogh B, Rajhans A, Schmerl B (2010) Augmenting software architectures with physical components. In: Proceedings of the embedded real time software and systems conference, pp 19–21, 2010
17. Marwedel P (2001) Embedded and cyber-physical systems in a nutshell, Knowledge center article, DAC.COM, 2001, pp 1–21
18. Hewitt C (1985) The challenge of open systems: current logic programming methods may be insufficient for developing the intelligent systems of the future. Byte 10(4):223–242
19. Stankovic JA (1988) Misconceptions about real-time computing: a serious problem for next-generation systems. Comput 21(10):10–19
20. Benveniste A, Berry G (1991) The synchronous approach to reactive and real-time systems. Proc IEEE 79(9):1270–1282
21. Baillieul J, Antsaklis P (2007) Control and communication challenges in networked real-time systems. Proc IEEE 95(1):9–28

22. Aoyama M, Tanabe H (2011) A design methodology for real-time distributed software architecture based on the behavioral properties and its application to advanced automotive software. In: Proceedings of the 18th Asia-Pacific software engineering conference, IEEE, pp 211–218, 2011
23. Chong C-Y, Kumar SP (2003) Sensor networks: evolution, opportunities, and challenges. Proc IEEE 91(8):1247–1256
24. Sztipanovits J, Koutsoukos X, Karsai G, Kottenstette N, Antsaklis P, Gupta V, Goodwine B, Baras J, Wang S (2012) Toward a science of cyber–physical system integration. Proc IEEE 100(1):29–44
25. Kim K-D, Kumar PR (2012) Cyber-physical systems: a perspective at the Centennial. In: Proceedings of the IEEE, vol 100. pp 1287–1308, 2012
26. Ma J, Wen J, Huang R, Huang B (2011) Cyber-individual meets brain informatics. IEEE Intell Syst 26(5):30–37
27. Lee EA (2010) CPS Foundations. In: Proceedings of the 47th design automation conference, ACM, pp 737–742, 2010
28. Ren C-J, Huang H-B, Jin S (2008) Specification of agent in complex adaptive system. In: Proceedings of the international symposium on computer science and computational technology, 2, pp 210–216
29. Ning H, Liu H (2012) Cyber-physical-social based security architecture for future internet of things. Adv Internet Things 2:1–7
30. Lee EA (2007) Computing needs time. Commun ACM 52(5):70–79
31. Dillon TS, Zhuge H, Wu C, Singh J, Chang E (2011) Web-of-things framework for cyber–physical systems, Concurrency Comput: Pract Experience 23(9):905–923
32. Work D, Bayen A, Jacobson Q (2008) Automotive cyber physical systems in the context of human mobility. In: Proceedings of national workshop on high-confidence automotive cyber-physical systems, Troy, MI, 3–4 April 2008
33. Lee I, Sokolsky O (2010) Medical cyber physical systems. In: Proceedings of DAC '10, ACM, Anaheim, CA, pp. 743–748, 13–18 June 2010
34. Martín HJA, de Lope J, Maravall D (2009) Adaptation, anticipation and rationality in natural and artificial systems: computational paradigms mimicking nature. Nat Comput 8(4):757–775
35. Herrmann K, Mühl G, Geihs K (2005) Self-management: the solution to complexity or just another problem? IEEE Distrib Syst Online 6(1):1–17
36. Kay JJ (2000) Ecosystems as self-organizing holarchic open systems: narratives and the second law of thermodynamics. In: Handbook of ecosystem theories and management, CRC Press, Lewis Publishers, pp 135–160, 2000
37. Biamino G (2012) A semantic model for socially aware objects. Adv Internet Things 2(7):47–55
38. Arthur WB (1993) Why do things become more complex? Sci Am 268(5):92
39. Wolfram S (1986) Approaches to complexity engineering. Physica D(22):385–399
40. Amaral LAN, Ottino JM (2004) Complex networks: completing the framework for the study of complex systems. Eur Phys J 38:47–162
41. Fisk D (2004) Engineering complexity. Inter-disciplinary Science Reviews 29(2):151–161
42. Dent EB (1999) Complexity science: a worldview shift. Emergence 1:5–9
43. Delic K, Dum R (2006) On the emerging future of complexity sciences. ACM Ubiquity 7(10):1
44. Yaneer B-Y (2002) Multi-scale complex systems analysis and enlightened evolutionary engineering. NECSI, Cambridge, MA, pp 25–29
45. Saunders PT, Ho MW (1976) On the increase in complexity in evolution. J Theor Biol 63:375–384
46. Johnson C (2005) What are emergent properties and how do they affect the engineering of complex systems? Reliab Eng Syst Saf 91(12):1475–1481
47. Liu J, Hu BC (2008) On emergent complex behaviour, self-organised criticality and phase transitions in multi-agent systems: autonomy oriented computing perspectives. Int J Model Ident Control 3(1):3–16

48. Herrmann K, Mühl G, Geihs K (2005) Self-management: the solution to complexity or just another problem? IEEE Distrib Syst Online 6(1):1541–4922
49. Kramer J, Magee J (2007) Self-managed systems: an architectural challenge. In: Future of software engineering, IEEE Computer Society, Washington, DC, pp 259–268, 2007
50. Steels L (1991) Towards a theory of emergent functionality. In: From animals to animats: 1st international conference on simulation of adaptive behaviour, Paris, France, pp 451–461, 1991
51. Sole R, Ferrer-Cancho R, Montoya J, Valverde S (2002) Selection, tinkering and emergence in complex networks. Complexity 8(1):20–31
52. Gossler G, Sifakis J (2005) Composition for component-based modelling. Sci Comput Program 55(1–3):161–183
53. Brooks A, Kaupp T, Makarenko A, Williams S, Orebäck A (2005) Towards component-based robotics. In: Proceedings of the international conference intelligent robots and systems, IEEE, pp 163–168, 2005
54. Arzén K-E, Bicchi A, Dini G, Hailes S (2007) A component-based approach to the design of networked control systems. Eur J Control 13(2–3):261–279
55. Lee EA, Xiong Y (2001) System-level types for component-based design, In: Embedded software, Springer, Berlin, pp 237–253, 2001
56. Sztipanovits J (2007) Composition of cyber-physical systems, In: Proceedings of the 14th IEEE international conference and workshop on the engineering of computer-based systems, pp 3–6, 2007
57. Brazier FMT, Cornelissen F, Jonker CM, Treur J (2000) Compositional specification and reuse of a generic co-operative agent model. Int J Coop Inf Syst 9:171–207
58. Graham S, Baliga G, Kumar P (2009) Abstractions, architecture, mechanisms, and a middleware for networked control. IEEE Trans Autom Control 54(7):1490–1503
59. Willems JC (2007) The behavioral approach to open and interconnected systems. IEEE Control Syst Mag 27(6):46–99
60. Wan J, Suo H, Yan H, Liu J (2011) A general test platform for cyber-physical systems: unmanned vehicle with wireless sensor network navigation. In: Proceedings of the international conference on advances in engineering, Nanjing, China, pp 1–5, 2011
61. Gorissen D, Couckuyt I, Demeester P, Dhaene T, Crombecq K (2010) A surrogate modelling and adaptive sampling toolbox for computer based design. J Mach Learn Res 11(7):2051–2055
62. Atkinson C, Kühne T (2003) Model-driven development: a metamodeling foundation. IEEE Softw 20(5):36–41
63. Selic B (2003) The pragmatics of model-driven development. IEEE Softw 20(5):19–25
64. Thacker RA et al (2010) Automatic abstraction for verification of cyber-physical systems. In: Proceedings of international conference on cyber-physical systems, pp 12–21, 2010

Chapter 6
Models of Design

Udo Lindemann

6.1 Introduction

Talking about a or the "model of design" requires some basic discussions. Design may address different aspects such as the product as a result of a design process done within a design organization (Fig. 6.1).

Another model of design may be a Black Box with some input information (requirements, need,...) and some output information (BOM, CAD-models, computation, ...) and, in addition, resources and management information. This abstract model represents the process of design. The important discussion will be about the benefit supplied by a model like this.

The design of a product including its shape, color, surface, and user interface represents another perspective on "design."

Due to different views, the concepts of "design" and "models" have to be discussed.

6.2 Design

Design of financial products, fast food products, children's playgrounds, software, games, etc.; there are no limits to our imagination and in all areas people talk about design. There have been a number of attempts to build a common understanding of design. For example in 1998, a group of scientists discussed the possibilities to develop a "Universal Design Theory" [3] across all disciplines.

The understanding of the word and the concept of "design" may differ widely, depending on the situation and the group of people involved. The design of a Diesel engine may be seen from different perspectives: there is the engineer, the

U. Lindemann (✉)
Technische Universität München, Munich, Germany
e-mail: udo.lindemann@pe.mw.tum.de

A. Chakrabarti and L. T. M. Blessing (eds.), *An Anthology of Theories and Models of Design*, DOI: 10.1007/978-1-4471-6338-1_6,

Fig. 6.1 Model of design

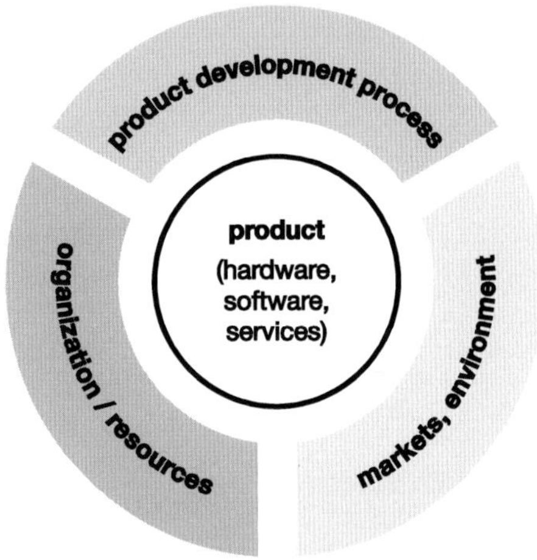

industrial designer, the software engineer—they all have different models and goals in mind when talking about "design." Even within engineering design one has to consider the role of the mechanical engineer, who usually thinks about stress and meshing of parts; the engineer in thermodynamics has a different idea of design as a problem of heat transfer; the production engineer may see the design of a shaft as a matter of handling and logistics. And there are many more aspects and perspectives.

There is a differentiation between the process of designing products and the design of a product. A product should be understood as an artifact we may be able to sell; it may be just a simple mechanical part, a piece of software, a mechatronic product, a solution for a customer including services, or large and complex systems. Regardless of the area of application we need to have a clear perception, if we discuss the matter of processes of design or the design of a product.

6.3 Design Process

Looking at the process of design we may state that there are several actions, executed in a sequential and/or parallel way. We can often observe small or large iterations. When we look at the results of these actions we can observe successful and unsuccessful actions and the whole range in-between. Quite often it takes a number of subsequent actions and a certain time gap before an action may be judged to be successful. And there is the question about the degree of details in the process we are looking at. We may look at the overall process of designing a complex system or at the "micro"-steps of thinking during design. We may (first) look at only one involved discipline like mechanical engineering or (second) as an intermediate stage at a set of disciplines involved in the design of a mechatronic

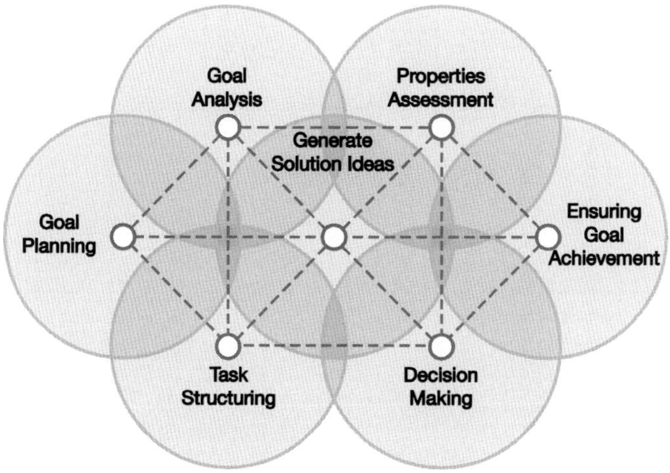

Fig. 6.2 Munich procedural model [7]

product or (third) at all importance aspects like engineering, finances, marketing, legal issues, and others. Based on this short discussion about the process point of view it seems that there is a variety of models merely for the processes of design.

Two examples may underline the discussion.

The Munich Procedural Model (MPM) (Fig. 6.2) is a generic model based on other preceding procedural models known in engineering design and in systems engineering. The model contains seven elements that represent tasks to be fulfilled during problem-solving processes. These elements are interlinked in different ways to indicate iterations and other "jumps" depending on the given situation. It focusses on the phases of understanding the problem and clarifying the task. The final stage of ensuring to achieve the goal is added; generating solution ideas builds the center. In the end, it is a generic type of a model supporting problem solving on all the different levels of abstraction and in the whole range of complexity. The main purpose of this model is strengthening the analysis of a given problem and supporting navigation through a set of actions, especially under stress.

The second example is the overall product development process model "FORFLOW," which has been developed within a team of researchers [11]. Figure 6.3 shows the upper level of the model with six major steps and indicates further levels of details and the overall model in the background. Even though there are several detailed steps it is still a generic model that has to be applied to given situations and be completed by iterations and additional parallel actions. A standard workflow is suggested indicating all activities in a general way. It has to be adapted to the actual project. This model was created to support process planning in product development projects for mechatronic products. It delivers guidelines for the initial planning but also for a detailed navigation during the project including continuous detail planning.

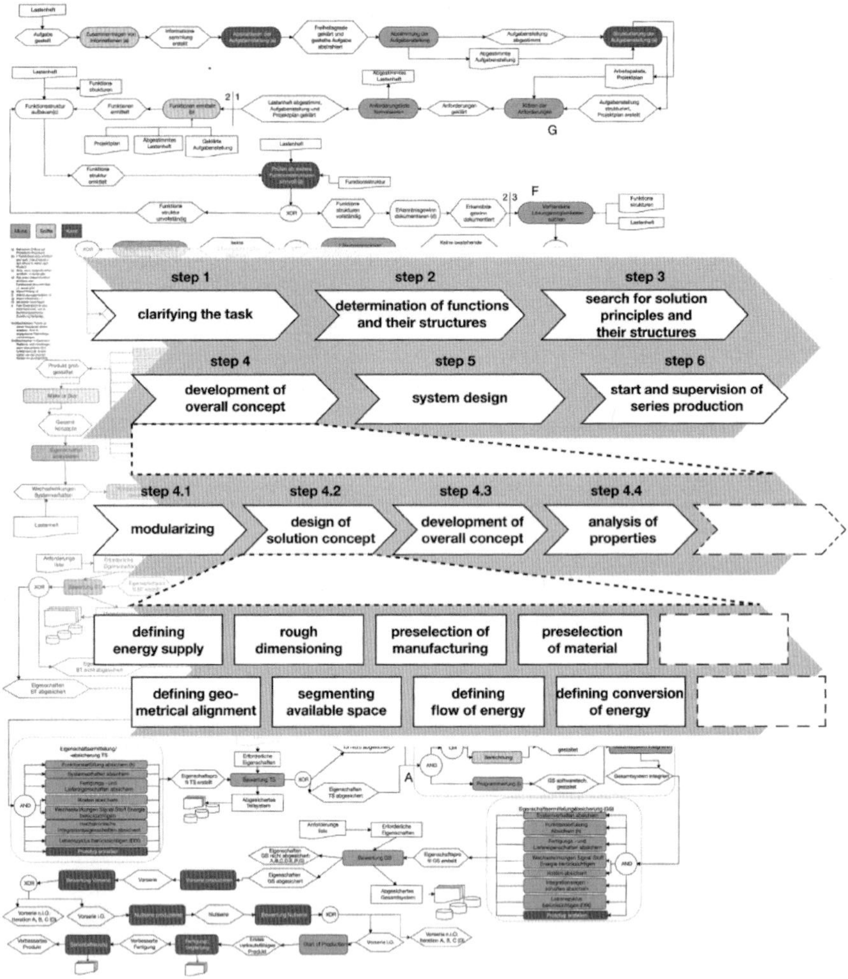

Fig. 6.3 FORFLOW process model [9]

Another concept is the product as a design.

6.4 Product Design

The Munich Model of Product Concretization (Fig. 6.4) shows an example of a product model composed of four partial models. It is based on a number of former models like the framework proposed by Grabowski [3]. There is a set of requirements collected, structured, and documented in a kind of requirement model. This model will be completed step by step during the whole process of

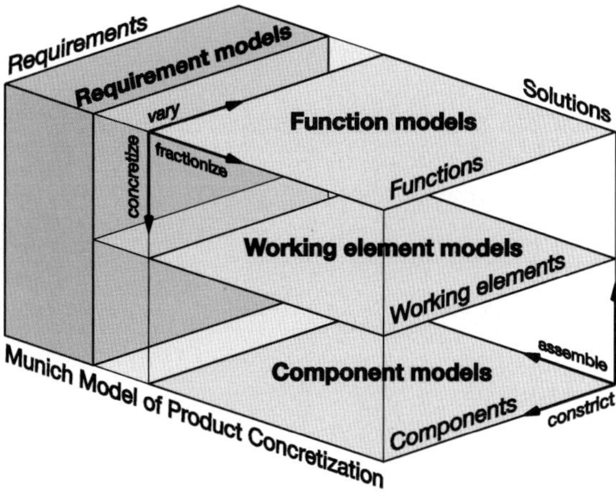

Fig. 6.4 Munich model of product concretization [8]

design. The solution orientation is supplied by the functional model on an abstract level. The next level of concretization is the working model (usually including physics), which may also be interpreted as a behavioral model. Finally, the component model has to be defined and documented, which may also be named the structural model. Depending on the point of view there are structures on all levels and within all partial models.

This model should not be compared with a process model. It may help to think about a sequence of process steps but it is independent of processes.

This is only one example of a large set of models used to describe or analyze products. CAD- models are used to document design-oriented details for production, usage, and recycling; the bill of material supports the material planning and cost calculation; finite element models allow numerical calculation of stress, deformation, heat flow, etc.

Lauer [6] identified a large number of different documents in product development in practice. The major part of these documents may be interpreted as partial model of the overall product model as described by the STEP standards [4]. And there are of course further categories like organizational or psychological models.

Only a few examples for models of design objects (products) and design activities (processes) have been discussed. These models are based on research and experience of the author and his team.

To underline the difficulty of the correct interpretation of models of other author's one short example will be discussed. This example is part of a publication Umeda et al. [15] about the FBS framework. They discuss (among other points) the purpose of Rodenacker's model of function and state that his intention was the support of novices in design. Looking at this statement one should consider that this interpretation is based on Rodenacker's teaching book published in 1971 [10], the origin of his model goes back to his dissertation in 1936 [9]. In the context of his book out of 1971 to be used primarily by students the purpose was the support for novices in new product development. The original purpose was about building some basic ways of abstract description of machines. On the other hand, Umeda et al. [15] looked in 1990 at computable (etc.) models, which up to a certain amount is a different set of purposes at another point of time and for a different group of people/users.

Some of the important points mentioned in this example like the purpose of the model or the users of the model (actors) will be addressed within the following section; here the general nature of "model" is discussed.

6.5 Model

There are numerous different definitions of the term "model"; a few will be discussed in this chapter.

The Oxford Dictionary [14], extract notes:

- three-dimensional representation of a person or thing or of a proposed structure, typically on a smaller scale than the original
- thing used as an example to follow or imitate
- simplified description, especially a mathematical one, of a system or process, to assist calculations and predictions
- person employed to display clothes by wearing them
- particular design or version of a product.

Shannon [12] wrote: "A model is a representation of an object, system or idea in some other form than itself". The definition by DIN 19226, [2] (German standard, translated by the author) is as follows: "A model is the image of a system or a process within another conceptual or representational system." VDI 3633, [16] (translated by the author) is suggesting: "A model is the simplified reproduction of a planned or an existing system including its processes within another conceptual or representational system."

All these definitions leave room for interpretation.

Stachowiak [13] wrote a book about model theory and pointed out that there are three important characteristics of a model: transformation, reduction, and pragmatism (Fig. 6.5).

Modeling is always based on an original. It may be an existing product, the idea of a new product, a CAD-model, a process, or an organization. The original owns a (large) set of attributes describing it. When building a model, some of these attributes are selected and transformed to define the model. Thus, the model contains a subset of the attributes of the original which may have been modified by the transformation process. Model specific attributes (such as a coordinate system in CAD) may additionally be added.

Looking especially at pragmatism there are three aspects of high importance. The purpose, the users (actors), and the time frame of usage. This may be stated as

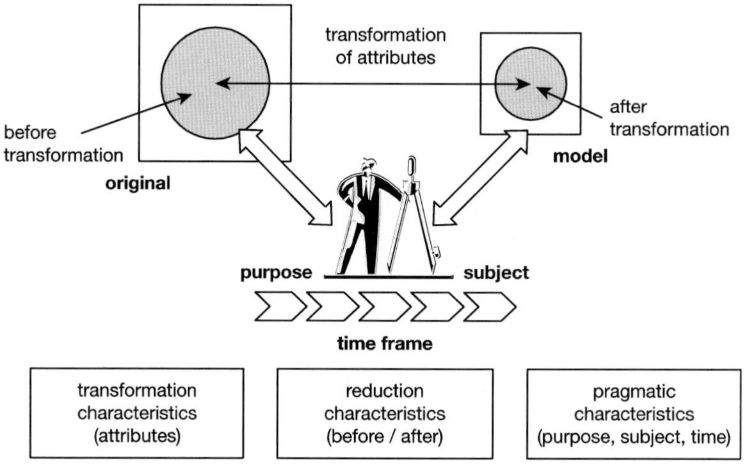

Fig. 6.5 Characteristics of modeling [according to 13]

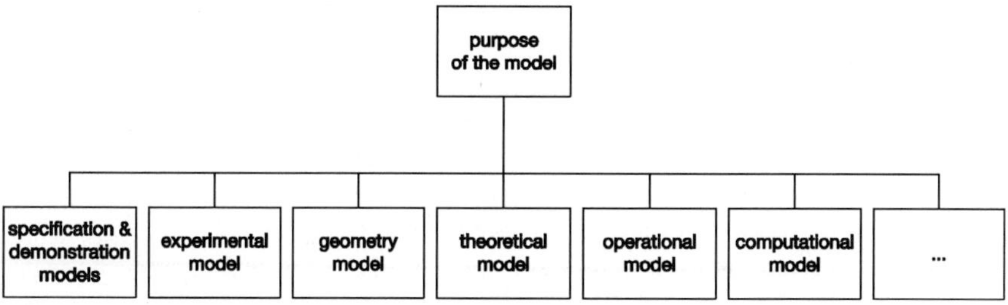

Fig. 6.6 Purpose of a model

why, for whom, and when. These characteristics serve to define the boundary of the system, its elements and the interdependencies.

Building a model has always a specific purpose. Based on this purpose the reduction of used attributes and the way of transformation has to be adapted. As there is a specific purpose guiding the process of modeling, there are of course limits of the validity of a model. This is why there has to be a clear statement about the purpose of a model. Some examples are shown in Fig. 6.6.

Further examples may help to understand the situation, communication, or documentation in general.

Another important aspect is that the modeling is done for a subject. A model may be built:

- to explain the function of an original to customers.
- to simulate the stress in a bolt by a computational engineer.
- to analyze the production cost by a person in the financial department.
- …

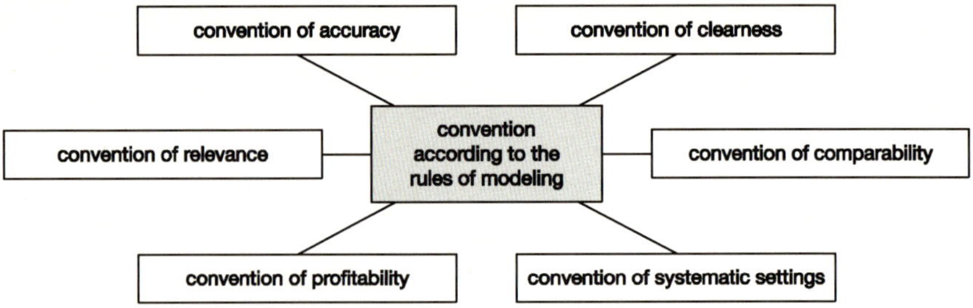

Fig. 6.7 Modeling conventions [1]

There is always a limiting time frame when the model is built. The limitation of capacity to be invested in building the model and the situation at the point of time when the model has been generated are important facts that should be known when using or discussing a model.

6.6 Quality of a Model

Starting with the purpose and the subject of the model, a number of different perspectives have to be kept in mind.

There is a set of different conventions (Fig. 6.7) that have to be considered during modeling. The interpretation of these conventions depends on the purpose (why), the subject (for whom), and the time frame (when).

The accuracy required is highly dependent on the purpose and the time frame and it represents the correspondence between original and model. In early phases of design, the cost calculation model will be good enough to estimate the range of future cost, in later phases the accuracy of the model has to be improved.

The clearness of the model is relevant for the subject, i.e., the user of the model. It is important to know the limits of a model and of course its purpose.

In a similar way, comparability, relevance, profitability, and systematic settings can be discussed.

These conventions provide a first set of requirements for a model. The list has to be completed by following a set of questions.

6.7 Requirements

Table 6.1 contains a systematic overview of the main questions to be aware of during collecting requirements in a generic way. The main focus has to be placed on the purpose of the model including the aspects of the subject and the time frame.

Table 6.1 Requirements [according to 5]

General model information	
Model name	A clear and unambiguous name should be chosen for each model
Subproject	Name of the subproject, which carries out the research
Modeling project purpose	
Purpose of the model	Which goal shall be achieved through the model? Are there any subgoals?
Original and system boundary	What is the original of the model (object of observation)? Can a model boundary be drawn already?
Classification of the model in the objective of the subproject	Which role plays the model in the objective of the subproject
User of the model	Who will use the developed model? (e.g., use in industry or science)
Relation to other models within the subproject	If other models exist or are planned within the subproject, which relations exist between this model and the other models?
Relation to models of other subprojects	Which relations between this model and existing/planned models of other subprojects already exist?
Requirements of the interfaces to other models	Which requirements for the interfaces to other models already exist?
Minimum requirements of the functionality of the model	When is the functionality of the model reached? What are the minimum requirements of the model concerning its functionality?
Timetable	When shall the model be completed? When shall the results of the model be used? What are important milestones during the development of the model?
Modeling project environment	
Institute/subproject	Which institute is responsible for the subproject/model?
Responsible person	Which persons are involved?
Previous knowledge	Which existing information/knowledge can be used for the modeling project? Hereby knowledge of the institute, previous projects, and personal knowledge should be considered
Modeling tool/language	Which modeling languages and tools are available and which one shall be used? Are certain languages/tools already determined (reasons)?
Modeling project outcome	
Presentation of results	Where and in which way shall the results be presented? (Graphical/verbal…)
Documentation	In which way shall the model/its usage/experiments be documented?
Requirements for a further use	Which requirements have to be fulfilled if the model shall be used beyond its main purpose? (e.g., regular updates, transfer to other projects…)
Necessary effort for maintenance and care	Is a regular maintenance and care necessary for the use of the model?
Visualization	

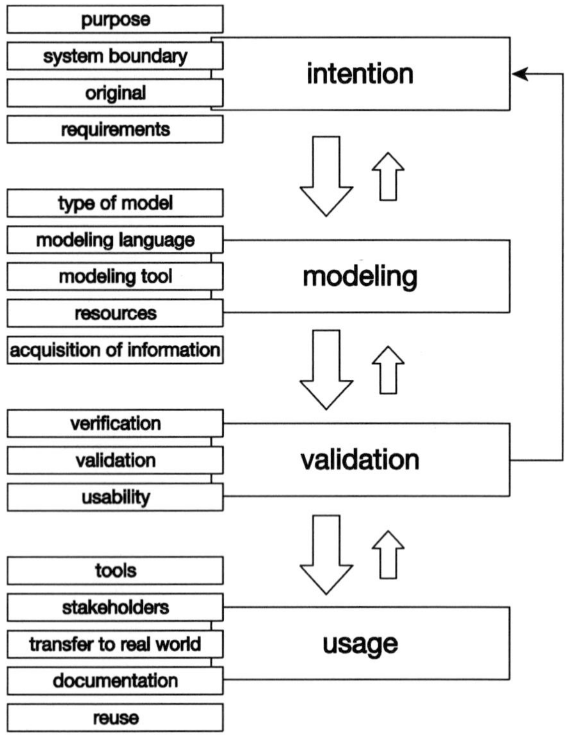

Fig. 6.8 Process of modeling

6.8 Process of Modeling

The generation and use of a model may follow the procedure shown in Fig. 6.8. Based on the discussion of purpose and requirements there are four major steps when working with models: Start with the definition of the intention (pragmatism including why, for whom, when), then build the model (reproduction and reduction based on pragmatism), then verify and validate the model and finally use the model. In addition, there are iterations especially after validation.

The intention includes the purpose, the definition of the system boundaries, the original, and the specific requirements.

In a second major step, the type of the model, the modeling language to be used, the modeling tool, the required resources, and the acquisition of necessary information have to be defined. The type of the model depends on the content, the purpose, and other characteristics. In processes, this may be the model of information flow, the structure of responsibilities, activities, and others. As for the product, this may be a list of requirements, the structure of functions, the use of a 3D-CAD-model, a video-sequence of a simulation tool, or several different forms of models. All possibilities require decisions about the language of modeling and the adequate tool. Data acquisition and achieving high data quality is often challenging.

During verification and validation the quality of the model has to be satisfactorily shown with regard to the given intention. Validation is important for quality assurance. Verification has to guarantee that all requirements are fulfilled in a correct way and validation has to show that the purpose of the model will be fulfilled. Usability checks should ensure that the subject (the user of the model) will be able to use the model in a correct way.

Based on the short discussion about modeling and design, findings and points of view about models of design will be pointed out as a conclusion.

6.9 Models of Design

A small number of models of design have been mentioned before. The general discussion about modeling has shown that models have to follow their purpose; a specific subject (user of the model) and that they are related to the given time frame. Due to these aspects we always have to accept that there will be a large number of models of design.

There are driving forces for an expansion of the family of models, as products are moving to a more complex level via mechatronic/adaptronic products toward PSS (Product Service Systems). The processes become more complex, too, due to trends such as globalization, a number of diversity issues in our societies, legislation, etc.

The large or even increasing number of models has to be accepted. But recognizing this and the above-discussed issues, all authors of new or modified models should clearly note what their model was made for. In the literature we can find a lot of models without a clear statement of their purpose, their subject, and the time frame. It should be a rule that authors not only present a model explaining design (or a process or a product ...) but indicate the reasons for this specific kind of a model. This kind of additional specification will help to recognize the limits of its validity.

There are several problems to be solved in design research. The number of different types of models and the number of different modeling languages are some of the key problems that have to be addressed. The modeling types and different languages have to meet the requirements of usability and purpose orientation.

References

1. Becker J (1998) Die Grundsätze ordnungsgemäßer Modellierung. Conference Modellierung betrieblicher Informationssysteme, Koblenz
2. DIN 19226 (1994) Beuth Berlin
3. Grabowski H, Rude S, Grein G (1998) Universal design theory. Shaker Aachen
4. ISO 10303

5. Kohn A et al. (2013) Improving common model understanding within collaborative engineering design research projects. International conference on research into design ICoRD'13. IIT Madras Chennai
6. Lauer M (2010) Integrative Dokumenten—und Prozessbeschreibung in dynamischen Produktentwicklungsprozessen. Dissertation, Technische Universität München
7. Lindemann U (2009) Methodische Entwicklung technischer Produkte. Springer, Berlin
8. Ponn J, Lindemann U (2011) Konzeptentwicklung und Gestaltung technischer Produkte. Springer, Berlin
9. Rodenacker W (1936) Anwendung der vergleichenden Getriebelehre nach Franke auf hydraulische Messapparate und Arbeitsmaschinen. Dissertationsdruckerei und Verlag Konrad Triltsch Würzburg
10. Rodenacker W (1971) Methodisches Konstruieren. Springer, Berlin
11. Roelofsen J, Lindemann U (2010) A framework for situation specific planning of product development processes. International design conference—DESIGN 2010, Dubrovnik
12. Shanon RE (1975) Systems simulation: the Art and science, Prentice-Hall,Englewood
13. Stachowiak H (1973) Allgemeine modelltheorie, Springer, Wien
14. The Oxford Dictionary (2012) http://oxforddictionaries.com/definition/english/-model. on 7 Oct 2012
15. Umeda Y, Takeda H, Tomiyama T, Yoshikawa H (1990) Function, behaviour, and structure. Applications of artificial intelligence in engineering V 1:177–194
16. VDI 3633 (2010) Beuth Berlin

Chapter 7
Perceiving Design as Modelling: A Cybernetic Systems Perspective

Anja M. Maier, David C. Wynn, Thomas J. Howard and Mogens Myrup Andreasen

7.1 Introduction

In many respects, design practice varies from person to person and from design project to design project. Based on the observation that creation and use of models is central to engineering design, to the extent that designing might be perceived as a propagation from model to model [37] and modelling can be viewed as 'the language of the designer' [2], in this chapter we propose to put the creation and use of models at the centre of an analysis of designing. Following Roozenburg and Eekels [36], we view 'modelling' as a heuristic process of creating and manipulating models, or generating and executing (or 'simulating') models (Fig. 7.1). Modelling is also a system of interactions between models, the modeller, and the thing or idea that is modelled. In overview, and prior to further elaborations later in this chapter, we consider the purpose of a model to be explaining or predicting behaviour, or articulating and realising something new. A model is a simplified and therefore to a certain extent a fictional or idealised representation. Interestingly, the term 'model' is rarely seen in dictionaries of engineering or architecture [33, p. 8].

Design engineers create and use a variety of models of the not-yet-finished product, for instance, to express the end products' forms and functions, e.g. through mathematical models, geometry models, parts drawings and function

A. M. Maier (✉) · T. J. Howard · M. M. Andreasen
The Department of Management Engineering, Technical University of Denmark, Lyngby, Denmark
e-mail: anja.maier@cantab.net

T. J. Howard
e-mail: thow@mek.dtu.dk

M. M. Andreasen
e-mail: mmya@mek.dtu.dk

D. C. Wynn
Department of Engineering, University of Cambridge, Cambridge, UK
e-mail: davidwynnemail@gmail.com

A. Chakrabarti and L. T. M. Blessing (eds.), *An Anthology of Theories and Models of Design*, DOI: 10.1007/978-1-4471-6338-1_7,

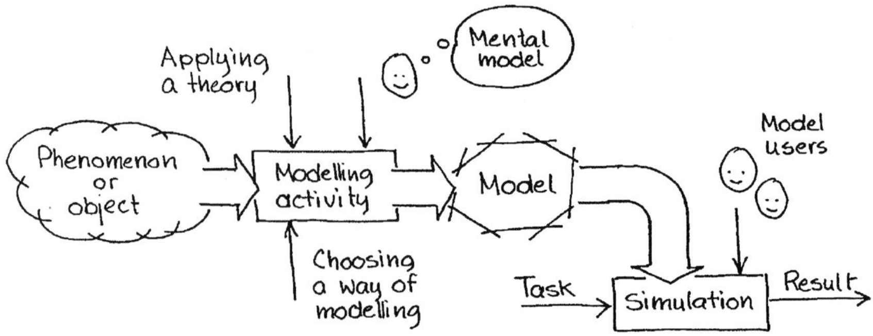

Fig. 7.1 The modelling process

models (e.g. Buur and Andreasen [8]). Models are also used to plan and execute the design process and understand who does what, along with time and resource constraints (e.g. Engwall et al. [15]). Such models can take the form of process flowcharts, formal procedures, Gantt charts, work breakdown structures and formal procedures [7]. Considering the latest developments in computer calculation speed and digitisation of design, development and manufacturing and digital fabrication devices with a focus on model-based or model-centric development, model 'manufacture' might become even more pervasive and the role of both product and process models in engineering design practice may become even more pronounced in future.

In this chapter, we will discuss what we understand by models and modelling, the purposes of modelling in designing, and the characteristics of a good model—or good modelling—that make it appropriate to enable design cognition and collaboration. We suggest that a cybernetic perspective may help to understand designing as a self-regulating modelling system and indicate some likely implications that could guide effective modelling in design.

7.2 Aspects of Modelling in Design

This section describes applications of product and process models in related design fields, discusses some of the purposes of modelling, and provides a brief review of perspectives on process models in engineering design literature.

7.2.1 Applications of Models in Designing

Models support, guide, and embody progress of the design process from the initial recognition of a market need or idea until the product is sold and recycled. Modelling thus plays a central role in most (if not all) design fields. For example:

An *engineering design* process may flow from mental exercises, impressions or thought models, turning something over in the mind's eye [16], to devising artefact and process models such as flowcharts, technical drawings [20], design sketches [24], function models, virtual models (such as CAD, FEA and PDM) and physical prototypes (from breadboard mock-up to 3D printing). To give an example from a situation the first author observed: A design engineer working in a medium-sized aerospace component supplier, walks with a piece of hardware in his hand across the hallway and drops it on his colleague's desk with the words 'This model is broken. Fix it'. He uses the term 'model' to refer to a full-scale prototype of an actuator for a jet engine fuel injection system. The colleague inspects the piece from all sides, finds some pressure points and replies by saying 'can I see the results from your stress and heat simulation models'?

In *product development*, design managers create models for many purposes—to explore, create, understand, rationalise, communicate and regulate. By making and using models, the designer is able to gain information about the relations between decisions and consequences. Models in product development concern processes as well as products; indeed the term process is often used interchangeably with 'process model'. Through a series of interviews with design managers, Engwall et al. [15] list five different conceptions of process model use: administrating, organising, sense-giving, team-building and engineering, i.e. solving technical problems. The models discussed in the interviews ranged from rough outlines of the workflow to more sophisticated versions of comprehensive management systems used in design and product development (see also Browning [7]).

In industrial design and architecture, (physical) design modelling is most commonly referred to as conceptualising and materialising design intent, as a way in which designers realise mental concepts [12, 38]. There are many steps from an idea to a CAD model to a physical part. A designer might start with a sketch or drawings to work out the dimensions and how things are going to go together, building up structure, using cardboard as structural element or other materials such as clay, wood, foam, plastic or foil and tools such as cutters and glue guns and other workshop tools or more advanced additive manufacturing techniques. By working with their hands, designers develop an understanding and appreciation for instance for the form, the function, and the usability of the final product to be designed. The models are not just a way to externalise the designers' thoughts; the techniques have some emergent properties that surprise and inspire the designers.

In *interaction, ergonomic and user-centered design*, modelling practice might typically be thought of as creating and working with personas as prototypical users having goals, desired experiences, backgrounds and mental models. Further, modelling practice in these domains includes devising storyboards, customer journey mapping to define the different touchpoints that characterise the interaction of the user with the product or service, and use cases and scenarios which product users are engaged in or want to engage in (e.g. Buxton [9]). All these activities are part of designing user experience as modelling of products and services.

instance a freehand sketch, a bold freehand drawing, straight-edge drawing, a drafting machine, templates or a plotter.

Specific drawings in design may be viewed as the combination of the selections within each of the four characteristics described above [42]. This illustrates how the process of modelling may be viewed as communication—selecting and encoding the content of the 'message'. The code used (e.g. notation, drafting standards, projections) must be known to the participants in the modelling process. Put another way, a model may be viewed as a communicative medium between different stakeholders. This use of the term 'medium' assumes a broader definition than conventional use of the term might permit, but it is not unprecedented [11]. Luhmann [26, p. 220, 27, p. 160], for example, distinguishes between three different types of media that are complementary: language, which is visible in linguistic forms, such as sentences; media of dissemination, such as writing, printing and electronic broadcasting and symbolically generalised communication media, such as money as a medium for transaction (see also McLuhan [32]). Despite writing for a different audience, their line of argument could be extended to denote a product, or representations of it (such as a sketch, CAD model or physical prototype) as media (see also Maier et al. [28]).

7.2.4 Perspectives on Process Models in Design Research

Design researchers have long been concerned with developing models, procedures and methods to assist effective product development. Andreasen [2] explains a now-common view that methods and models allow the properties and characteristics of the design to be determined, and furthermore that design can be viewed as creating a sequence of models to obtain answers to queries raised while designing. According to Eckert and Stacey [14] and also in this volume, it is now 'established wisdom in the design community that models are a useful means of understanding and interacting with both products and processes'. Albers and Braun [1] and also in this volume write that process models 'become increasingly important to designers as complexity grows'. However, models are often difficult to work with in practice; Wynn and Clarkson [48] write that no method or process model provides a 'silver bullet solution'. Smith and Morrow [41] suggest that a process model should have four characteristics to be useful to support managerial decision-making: 'it addresses an important managerial issue; the decision-making is based on information that is available and accurate; the assumptions and simplifications of the model are reasonable; and the model is computationally tractable'.

Practitioners, researchers and educators alike are seeking to fully understand the multiplicity of model types and how they might be best deployed at appropriate stages in the design process [33]. Most analyses of models in design focus on developing classifications according to model characteristics, which vary largely due to their foci on different situations. For example, Wynn [47] develops a classification scheme in which design process models are viewed as abstract, procedural or analytical on the top level, where (for instance) procedural models are divided into those focusing on project guidance and those focusing on design activity guidance. In a literature review of over 100 design and creativity process models, Howard et al. [22] showed the gap between the procedural engineering process models and the iterative, flexible creative process models. In a similar study, Gericke and Blessing [18] conclude that comparison of existing process models in design is challenging due to the differing level of detail and specific application contexts of each.

The prevailing conclusion from authors writing on models and methods in design seems to be that too much heterogeneity is problematic and that design research should aim towards rationalisation, consolidation and integration of the ideas. Thus, for instance, Albers and Braun aim to '[en]compass entire engineering processes in one consistent modelling framework' [1]. Browning et al. [6] argue that many benefits could be gained if an organisation held a single, consistent model from which different views could be generated on demand.

7.3 Philosophical Perspectives on Models in Design

The previous section summarised some ways that models are used in design practice. On a philosophical level, questions of what a model is, what it is used for and how modelling functions, are often debated. There is no uniform terminology used—be it by (natural) scientists, designers or philosophers of science to name a few disciplines engaged in the debate. Despite the fact that discussion around the role of models has generated considerable interest, e.g. among logicians and philosophers of science (e.g. [17, 43, 45]) and among engineering design researchers (e.g. [8, 14, 30, 49]), there remains no clear consensus regarding what models are, how they work in engineering design, and how a conceptualisation of model creation and use might function as a platform for design theory.

7.3.1 Explanatory, Predictive and Synthetic Models in Design

In many respects designing is different from some other uses of models studied in the literature. In particular, much of the philosophy of science literature focuses on a specific kind of model that aims to describe a situation as it really exists. In designing (and many other organisational situations) models are often not used in this way. Models of a product (synonymously used with artefact models, product models or designs as all representations leading up to the final product), of the design process and of an organisation are all used to represent a situation in mind.

This difference has important consequences for the way models and modelling should be considered. By focusing on the claimed relationship between a model and the 'real world', modelling can be perceived as either explanatory, predictive or synthetic:

Explanatory, where a claim is made, or belief is held, that the workings of a model map directly onto, or truly explain, 'real-world' mechanisms that 'cause' observable behaviour. Whether or not this is a sensible perspective to take is philosophically controversial, although in practice is generally held with respect to some established theories such as evolution by natural selection. An example of design modelling explaining real-world behaviour is the use of kinematic analysis [10] to understand the behaviour of a mechanism. By sketching a mechanism and through simple analysis of the degrees of freedom of the members, it is possible to understand why a mechanism behaves as it does. If it works as expected it will be correctly constrained, if it moves in an undesirable/unpredictable pattern it will be underconstrained, if it jams or does not move it will be overconstrained, and in special cases, if it works as desired but exhibits increased wear, friction and/or assembly issues, then it will be over constrained by one degree.

Predictive, where a claim is made that a model can predict phenomena, but it is acknowledged that underlying real-world mechanisms may not exist in the form the model suggests, or the issue is viewed as unimportant (such as the Newtonian description of the motion of bodies under gravity). Good examples of models in designing that are acknowledged to be predictive, but not explanatory, are correlations based on past designs that are used to make early design decisions. Another set of predictive models are those based on scientific computation, in which many of the relevant issues are illustrated by computational fluid dynamics (CFD). CFD can be applied in many ways. Numerical models based on potential flow equations include issues such as statistical characterisation of numerical data, estimating the probabilistic future behaviour of a system based on past behaviour. This can include prediction outside the data range (extrapolation) or within the data range (interpolation) of data based on best fit, error estimates of observations, spectral analysis of data or model-generated output. The example of CFD illustrates how scientific computations are based on simplifications to ensure tractability, and are explicitly not explanatory.

Synthetic, where a model is explicitly recognised to not represent a real situation, but rather to represent an idea and thus to bring a situation into being. In synthetic terms, a model is executed in order to gain knowledge about actions that are possibly required today in order to guarantee or even generate a satisfactory future [40]. Stepping back from individual modelling actions that might be viewed as explanatory or predictive, it seems that most models in design fulfil a synthetic role. For instance, when designers sketch a mechanism, then formalise and analyse it, they are on one level analysing, but stepping back they are synthesising something that did not previously exist.

The difference between these categories lies mainly in the philosophical viewpoint, although some models seem to be best viewed in different ways depending on their use, (for example, Finite Element Analysis (FEA) can be used to explain the stress in a loaded joint or predict it). In the context of designing, pragmatically speaking, there may be little significant difference between an explanatory and predictive outlook. Eckert and Stacey argue for such a pragmatic perspective: 'It is important that designers and managers understand that models can be interpreted in different ways and that people have different expectations of them. Rather than prescribing an interpretation of models, the understanding of the role that models can play in an organisation needs to be negotiated within a team and an organisation. An understanding of a model is a cognitive construct rather than an inherent property of the model, and a shared understanding is constructed through social processes of discussion and clarification' [14, p. 11].

7.4 Effectiveness of a Model

All models have limitations. No model can possibly describe, explain and predict every detail of a phenomenon—and nor should it. An important question when considering the role of models and modelling is to understand when a model is a 'good' model, i.e. what makes a model helpful to support understanding, analysis or design.

The issues are different depending on which perspective of 'model' is taken. From the explanatory and predictive viewpoints, one of the key criteria for a good model is that it should provide an accurate match onto its target. This might be interpreted as its ability to explain underlying mechanisms (for the explanatory viewpoint) and/or to accurately predict patterns in observations (for the predictive viewpoint). The basic assumption in both cases, and source of much philosophical difficulty, is that a model (more specifically, its interpretation) is internal to the modeller, while the target is 'outside' and can only be observed through imperfect 'sensors'. Some mechanism must thus allow observations of the target to be accumulated and processed to refine the internal model. This viewpoint raises many of the philosophical 'big problems' of induction, causality and so forth. Important questions here include: Should we believe that repeated observations give improved confidence in a model's accuracy? Is it meaningful to assume a cause or mechanism underlying observable phenomena? A great deal of philosophical effort has been dedicated to these kinds of problems; it has been said that the degree to which a model is taken as exactly and adequately representing reality is one of the most significant ways that claims about models in science can differ [25].

Although relevant, these questions might not be the best approach to understanding models in designing, because of the primarily synthetic nature of models in this context. Understanding the goodness of synthetic models raises a different set of issues, because the goodness of a model is not so immediately related to how well it represents a target. One key difficulty lies in apparent cyclic causality [44]. One might ask: is it meaningful to ask how well a model represents a situation, if that situation will be created or changed through the modelling? (Nevertheless as explained in the above example, synthetic models must have some degree of predictive power because they are able to bridge between a concept and its 'real world' implications, to allow judgements to be made and synthesis based on those models to progress). The use of models may be compared to children playing with toys. A few LEGO blocks put together, for example, may unfold a fantastic world in the child's mind. We cannot reason directly from the model to its effects, because the relationship also includes the modeller or interpreter.

What makes a model a good model thus lies not so much in goodness of fit, meaning how accurately it represents its 'target', but rather the degree to which it enables decision-making that turns out to add value given a certain purpose and context.

As Browning puts it, fitness of a process model depends on the alignment of its content and structure with what is appropriate to support a particular decision, purpose or use case [7, p. 75] as also argued earlier by Ozgur [34]. Some criteria that a good synthetic model of an artefact should fulfil might include: it should represent an imagined situation that is nevertheless plausible to transfer to the real world; studying the model should have potential to reveal problems that might arise in the idea if it were made real; the model should participate explicitly or implicitly in suggesting a next action, or at least narrow down the possibilities, for progressing the idea and it should represent an incremental step in detailing or concretion of the idea, in relation to its predecessors and successors in a chain of models fulfilling these same criteria.

Taking such issues into account, it seems that the problem of understanding how good a synthetic model is, could well require a different frame of reference than understanding how well a model can explain or predict observable phenomena if this is what it is expected to do.

7.5 Cybernetic Perspective for a Model-Centric View of Designing

The previous section highlights some of the issues relating to understanding the goodness, or usefulness, of a model viewed in the synthetic sense. Our final point is now made, namely that cybernetic thinking can help to unpack some of these issues and thus reach a better understanding of the effectiveness of models and modelling as used in design.

7.5.1 Understanding Models and Modelling in the Cybernetic Sense

The term 'cybernetics' is derived from the Greek word kybernetes meaning helmsman or cox, from which today's terms of governor, regulator, controller also originate. Cybernetics aims to provide a meta-language to describe different kinds

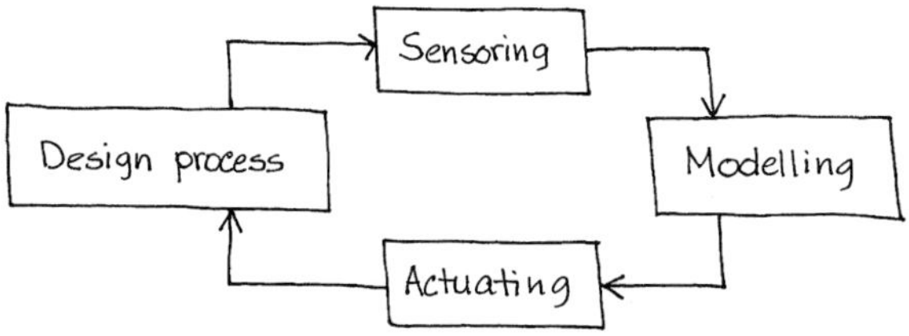

Fig. 7.3 The design process as a self-regulated system

of systems by focusing on information, control and circularity. It is concerned with understanding how systems are, or can be controlled through (self)regulation of their behaviour in the presence of uncertainty, disturbance and changing objectives [5, 46].

It has been proposed that the design process can be viewed as a cybernetic system and that this perspective can reveal the roles of models and modelling within it [30, 50]. According to the authors' interpretation, the participants in the process act as cybernetic 'controllers'. They 'sense' the state of the process from the viewpoint of their own interactions within it (see Fig. 7.3). They use process models and design methods to guide their response to this perceived state, thus acting as 'actuators' that influence the process according to their goals. In elaborating how a model affects an unfolding process, this viewpoint provides a framework to consider the synthetic role of models in designing.

As an illustrative example, consider following a road map. While travelling down a motorway you see junction numbers and road signs, with reference to the map (the model) you can then locate your current position and then decide which exit to take to reach your destination. Similarly in design: for example, a FEA model may show stress hotspots on a model. Based on these hotspots, the geometric computer-aided design (CAD) model may be adjusted to compensate for these hotspots, placing extra material around areas of high stress concentration. Notice here the cybernetic system is between a model of the product, a model or the situation and the modeller. This occurs up until manufacture or prototyping where the actual produced parts and products can be inspected, replacing the virtual situation with a real situation.

The understanding of 'model' in the cybernetic sense is somewhat different to the colloquial use and the situations outlined earlier in this chapter. In the cybernetic sense, a model must be a description or conception of a situation that is used to guide or influence the response to that situation. Thus, a process model, for instance, must not only represent the process, but also be 'brought to life' by interpreting it in the context of a given situation and with respect to a goal, and the resulting insights must be used to take action. To 'bring a model to life' requires that the modeller interprets and incorporates aspects of the situation to be regulated into the model as they understand it. Interpreting a model in this way to form an actionable mental model is also a form of modelling. A product model, by highlighting features of an emerging artefact, can similarly be viewed as guiding the designer's response to the current design state. The product as the emerging design object is regulated with respect to meeting the design objectives and requirements, while the design process is regulated with respect to project objectives, such as time, cost and conformance to company procedures.

To summarise, cybernetics views a model not primarily as a representation of a target, but rather as a tool for making decisions (even if a model may not be directly used to take decisions, it may inform decision policies). According to this perspective, a good model is one that helps to elicit and/or generate information that make a good sequence of decisions—this does not necessarily require a good or accurate representation, but an ability to connect cause and effect within the certain specific context, in which only a certain range of actions may be possible and only a certain range of effects may be observable. In turn, a good sequence of decisions is one that results in a good result, accounting for the cyclic causality that results from the model playing a synthetic role. A good result is a design process that delivers a suitable product, meeting foreseen and perhaps unforeseen customer requirements, within a suitable time and cost.

Considering the principles of cybernetics reveals a number of suggestions for organisational behaviours to improve modelling in design from this perspective. These implications are summarised below (for further discussion, refer to Wynn et al. [50], Maier et al. [30]).

Key cybernetic principles pertaining to the effectiveness of a regulated system are the principles of requisite knowledge [21] which is related to Ashby's well-known law of requisite variety [4]. In effect, these respectively state that effective regulation requires a suitable model of the effects of one's actions, and second, that these actions must also be carried out. Selecting an action that is exactly optimal would require that the model used to make these predictions has a level of complexity requisite to that of the system under regulation. Consideration of these principles highlights that, an effective modelling system must detect deviation, must possess suitable models to decide what action to take, and must be able to carry out those actions. A modelling system, to reiterate, includes in our view the modeller, the model and the modelled object. Of course, in a complex system such as the design process, requisite knowledge and variety are not usually possible, since models are by their nature much simpler than the processes they help to regulate. Thus, one might think of regulation as influence, rather than as control.

7.5.2 Cybernetic Systems and Learning

As modelling systems seek to adapt to an ever-changing environment they can be said to learn. Learning uses feedback about system performance to improve the model that governs response to stimuli. Argyris and Schön [3] distinguish between single-loop and double-loop learning. Single-loop learning corresponds to changes to strategies of action (we can interpret this as the model used to decide how to respond to observations and changes). In terms of process operation and improvement, a model might be updated when advice derived through that model, does not yield the expected benefits. The change is in the model only. In double-loop learning, a connection is made at a higher level among (1) the observed effects of actions; (2) the models that were used to guide action and (3) the values and norms by which action is deemed successful. Here, an example might be that the goal of modelling was recognised to be inappropriate to yield overall objectives. A focus on production cost might then be augmented with greater consideration of lifecycle costs, and this in turn would require changes to the models used to guide decision-making. The change is in the model and in the system objectives.

Consideration of these principles suggests that an effective modelling system should reflect on the consequences of its actions, and should align the objectives of its modelling parts to minimise conflicting actions.

Another important aspect of modelling is abstracting the complexity of a real system to highlight certain factors which are most pertinent to decision-making according to the pertinent objectives. In the context of mathematical or simulation modelling, for instance, it is necessary to determine a small set of assumptions and variables in order to render analysis tractable. Finding an appropriate way to do this is often not obvious when a modeller is faced by complex, ambiguous situations such as in design processes.

Finally, the ability of a dynamic system to remain stable under changing conditions may assist learning by making it easier to identify whether modelling interventions actually result in improved performance. This is especially important when the system and its environment are continuously changing and when many models are in operation concurrently. In practice, stability may often be enhanced by minimising information flow delays and enabling rapid feedback in response to changes.

7.5.3 Implications for Modelling in Design

A key characteristic of design processes is that they are not concerned with simply processing information relative to a fixed goal, or a goal that varies along fixed dimensions. Rather, design is about doing something new or at least on some level in a new way. The design process creates knowledge that in turn leads to refinement or even redefinition of the design objectives. The design process thus redirects itself in unforeseeable ways as it progresses. It is not only goal-directed, but also goal-directing and even goal-defining. Each step in the design process, as it unfolds, creates possible options for progression while closing off others. This leads to a number of behaviours, such as being iterative (e.g. Wynn et al. [49]. For an overview of characteristics of engineering design processes, see Maier and Störrle [29]).

A key point highlighted at several places in the earlier discussion is that the design process constructs and maintains models as well as using them to regulate itself and its interactions with its environment. The modelling processes within such a system may be characterised as incorporating all activities that form a part of developing models, including the development of the modellers' perception and imagination [23, p. 101].

To be useful in guiding a process given the dynamic nature of the process and possibility for multiple directions, the system of models used to regulate a designing organisation must exhibit variety requisite to that of the process. Put another way a method or process model to support designing should not be too detailed and prescriptive, because it could then only provide guidance for very specific situations—whereas designing is dynamic and emergent in nature. In practice an organisation has many process models (interpreted broadly, many ideas of the best way to approach certain situations), and possible interpretations thereof, thus allowing selection of guidance based on apparent best fit of each model to the ambiguous information available at each point in time [35].

The population of process models and interpretations they afford are continuously refined, developed and discarded as knowledge is created and accumulated regarding the design context, and as particular models prove to be useful, or not, within that context. The variety of models and interpretations allows the emerging complexity to be managed. This seems contradictory to prevailing wisdom that process models in an organisation should be rationalised and simplified; a variety of models may indeed cause difficulties in coordination and control—but it also embodies the capability to adapt to change.

7.6 Concluding Remarks

A model-centric view of designing has potential to be considered as an integrative view of designing. Taking models as a starting point for analysing design activity takes a systems view and integrates the modeller, the model user and the object it is representing. A model-centric view of designing through the lens of cybernetics helps in becoming cognisant of the role of models in the way designing unfolds. Like other knowledge-intensive activities, designing depends on reasoning about, and with, models (construed broadly). To support designing it may help to focus on improving models. But what are the criteria for good models? The major implication of the cybernetic view suggested here is that we need to focus not on the content of a model, meaning how well it represents, but how it fits into the context of use. In other words, the goodness of a model depends on the context of its application.

Models, e.g. process models, need to be matched to the ways of working of the individuals and the characteristics of the problem they are solving. Some groups need well-defined prescriptive process models (how to do things, how to coordinate themselves) while others seem to be very effective with an implicit shared understanding of the work steps and coordination requirements. But without an effective process model, even an implicit one, it is difficult to coordinate effectively. Using cybernetics to explain modelling in design emphasises a systems perspective, where the mode of inquiry is synthesis. It asks us to think of the parts and their interactions; the modellers, the models, the modelled things and their compound influence throughout the process of design. Viewing designing in this systemic way may help to suggest criteria that enable design teams and organisations achieve their goals in different situations. These goals may include stability, flexibility, adaptability, reliability and scalability to name a few.

Acknowledgments The authors wish thank the two anonymous reviewers for their suggestions and the following people for stimulating discussions on modelling and models in designing which influenced the thoughts and arguments proposed in this chapter: Tyson Browning, P. John Clarkson, Jaap Daalhuizen, Claudia Eckert, Boris Eisenbart, Kilian Gericke, Khadidja Grebici, Ralf Stetter, Harald Störrle and Ken Wallace.

References

1. Albers A, Braun A (2011) A generalized framework to compass and to support complex product engineering processes. Int J of Prod Dev 15(1):6–25
2. Andreasen MM (1994) Modelling—the language of the designer. J Eng Des 5(2):103–115
3. Argyris C, Schön DA (1978) Organisational learning. a theory of action perspective. Addison-Wesley, Massachusetts
4. Ashby WR (1956) Principles of self organizing systems. Aldine, Chicago
5. Ashby WR (1965) An introduction to cybernetics. Chapman and Hall Ltd, London
6. Browning TR, Fricke E, Negele H (2006) Key concepts in modeling product development processes. Syst Eng 9(2):104–128
7. Browning TR (2009) The many views of a process: toward a process architecture framework for product development processes. Syst Eng 12(1):69–90
8. Buur J, Andreasen MM (1989) Design models in mechatronic product development. Des Stud 10(3):155–162
9. Buxton B (2007) Sketching user experiences: getting the design right and the right design (interactive technologies). Morgan Kaufmann, Massachusetts
10. Christensen ME, Howard TJ, Rasmussen JJ (2012) The foundation for robust design: enabling robustness through kinematic design and design clarity. In: Proceedings of the 12th international design conference DESIGN 2012, pp 817–826)
11. Crilly N, Maier AM, Clarkson PJ (2008) Representing artefacts as media: modelling the relationship between designer intent and consumer experience. Int J Des 2(3):15–27
12. Cuff D (1992) Architecture: the story of practice. MIT Press, Boston
13. Deutsch KW (1952) On communication models in the social sciences. Pub Opin Quart 16(3):356–380
14. Eckert CM, Stacey MK (2010) What is a Process Model? Reflections on the epistemology of process models. In: Heisig P, Clarkson PJ, Vajna S (eds) Modelling and management of engineering processes. Springer, New York, pp 3–14
15. Engwall M, Kling R, Werr A (2005) Models in action: how management models are interpreted in new product development. R&D Manag 35(4):427–439
16. Ferguson ES (2001) Engineering and the Mind's Eye. The MIT Press, Cambridge
17. Frigg R, Hartmann S (2012) Models in Science. In: Edward NZ (ed) The stanford encyclopedia of philosophy (Fall 2012 Edition). http://plato.stanford.edu/cgi-bin/encyclopedia/archinfo.cgi?entry=models-science
18. Gericke K, Blessing L (2011) Comparisons of design methodologies and process models across disciplines: a literature review. In: ICED 11–18th international conference on engineering design—impacting society through engineering design vol 1, pp. 393–404

19. Goldschmidt G (1995) The designer as a team of one. Des Stud 16(2):189–209
20. Henderson K (1999) On line and on paper: visual representations, visual culture, and computer graphics in design engineering. MIT Press, Cambridge
21. Heylighen F (1992) Principles of Systems and Cybernetics: an evolutionary perspective. In: Trappl R (ed) Cybernetics and systems'92. World Science, Singapore, pp 3–10
22. Howard TJ, Culley SJ, Dekoninck E (2008) Describing the creative design process by the integration of engineering design and cognitive psychology literature. Des Stud 29(2):160–180
23. Hubka V, Eder EW (1987) A scientific approach to engineering design. Des Stud 8:123–137
24. Kavakli M, Gero JS (2001) Sketching as mental imagery processing. Des Stud 22(4):347–364
25. Lloyd EA (1998) Routledge encyclopedia of philosophy. Routledge, London
26. Luhmann N (1984) Soziale Systeme. Suhrkamp, Frankfurt am Main
27. Luhmann N (1995) Social systems. Stanford University Press, Stanford
28. Maier AM, Eckert CM, Clarkson PJ (2005) A meta-model for communication in engineering design. CoDesign 1(4):243–254
29. Maier A, Störrle H (2011) What are the characteristics of engineering design processes? In: Proceedings of the 18th international conference on engineering design, vol 1 Design P, pp 188–198)
30. Maier A, Wynn DC, Andreasen MM, Clarkson PJ (2012) A cybernetic perspective on methods and process models in collaborative designing. In: Marjanovic D, Storga M, Pavkovic N, Bojcetic N (eds) Proceedings of the 12th international design conference DESIGN 2012, pp 233–240
31. Maier MW, Rechtin E (2002) The art of systems architecting, 2nd edn. CRC Press, Boca Raton
32. McLuhan M (1964) Understanding media: the extensions of man. McGraw-Hill, New York
33. Morris M (2006) Models: architecture and the miniature. Wiley, Chichester
34. Ozgur E (2004) Effective inquiry for innovative engineering design. Kluwer Acadmic Publishers, Boston
35. Pask G (1975) The cybernetics of human learning and performance. Hutchinson, Paris
36. Roozenburg NFM, Eekels J (1995) Product design, fundamentals and methods. Wiley, Chichester
37. Roth K (1986) Modellbildung fuer das methodische Konstruieren ohne und mit Rechnerunterstuetzung. VDI Zeitschrift, 128
38. Sass L, Oxman R (2006) Materializing design: the implications of rapid prototyping in digital design. Des Stud 27(3):325–355
39. Schön DA (1992) Designing as reflective conversation with the materials of a design situation. Knowl-Based Syst 5(1):3–14
40. Simon HA (1969) The sciences of the artificial. Institue of Technology Press, Cambridge
41. Smith RP, Morrow JA (1999) Product development process modeling. Des Stud 20(3):237–261
42. Tjalve E, Andreasen MM, Schmidt FF (1979) Engineering graphic modelling: a practical guide to drawing and dseign. Newness Butterworths, London
43. Van Fraassen B (2008) Scientific representation: paradoxes of perspective. Oxford University Press, Oxford
44. Von Foerster H (2002) Understanding understanding: essays on cybernetics and cognition. Springer, New York
45. Wartofsky MW (1979) Models: representation and scientific understanding. Macmillan, Basingstoke
46. Wiener N (1948) Cybernetics, or Communication and Control in the Animal and the Machine. Cambridge: MIT Press
47. Wynn DC (2007) Model-based approaches to support process improvement in complex product development. University of Cambridge: Cambridge
48. Wynn DC, Clarkson PJ (2005) Models of designing. In: Clarkson PJ, Eckert C (eds) Design process improvement—a review of current practice. Springer, London, pp 34–59

49. Wynn DC, Eckert CM, Clarkson PJ (2007) Modelling iteration in engineering design. In: Bocquet J.-C (ed) Proceedings of 16th international conference on engineering design, Design Society, Paris (paper id: 561)
50. Wynn DC, Maier AM, Clarkson PJ (2010) How can PD process modelling be made more useful? In: Marjanovic D, Storga M, Pavkovic N, Bojcetic N (eds) Proceedings of the 11th international design conference DESIGN 2010, pp 511–522

Chapter 8
The Contact and Channel Approach ($C\&C^2$-A): Relating a System's Physical Structure to Its Functionality

Albert Albers and Eike Wintergerst

8.1 Introduction

Research on theories and models of design is often motivated from observations in designing, i.e. they address a specific purpose and are intended to describe, explain or predict certain phenomena that pose an unsolved challenge both for the research community and for design practitioners. This chapter is dedicated to one of these present challenges in engineering design: to explain and to provide efficient means to describe how the quality of a product's technical functions and properties depend on the design of its physical characteristics.

The following sections deal with theories and models that address this topic and explain how analysis and synthesis of functions can be facilitated in complex systems engineering projects. Based on theoretical considerations and empirical research results, conclusions are drawn that motivate further research on an integrated modelling approach.

In particular, the Contact and Channel Approach ($C\&C^2$-A) is introduced which addresses the need to associate a product's functions with its physical structure and embodiment. Strengths and limitations are compared and discussed based on observations from recent research projects that contribute to its evaluation and further development in design practice.

A. Albers (✉) · E. Wintergerst
IPEK—Institute of Product Engineering, Karlsruhe Institute of Technology (KIT), Karlsruhe, Germany
e-mail: albert.albers@kit.edu

E. Wintergerst
e-mail: eike.wintergerst@kit.edu

A. Chakrabarti and L. T. M. Blessing (eds.), *An Anthology of Theories and Models of Design*, DOI: 10.1007/978-1-4471-6338-1_8,

8.2 Analysis and Synthesis in Engineering Design

Activities of engineering designers are generally characterised by the task to find solutions for complex technical problems: 'Engineering designers solve design problems related to product functionality' [1]. Changing customer demands and improved solutions from competitors are generating pressure on companies to optimise their products under technical and economic conditions. The majority of product engineering projects is therefore focused on Product Generation Development [2]: optimising existing products as well as developing new product generations with improved and extended functionality. Objectives like these require systematic changes on a product's physical structure. This chapter is therefore dedicated to models that facilitate designers in describing how the embodiment design of a product determines its functions and properties (e.g. its performance, reliability or safety).

From a psychology perspective, product engineering can be described as a collection of both thinking and explicating activities with the purpose of developing models of a future object which currently does not yet exist [3]. These activities are composed of both opportunistic and systematic periods [4]. Engineering designers usually do not follow a strictly formal procedure, but rather progress in iterations on different levels of abstraction. While doing so, they combine both mental and explicit models and quickly switch between analysis and synthesis activities. Studies show that designers even in early stages of conceptual design work with component geometry as the primary representation of products [5–7]. Especially for analysis activities, designers are more confident and correct in making conclusions when using high-fidelity and physical representations [7]. While synthesis is a strongly mental endeavour because of the affordance of creativity and cross-linking thinking, analysis activities are grounded on the possibility to use explicit models and tools. Analysis is especially important to focus a design problem, to specify its characteristics and to set a basic understanding about what kind of solution principles might address it. 'Many products carry components or solutions principles across from their predecessor designs. It is therefore important for designers to understand how the original product works, what its strengths and weaknesses are and what problems have occurred with it' [5]. Besides the clarification of a design problem and related objectives, analysis plays a major role in validating system properties according to their desired quality. These validation activities proceed in parallel to synthesis activities where the newly gained knowledge is transferred into adjusted specifications for embodiment design characteristics. One aim of engineering design research is therefore to develop theories, methods and models that facilitate designers to analyse design problems and to create appropriate solutions [8].

Manifold purposes and perspectives may motivate designers to make use of theories and product models. In general, theories and product models provide a framework for making information accessible (analysis) as well as for expressing design concepts and decisions (synthesis). They serve designers to capture, to focus, to structure, to make explicit and to simplify the complex relationships of a system's properties and characteristics. Thus, they serve as a means to overview, explore, understand and communicate such relationships at a systems level. It can be argued that designs evolve through abstraction and refinement, which occur in the framework of those product models [5]. Product models may consequently provide information that supports both creative thinking as well as deductive reasoning.

Since the primary task of designing is to define a system's characteristics in a way that it may perform its functions in a defined quality, product models should refer to physical characteristics and the related functional properties of a system. However, in design practice product models often describe just the physical structure of a product. The focus of the following sections will thus be set on findings from design research that recognise the importance of modelling embodiment design characteristics in relation to corresponding functions and properties of a system for the purpose of functional analysis and systems engineering design.

8.3 Functions in Design Research and Practice

The market value of a product often refers to the quality of its functions. Creating these functions by defining appropriate design parameters under technical and economical boundary conditions is a major challenge in the product development process. The relationship between a product's physical embodiment and its functions has thus been studied extensively and a number of related concepts are being proposed in the literature. Many of them focus on the term 'function' to articulate a purpose, how a product is intended to behave and how it might achieve this, as well as what it definitely should not do, and how to prevent this from occurring.

8.3.1 Modelling Functions in Relation to a Product's Physical Structure

The final result of a product development process is usually the documentation of its physical embodiment in technical drawings, virtual models and real prototypes. They specify design characteristics of systems, subsystems and single components that interact with each other in order to fulfil functions in a desired sequence, quality and within a framework of boundary conditions (e.g. influences from remaining systems and the environment). Except from main functions of a product, designers in practice often deal with sub-functions and properties only mentally, without explicitly noting them down or stating their relation to the respective embodiment design [9]. As a consequence, neither these functions nor the relation of functions or boundary conditions to the product's embodiment design are usually documented in any way—only the knowledge and mental models of the participating designers are complementing the available explicit documents. Based on a study in industrial enterprises, Matthiesen reports that although 'most design engineers are familiar with function structures and appreciate them as sensible and profitable, [...] it proved to be impossible to find a single distinct function structure in any case'. He concludes that in design practice, 'functions are not very unambiguous, if they are documented at all, and are not associated with the embodiment documentation' [9]. This may lead to various kinds of problems and inefficiency in design processes: in discussions of functional problems with other designers, negotiations on tolerances versus production costs or simply to find out which design parameters should be modified in order to optimise a product's functionality.

Engineering design research is drawing its motivation from this situation. Various approaches contribute to this scientific discussion and provide means to improve this situation. For instance, CPM [10] describes a generalised framework to express properties of a product in relation to its physical embodiment and design characteristics. Accordingly, design characteristics can be directly specified by designers, expressing, e.g. the shape and the structure of a product. In contrast, functions and properties can only be indirectly implemented into a product. They directly depend on the values and configuration of design characteristics and can only be analysed through testing/usage of a product. Thus, designers are supposed to implement functions and properties by determining and iteratively specifying appropriate design characteristics (synthesis). The Axiomatic Design Theory [11] describes this iterative behaviour of designers as a sequence of 'zig-zagging' modelling activities. In this process, the embodiment of functional requirements into components generates new sub-requirements on both function and form which must be addressed by further detailing the component designs. Another model that implements both views on functions and embodiment design is introduced by Lindeman et al. [12]. Based on design structure matrices (DSM), relations between functions and systems or components can be displayed and used for manifold purposes, e.g. optimal configuration of system architectures.

For the purpose of introducing new solution ideas, various engineering design methodologies (e.g. [8, 13, 14]) advocate abstract function structures that 'represent the desired functional behaviour of products and their elements' [15] without reference to embodiment design. They recommend describing functions in a solution-neutral way in order to overcome mental fixations on existing solution principles when looking for new and creative ideas. This process of generating new solutions is thus characterised by incremental steps between abstract models that describe desired functions and (stepwise becoming more concrete) models that specify the physical embodiment and configuration of a system. However, Lindemann et al. point out that the quality of a solution is 'not determined by the result of a function analysis method, e.g., a function structure, but by the intellectual process of abstraction and structuring which leads to a better understanding of a design problem. […] This implies that the main purpose of function structures is to inspire thinking about the technical context' of a design problem [16].

Studies (e.g. [17]) show that the notion of function plays a key part in people's ability to analyse and solve a design problem. However, individual designers cannot be assumed to have similar understandings of a product's functions. The purpose of modelling functions, the notion of what a function might express and how it should be described depends on an individual perspective of a person. Even in design research, the meaning of the term 'function' and the usefulness of respective models is subject to controversial debates. As an example, Erden lists 18 different notions of function in engineering domains [18]. A brief review of the literature reveals a large variety of notions and concepts [19] that reflect the manifold perspectives of product design. In engineering design practice, functions are often described pragmatically as abstract affordances to a system [20]. They are used as a subjective concept of what a system is intended to do. Consequently, discussions in the design research community have brought up the understanding that functions can have several different meanings and expressions, but should refer to purposeful interactions between systems and their environment [17]. Functions and properties, however, refer to the condition and the capability of an assembly of (statically or dynamically) interacting components.

At least in mechanical design practice, when somebody talks about the function of a technical artefact they always have some kind of physical representation that could realise this function in mind [17]. Hence, it is not surprising that product models in engineering practice often describe artefacts that contribute to the purpose of a system. However, they do not express how and why these artefacts interact with each other to fulfil this purpose in a specific application situation [21]. Functions are often described only implicitly, e.g. by specific technical terms or concrete visualisations of a (sub-)system's physical embodiment [9]. Additional oral or written explanations and the name of the assemblies and elements may explain the system's interactions and functions to a person that is educated in reading this model. Araujo explains these circumstances with a human's preference for concrete descriptions of a problem, making it easier to understand and develop (not necessarily optimal) solutions [22]. Another reason why designers settle early on concrete solution representations is the lack of intermediate models that describe designs on different levels of abstraction (e.g. abstract functional representations and concrete form representations, see [23]). This natural human behaviour can be seen in contrast to prescriptive approaches that advocate a solely abstract, solution-neutral description of functions in order to overcome early solution convergence and to open-up new solution spaces.

Studies on acceptance of design methods in practice show that abstract models do not match a designer's intuitive thinking and expectations for efficient work support [24]. Hacker [3] emphasises that conceptual and embodiment design activities depend on creative thinking of physical structures that may fulfil functions in a desired quality. Thus, functions must be mentally 'translated' into physical structures and physical structures must be continuously related to their functions. The mental challenge of creative product development may thus be

relevant structure and design characteristics that influence the quality of a function. This problem is enforced by a lack of documentation, e.g. in CAD-models and technical drawings which are not applied as intermediate representations. As a consequence, a lack of understanding about these basic correlations might hinder them to effectively create an optimised design solution. Research on the Contact and Channel Approach is thus driven from the question how to isolate the relevant physical structure of a product (both virtual and real) and how to consider both functions and embodiment design in a shared representation, using product models that are widely spread in engineering design practice—e.g. early sketches of principal solutions and principal ideas, drawings, CAD-Models, etc.

The initial situation in most product development projects is characterised by the existence of a previous product or product portfolio: 'only a small percentage of product design tasks start from scratch' [27]. The scope of the Contact and Channel Approach is therefore to support deductive reasoning about improvements of design solutions in terms of meeting more demanding design objectives in Product Generation Development projects:

In analysis activities: Recognise functions and their physical structures according to interactions of systems and components; identify design characteristics that are relevant for the quality of functions and properties.

In synthesis activities: Optimise the relevant design characteristics in order to improve the quality of functions and properties. Observe alternative solution principles by extending or reducing a product's Wirk-Structure.

For documentation and communication purposes: Store, retrieve and communicate information about functions and properties of interacting systems or components within commonly applicable product models.

The presented approach is based on previous works from Rodenacker, Roth, Hubka and others, who describe functions as the relation of input and output parameters of interacting technical (sub)systems and components. This abstraction of the term 'function' provides a basis for a research hypothesis:

Interactions and thus the exchange of input and output parameters take place in interfaces and physical structures which connect these interfaces. The values of the relevant design characteristics that specify these interfaces and structures define the quality of a function.

To investigate this hypothesis, a product modelling approach was established by Albers et al. that provides a formalised modelling language from which individual models can be derived. In order to address the need of designers to stick with concrete representations of problems and solutions, these models are not composed of abstract boxes and connecting arrows, but they display a concrete representation of the physical structure of a technical system. A Contact and Channel Model includes representations of the physical structures and the Wirk-Structure as well as functional descriptions of a technical system. These representations can be drawings, sketches of a conceptual physical structure, screenshots from a 2d- or 3d-product model or other graphical representations. These product models are enhanced with information to describe functions in terms of static and dynamic interactions between components. It thus addresses the lack of intermediate representations in design [5]. Formalised modelling elements are provided which are can be used to highlight the relevant structures and interfaces of the physical embodiment that contribute to the fulfilment of a function, as well as to integrate the relevant surrounding influences into the scope of a model. It is intended to assist engineering designers in making abstract functional descriptions more tangible.

The presented research results reach beyond theoretical discussions about definitions e.g. of the term 'function'. The purpose of this work is to provide a means to analyse engineering design problems and to synthesise (e.g. to optimise) technical solutions to these problems. It builds on the firm conviction of the authors that one component on its own cannot fulfil a function. Technical functions can be seen as a result of interactions between components (of a product) and systems (e.g. a product, its user and the environment). The realisation of a technical function thus depends on the selection and combination of appropriate solution principles (*conceptual design*). This in turn determines the configuration and coordination of interacting structures and interfaces. However, the quality of a technical function is not only dependent on the chosen solution principle, but is strongly influenced by the specification of relevant design characteristics (*embodiment design*). In mechanical design, these design characteristics define the constitution and alignment of structures and interfaces of interacting components. Their specific values determine the quality of functions and properties. In general, design characteristics denote physical structures and information parameters of interacting components or systems. Functions and properties determine the available performance, ergonomics, safety, etc. of a technical system. They are directly affected by changes on its design characteristics which might appear due to 'normal' wear and deterioration, but also due to various unintended external or internal influences, e.g. variation of specifications during design, deviations in production processes or unforeseen impacts in the application.

In conclusion, the quality of functions and properties depends on the specification of design characteristics of a product. Both emerge in iterations, since designing a product requires a simultaneous consideration of both functions and design characteristics. A designer's main task is thus to design the relevant characteristics of interacting components instead of volumes and surfaces of single parts. Previous research works refer these interfaces to the concept of working surfaces, their corresponding physical structures to the term working volumes [14, 28, 29].

8.4.2 Modelling Elements and Their Relationships

Conceptual and embodiment design comprises—on different levels of abstraction—the specification of all geometrical, material and procedural design characteristics that are relevant for function fulfilment. Depending on the condition of these design characteristics, the quality of a function can be varied accordingly.

Conversely, the desired quality of functions and properties determine the required specification of relevant design characteristics.

Product models can only be effectively used to communicate information about these complex interrelations if they can be represented sufficiently unambiguously to be interpreted in the same way. It is therefore necessary to define a common modelling language that allows transforming different mental models into explicit models that can be shared among designers. For this purpose, the Contact and Channel Approach introduces modelling elements and rules how to arrange them for an integrated description of functions and the physical design of a product.

The underlying idea is that a product cannot perform a function without interactions between its components and with its environment—one component itself cannot perform a function. For engineering designers, it is important to focus on the interfaces and physical structures that are relevant to perform a function. They belong to both the interacting components of a product as well as its interacting environmental systems. Depending on the operation mode of a technical system, only a particular number of interfaces and physical structures actively performs a function. They compose a *Wirk-Net* which stores, transforms and exchanges inputs and outputs, e.g. energy, material and information flows. The input and output flows that result from the performance of a function are called the 'Wirkung' of a function. In Contact and Channel Models, the Wirk-Net of a function is composed of the following modelling elements:

- *Channel and Support Structures* (*CSS*), which denote permanently or occasionally interacting *physical structures* of solid bodies, liquids, gases or fields,
- *Working Surface Pairs* (*WSP*), which represent *interfaces* between these physical structures and
- *Connector* (*C*)[1] modelling elements, which represent the 'Wirkung' and the state properties of the environment that is relevant for the function of a system.

Figure 8.1 shows the minimal configuration of WSP, CSS and Connectors that compose a Wirk-Net, using the example of a screw that is drilled into a wall. A Wirk-Net of the screw is composed of at least two Working Surfaces that are interconnected by a CSS and coupled to one Connector each. Every Connector comprises models of the relevant system environment as well as one interface (Working Surface) to the associated Working Surface of the product. Depending on the scope of modelling, this Wirk-Net could be decomposed in various other WSP and CSS, e.g. at the thread or the head of the screw in Fig. 8.1.

Depending on the operation mode of a technical system, different functions may be performed on different Wirk-Nets. The task of an engineering designer is to anticipate and to dimension all necessary Wirk-Nets to realise all desired functions within the technical system (according to the purpose of the System). This

[1] Examples can be found e.g. in literature about the XiL-framework [30], where Connectors are used in automotive validation scenarios to describe interfaces and models of the external systems 'street', 'environment', 'driver' and 'remaining drive-train', e.g. for validating powertrains on a test bench that is implemented into a virtual reality.

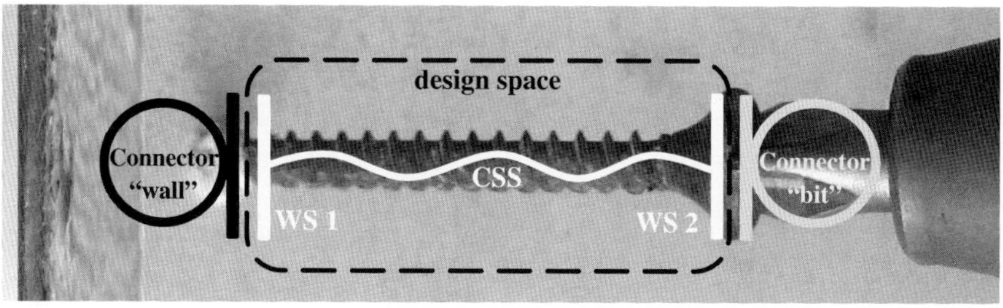

Fig. 8.1 Contact and channel model, including the wirk-net and the corresponding detailed physical structure of a screw

composition of all Wirk-Nets is called the *Wirk-Structure* of a technical system. In Fig. 8.2, the composition of Wirk-Nets within a Wirk-Structure of a hybrid powertrain is shown. It comprises all WSP, CSS and Connectors that are relevant for the desired functions, e.g. the drive-operation-modes of the car. Depending on a specific operation mode (e.g. recuperation, electric-, combustion-, boost-mode), only a part of the entire Wirk-Structure, the Wirk-Nets, are performing the particular functions.

Operation modes are characterised by different input parameter characteristics (the acceleration pedal position, position change speed like kick-down), environmental conditions (road gradient, headwind, temperature, etc.) as well as system state properties (state of charge of the high-voltage battery, ICE temperature, etc.). Examples can be observed in all kinds of products: a multi-plate clutch may transmit very low torques using only a part of all discs; a screw only uses a share of its thread to transmit low tensile loads; software only uses one of multiple CPUs for simple algorithms.

Albers et al. define further modelling elements such as Limiting Surfaces (LS) and Remaining Structures (RS) that do not contribute to a product's functions, but might be required due to boundary conditions of the design space (e.g. manufacturing conditions [31]).

All above-named modelling elements are fractal, i.e. they can be similarly modelled at different levels of abstraction. According to Fig. 8.3, the Wirk-Structure of a clutch, which is a sub-system of the hybrid powertrain in Fig. 8.2, can be modelled with the same modelling elements on a higher level of detail.

As shown in the examples above, the modelling language of the Contact and Channel Approach serves two purposes. First, the terms 'Working Surface', 'Working Surface Pair' and 'Channel and Support Structure' can be used as a common language to communicate about the Wirk-Structure of a product. Second, these modelling elements can be used as symbols in product models to abstract, to highlight and to connect the relevant design characteristics to functional descriptions of a product.

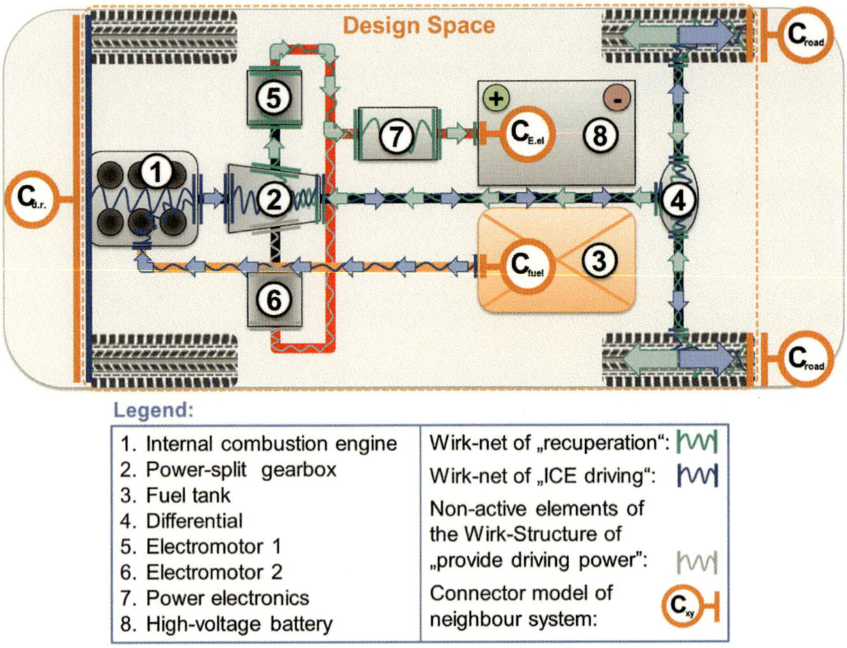

Fig. 8.2 Contact and channel model, including the wirk-structure and corresponding conceptual physical structure of a hybrid powertrain

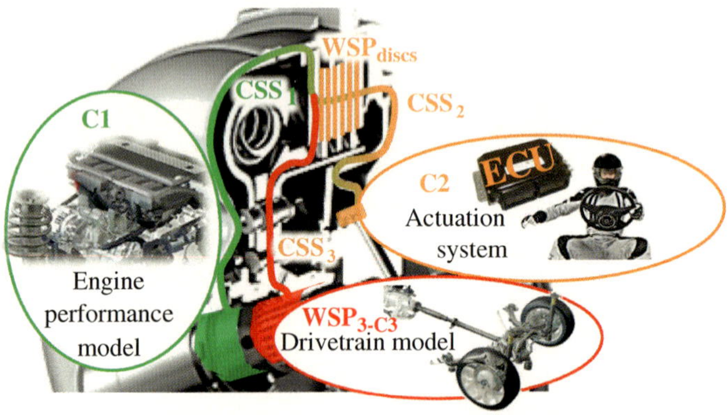

Fig. 8.3 Contact and channel model, including wirk-nets and a corresponding detailed physical structure of a multi-disc clutch

8.4.3 Guidelines for Application in Design Practice

Working with Contact and Channel Models serves the purpose to analyse design problems concerning product functionality and to specify appropriate solutions for an optimised Wirk-Structure. Both during analysis and synthesis activities, the representations can be progressively detailed through a repeating process of abstraction, creation, evaluation and modification. Abstractions enable designers to simplify problems and concentrate on particular aspects of a problem [5]. The modelling elements WSP, CSS and C can be applied independently from the level of abstraction that is selected for a representation (fractal character). A product's main function can be described with at least two Connectors, each including one Working Surface that form a WSP with the Working Surfaces of the product. The Working Surface Pairs of the product are connected by at least one CSS. However, this Wirk-Net can be decomposed in various other WSP and CSS that represent the Wirk-Net of sub-functions on a higher level of detail. For analysing functions and operation sequences of a system, it may be useful to create Contact and Channel Models in a top-down process to gain an overview on relevant influences and interactions of contributing (sub-)systems and components. If relevant aspects cannot be explained or turn out to be incomplete or wrong, it may be necessary to include additional influencing factors from the surrounding systems, to move back to a more abstract view or to move forth or back to another time or logical state.

Application of the Contact and Channel Approach in multiple case studies in design practice revealed that identification of the relevant Wirk-Nets is an easy first step towards a more comprehensive knowledge about a design problem. In contrast, reasoning about detailed design characteristics of WSP and CSS and their relation to functions and properties is often a non-trivial issue. Besides knowledge that can be taken as facts (e.g. from observations and experiences), speculations about cause and effect relations of design problems may arise during modelling. This uncertain information can be described in hypotheses and measures to test them. Application of the Contact and Channel Approach in multiple case studies in design practice provided opportunities to develop guidelines that are intended (a) to stimulate ideation processes for concrete solution principles and (b) to assist in deductive problem analysis.

The best-practice guidelines listed in Table 8.1 were derived from observations in design projects that were performed together with OEM and supplier companies within the engineering and construction, automotive and power-tool industry (e.g. [26]). Based on Contact and Channel Models, illustrations of important design characteristics and investigations about their influences on a design problem (e.g. to improve the quality of a function) may lead to a proposal of amendatory design measures. Depending on the type of problem, they can reach from suggestions to extend or replace sections within a Wirk-Structure (e.g. by introducing new WSP) to variations of its design characteristics.

Since design characteristics often depend on each other, it is not always possible to optimise a single design characteristic in order to improve the quality of a

Table 8.1 Guidelines to use contact and channel models in practice

	Guidelines	Methods, Models
Preparation	Define the purpose of modelling and select an adequate representation	Text, tables, lists, mind maps, presentations, etc.
	Determine the focus of modelling in time and space; only the relevant states, systems and external super-systems (environment) should be comprised	
Problem clarification, objective specification	Specify *functions* and *properties* that shall be observed and improved. Specify dependencies between functions and relevant properties	*Contact and Channel Models* (including text, tables, lists, images, real and virtual models)
	Determine the *Wirk-Nets* including the Connectors based on interactions that can be observed. Determine which *design characteristics* of WSP and CSS affect the quality of functions and properties	
	Describe functions and properties in reference to the Wirk-Nets that are affected, using the modelling elements WSP, CSS and C	
	Establish linkages between models that describe sequential functions in separate time or logical states	
Design specification	Switch between detailed and abstract views and move from qualitative to quantitative specifications to explore and to develop a design in more detail	
	Select those design characteristics that can be modified with reasonable efforts. Introduce new WSP and CSS if necessary	
	Determine how variations on design characteristics affect the quality of functions and properties	QFD, DSM, DoE, FMEA, etc.
	Derive, apply and validate design measures that may improve the quality of functions	Sketches, CAD, prototypes, etc.

function. Variations may afford extensive changes on related characteristics as well, imposing a rising complexity of the design problem. Moreover, variations on design characteristics may impose conflictive impacts on several properties—effects on the quality of functions are thus not always immediately evident. Despite this complexity, optimising the quality of a function must be accomplished effectively, i.e. designers must accomplish a maximum performance from a minimum design effort. The Contact and Channel Approach addresses this

Fig. 8.4 Modification of working surfaces in multi-disc clutch

challenge to identify the smallest possible set of design characteristics that impose a maximum impact on the quality of a function. Considering costs in design and manufacturing processes, the objective is to find design specifications that allow a reliable fulfilment of a function in a just acceptable quality. Besides modelling the Wirk-Structure of a product, further calculation, ideation, specification and evaluation purposes, additional methods and models must be taken into consideration that complement available design experiences and insights from Contact and Channel Models: e.g. Design Structure Matrices, spread sheet tools, simulation and experimental testing (see Table 8.1 for further recommendations).

Figure 8.4 shows an example for the application of a combination of methods and models. In order to optimise the mechanical and thermal strength of a lubricated multi-disc clutch, engineers were looking for matching sets of design characteristics that reduce the thermal load and therefore allow to improve the power density of the clutch [32]. For this purpose, a qualitative Contact and Channel Model was established, denoting potential Working Surface Pairs on the friction areas and their corresponding design characteristics that were assumed to be relevant. Quantitative models which were derived from simulations and measurements on sample clutch discs revealed that the Working Surface Pairs in friction areas on the clutch discs could be reduced in size when changing their material characteristics from steel to ceramics. However, modified objectives for the heat transfer to the surrounding lubrication (C_{Oil}) and unintended side effects such as increased oil ageing had to be considered in the following development process. Based on iterative simulation, testing and analytical calculations in Design Structure Matrices and graphs, detailed specifications for the relevant design characteristics of the relevant Working Surface Pairs and Channel and Support Structures could be derived that improve, e.g. specific friction power of the clutch for over 100 % [32].

8.5 Recent Studies and Results

Research on the Contact and Channel Approach focuses on two aspects: first, to develop a terminology that helps to make abstract functional descriptions more tangible. The second concern is the advancement of the applicability of Contact and Channel Models in design practice. Therefore, research on the Contact and Channel Approach is constantly evaluated in industrial projects. In the recent past, a strong focus was set on the cognition and articulation of functions and Wirk-Structures [5], the handling of Contact and Channel Models depending on applied tools and media [25], the formalisation and articulation of the terminology [25] as well as the transformation of analysis results into successful design specifications [26]. The results of these studies are summarised in the previous and the following sections.

8.5.1 Empirical Research and Case Studies: Evaluation in Design Practice

In the past 10 years, the Contact and Channel Approach was developed and evaluated in many research discussions, but also in multiple industrial design projects and workshops that were executed in order to solve real-design problems. More than 20 published and unpublished empirical studies were conducted in design practice to survey its applicability, usefulness and to draw further research potential from these findings. The main application of the Contact and Channel Approach in design practice has been so far for the purpose of analysing design problems on products in order to deductively derive specifications for improvements. Contact and Channel Models assist in problem clarification, e.g. to identify unintended physical effects, interactions between components or influences from the environment as well as unsuitable specifications of design characteristics.

Empirical research was conducted both in university lectures, in laboratory studies as well as in industrial practice. Due to the large number of application scenarios, only a brief introduction can be given in this section, demonstrating the scope and the range of applicability of the Contact and Channel Approach.

Examinations on student projects have been indicating that the student's cognitive ability to understand technical systems and to carry cognition forward to unknown systems has considerably increased since introducing the Contact and Channel Approach in teaching [23]. The approach is meeting cognitive needs of both students and designers to think visually about abstract functions by linking them to concrete locations in a product.

The Contact and Channel Approach has also been applied in multiple recent research projects on new and advanced technologies, e.g. to describe tribological systems behaviour of multi-disc clutches [32] or to facilitate micro-system development [33].

In the context of industrial design projects, this modelling approach could be successfully applied to develop conceptual design solutions for various problems and domains. Analysis of potential failure modes of a power tool was conducted based on modelling different logical function states [34]. It was observed that the terminology to describe Channel and Support Structures provides a logical structure for precautious examinations on the reliability of technical systems. Further empirical studies are currently ongoing, focussing on quantitative modelling of correlations between the observed or desired quality of functions and the specification of relevant design characteristics. They are driven by the intention to support deductive conclusions about design specifications that improve a system's properties, e.g. performance, safety, ergonomics, etc. These properties directly affect the quality of functions on different levels of abstraction: both in direct interaction of components as well as on a system's level.

Further insights about quantitative modelling opportunities, e.g. for configuration and evaluation of components according to their functional affordances and requirements, were derived from a software implementation of the Contact and Channel Approach for purposes of system architecture modelling [25]. The abstract concept of requirements can be decomposed into functions and properties, which in turn can be assigned to concrete locations on physical artefacts (WSP and CSS). This provides means to keep functions, functional requirements and objectives more transparent throughout a development process [35].

8.5.2 Strengths and Limitations of the Contact and Channel Approach

Despite of over one decade of research on the Contact and Channel Approach and multiple successful design projects in practice, several unexplored opportunities remain for future research. They can be derived from integrating quantitative modelling techniques as well as from unsolved limitations that are mainly imposed by a lack of tool support for integrated product models on a system's level.

With a growing number of associated empirical studies, both the terminology as well as guidelines for establishing and working with Contact and Channel Models were continuously refined. It could be observed that its strength can be seen in the small number of modelling elements (WS, WSP, CSS, Connectors) that must be applied to describe the Wirk-Nets and the Wirk-Structure of both simple and complex products on different levels of abstraction. Moreover, it is based on the simple hypothesis that any design can be adapted to functional requirements according to three basic operations: integration or elimination of Working Surface Pairs or Channel and Support Structures as well as modification of their design characteristics.

However, appropriate tools and media are required to effectively use the Contact and Channel Approach in complex product models. Concerning the tool support based on analogue media, complex analyses can be difficult because a great many images, sketches or drawings are required to progressively develop models for different time and logical states as well as for different abstraction views. Analogue models derived from the Contact and Channel Approach always represent a snap-shot in time, since sequences of a dynamic system's functions and operation modes cannot be integrated within the same model or dataset. Analogue models are unsuited to support design iterations for even a moderately complex system, as they usually afford modifications on multiple elements and levels of detail. They also often restrain designers to two-dimensional views, but offer a great variety and combination of different abstractions within a single model. However, it cannot be assured that analogue models are internally consistent. Working with large models is often challenging because the pen-and-paper model does not allow choosing which information is hided or displayed, e.g. by 'zooming' in and out to different levels of detail. Finally, analogue media do not lend themselves to global storage and retrieval of information, which is useful to support communication and distributed teamwork [25].

Digital media provide a huge variety of opportunities to establish models and to visualise or communicate relevant information in different views and formats. Not only internal consistency could be controlled within a Contact and Channel Model, but also relevant information could be shared between different models of engineering design practice.

Both abstract and detailed views could be derived from a common data set, e.g. in parametric CAD-models. This affords a detailed formalisation of the modelling terminology. In formalising the approach, the aim is rather to facilitate the construction, use and handling of large models than to make engineering design fully computable. First attempts have been made to implement the Contact and Channel Approach in a software tool to establish and to evaluate system architectures of mechatronic systems [25]. In another attempt, concepts of the Contact and Channel Approach were integrated into Model-Based System Engineering tools (e.g. [36]). However, only isolated software tools exist so far which are—due to their low degree of maturity—only sparely applied in modelling practice [35]. Current as well as future investigations will thus be applied on implementing the Contact and Channel Approach into commonly applied software tools in engineering practice (e.g. CAD systems) which allow the construction and manipulation of complex, hierarchical models.

Theoretical and empirical research works that provided contributions to the Contact and Channel Approach as well as other design methodologies so far have been mainly concerned with a qualitative modelling of functions. However, design practice calls for tools and methods that support modelling techniques which are capable of exploiting, representing and integrating quantitative information about specifications and conditions of both functions and embodiment design. For this purpose, recent works of the authors focus on methods and tools that facilitate quantification of models that link both functional and embodiment design descriptions.

8.6 Summary and Outlook

There is a general consensus among engineering design researchers that the relationship between functions and the physical structure of a technical system is a key issue of product development. A product's design can be clearly associated with its functions, functional capabilities and properties. However, design practice is still focused on modelling single parts without explicit documentation of interactions and external influences that impact on the quality of a product's functionality. Empirical research has thus been conducted in university lectures, in laboratory studies as well as in industrial practice to find out how to facilitate designers in understanding and communicating the complex relationship between abstract functions and the relevant physical structure of a product. This research is founded on the Systems Theory as well as previous work from Rodenacker, Roth, Hubka and others, who refer functional interactions between components to their constitution and design characteristics.

Results were summarised in the Contact and Channel Approach that advocates an integrated modelling of functional and physical descriptions of a product. Providing a formalised modelling language that can be used among designers to communicate about the Wirk-Structure of a technical system, it is intended to assist in making abstract functional descriptions more tangible.

Constant evaluation in industrial projects will remain a driving factor for future research, which will be focused on the advancement of the applicability of Contact and Channel Models in design practice. A strong focus in these works is set on improvements in tool support for integrated product models on a system's level. Another focus will be set on quantitative modelling, e.g. to specify and negotiate between design characteristics that influence the quality of functions and properties. However, contributions that can be expected from future research on the Contact and Channel Approach are not only based on advancements in its performance and ergonomics (e.g. plausibility of its theory and applicability in practice), but also on considerations of the human factors of applied methodology, e.g. individual experiences and creativity of designers.

References

1. Kim KM, Lee KP (2010) Two types of design approaches; regarding Industrial design and engineering design in product design. In: Proceedings of DESIGN 2010, Dubrovnik
2. Albers A, Ebel B, Lohmeyer Q (2012) System of objectives in complex product development. In: Proceedings of TMCE2012, Karlsruhe
3. Hacker W (1997) Improving engineering design—contributions of cognitive ergonomics, Ergonomics, 40(10):1088–1096
4. Israel JH (2009) Hybride Interaktionstechniken des immersiven Skizzierens in frühen Phasen der Produktentwicklung, PhD Thesis, Berlin
5. Eckert C, Alink T, Albers A (2010) Issue driven analysis of an existing product at different levels of abstraction. In: Proceedings of the DESIGN 2010, Dubrovnik

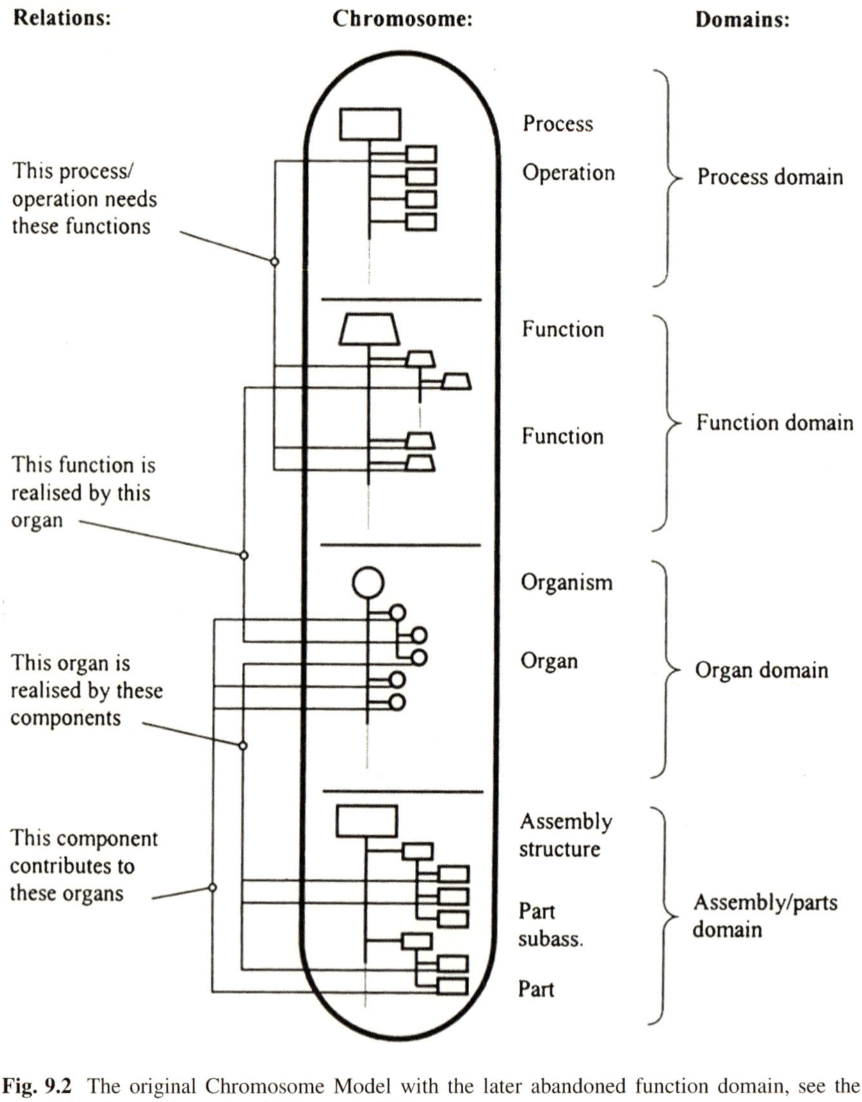

Fig. 9.2 The original Chromosome Model with the later abandoned function domain, see the text ([13] adapted from Ferreirinha et al. [12])

In the class of properties, we find the sub-class *functions*, characterised by being active effects. Realising the functions is a behavioural, not structural aspect of a system; therefore it becomes meaningless to articulate a function domain in the Chromosome Model and as a result the function domain was later abandoned; we find functions related to all three structures. Below we take a closer look upon functions. The purpose of a product is to establish a transformation and to deliver necessary effects for this transformation by its functionality. This relationship is

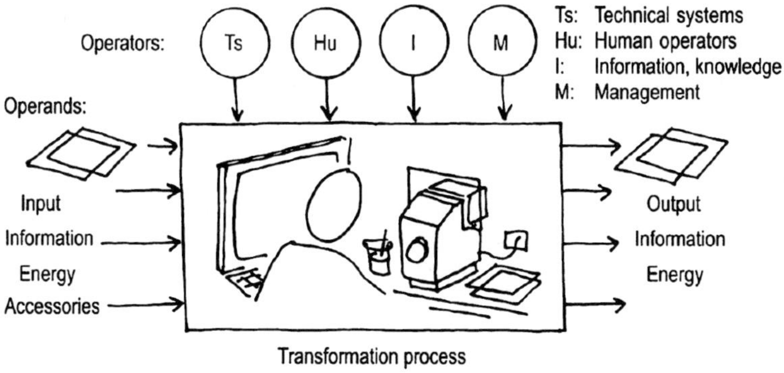

Fig. 9.3 Model of the Transformation System, Hubka and Eder [7]

articulated in the very fundamental Model of the Transformation System (Fig. 9.3).

The Domain Theory claims that the Transformation System actually may be seen as different views upon a product, namely the activity view and two fundamentally different views upon the product: an organ view and a part view. The Domain Theory also claims the existence of three types of systems related to the product; this may be articulated in the question: What are the characteristics and properties of the three systems? We will answer the question in the following.

9.2.1 The Activity Domain: How the Product Is Used

What Hubka call a transformation process in Fig. 9.3 we call a *technical activity*, to underline its relation to a technical product; it is a single or sequence of transformations in which the product is utilised (for example, see Fig. 9.8 for coffee brewing activity), or transformed (for instance the coffee pot being assembled). The very interesting aspect of the Transformation System Model in Fig. 9.3 is that it relates products, activities, technology and need satisfaction:

- The *technical activity* is determined by the user's application of the product. Together with the user experience and action with the result of the technical activity it satisfies the initially unsatisfied *need.*

When the product is used it contributes to the transformation of operands of the classes: material, energy, information and/or biological objects.

What are the characteristics of an activity? Because of the big variety in the state transformations or the technologies' nature we can only sketch some of the core *characteristics*:

- The *operands* being changed in terms of: material, energy, information or biological nature, characterised by their input and output state.

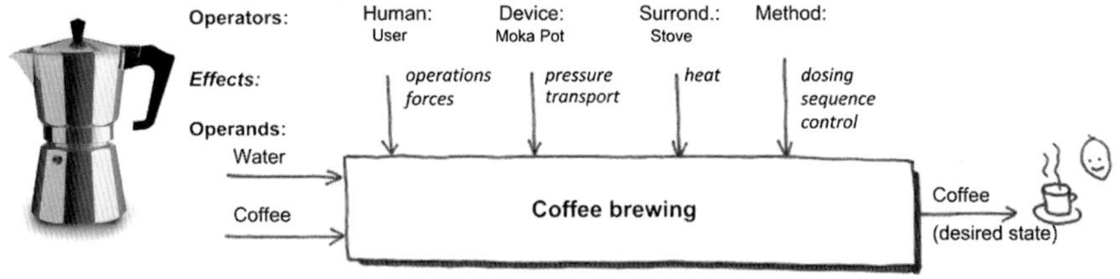

Fig. 9.4 Moka Pot and the coffee brewing activity showing operands, effects and operators

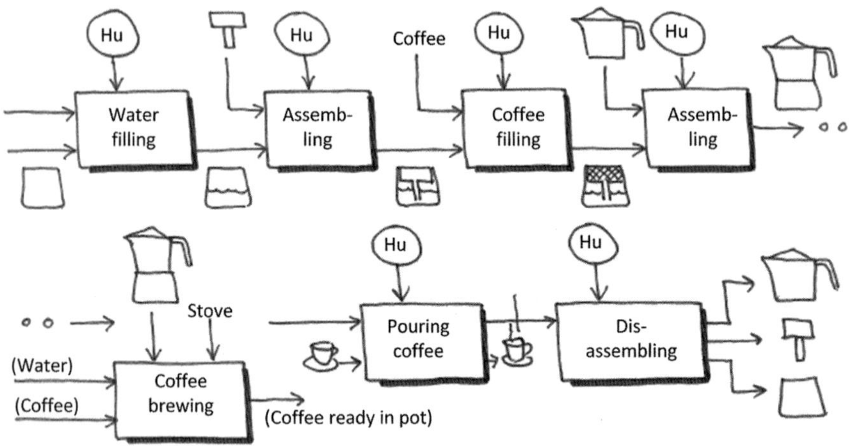

Fig. 9.5 The use activity related to the Moka Pot, showing the many human operations (Hu) and the central brewing process

- The necessary *effects* from the operators, their nature, their state and how they are carried or lead into contact with the operands.
- The conditions of the active *surroundings* necessary for the transformation to take place.

Example: Moka Pot. A Moka Pot is a simple product for brewing an espresso-type coffee. Figure 9.4 shows the pot and a model of the brewing activity in the form of Hubka and Eder's Transformation System Model (Fig. 9.3) with the addition of the operator, effects and operand labels.

The activity domain, belonging to the use of the Moka Pot, is the sequence of activities showing its use as illustrated in Fig. 9.5.

A use activity may be viewed and described on all levels of concretisation from abstract name, black box identification and pictorial articulation, mathematical and quantitative model of the technology to the full, concrete activity where the product is physically present and the user is in action. The activity's functionality and required properties may be articulated for proper selection of best solution.

The separation of transformation and product, which we do not find in, for instance, German language literature on Design Methodology, is one of Hubka's original ideas. In the literature, we find authors aware of composed transformations inside the product. In his efforts to clarify the concept of functions, Vermaas makes a distinction between actions of the device and functions of the device, and Houkes and Vermaas [16] introduce 'use plans' as the articulation of activities with the product. We also find observations of use activities in software engineering's 'use case' modelling. Howard and Andreasen [17] see an activity as a dynamic phenomenon which carries properties and functions:

- A *use function* is an active effect created by the use activity of the product.

It is important that the use result may be seen as a use function. It is often unnoticed that the main function of a product in itself (for instance the rotation of the cutting tool of a drilling machine) is not identical with the actual use function: to create holes. So the use result shall be seen as a most important function, satisfying the need: holes in the wall.

9.2.2 The Organ Domain: How the Product Functions

Many authors in the area of Design Methodology have recognised the importance of understanding the entities or, in system language, the elements of a product which are a carrier of certain functions. In parallel to biology Hubka, we call these entities organs, i.e., a structural, anatomical description. In the literature, organs are called function carriers, function elements or simply functions, which unfortunately mix up the view of a function as seen as a behavioural aspect of an organ. We define an organ as follows:

- An *organ* is a function element (or 'means') of a product, displaying a mode of action and a behaviour, which realise its function and carry its properties.

The identification or understanding of organs needs a dynamic perception of the product, it means one has to imagine what happens over time (a state change):

- An *organ* is based upon physical, chemical or biological phenomena. When stimuli (external effects) act on the organ, the organ delivers an effect which interacts with the surroundings, namely the wirk function.

Stimuli and effects may be material, energy, information or biological. The word wirk function is selected to remind us that wirk functions relate to the product's mode of action (German: Wirkungsweise), to be distinguished from use functions related to its activities.

Example: Moka Pot. The Moka Pot's organs are shown in Fig. 9.6a. The cross section shows how it works: When the pot is placed on a stove, the water will boil and be pressed through the filters and the ground coffee and collect in the serving pot. The pot consists of boiler organ, brewer organ (including two filtering organs),

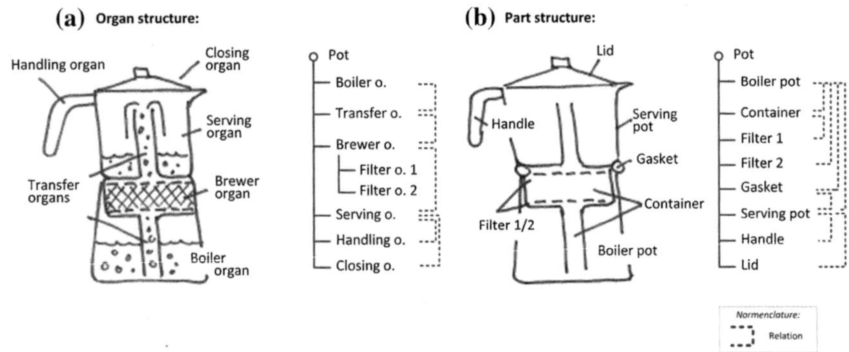

Fig. 9.6 The Moka Pot's organ structure (**a**) and part structure (**b**). The two structures are shown in a break down description similar to the Chromosome Model Fig. 9.3, and relations are added

transfer organs and a serving organ. Secondarily, we find also a closing organ (lid) and a handling organ.

The Moka Pot may be viewed as an organ system as described. Generally we may define:

- A product is a *system of organs*. The product's structure consists of its organs and their relations. The relations are active effects (Input/output).

The structure of organs explains how the composition of organs lead from the product's input to the effects needed for the transformation activity. In the Moka Pot example, we see how pressure and material flow become input or stimuli from the boiler organ to the brewing organ; another organ.

An organ may be viewed or described on all levels of concretisation as mentioned. Many scholars see the function carriers as an abstraction of the parts in a product; actually, we cannot reach any insight into the organs by abstracting a parts view. We see organs as just as concrete as parts; when the product is in action you can hear, see and smell the organs. Again referring to the example, the Moka Pot, a central organ is the organ creating pressure, realised by the closed chamber in which the boiling occurs. The closed vessel is created by the boiler cup's walls, the gasket and, surprisingly, the composed coffee powder kept in place by the filters. We return to the questions of organs' properties and modelling later in the chapter.

9.2.3 The Part Domain: How the Product is Build Up

In the mechanical area, products are specified by drawings showing parts and their assembly, where parts and the parts' assembly are primarily seen from a production viewpoint: What parts shall be produced? How shall they be assembled?

Second, the drawings allow skilled designers to read the organs and their functionality from the drawings. We define part in this way:

- A *part* is an elementary material element of a part system. Parts are building elements of an organ, realising the organ's mode of action by the part's physical states and interactions.

A system's elements may be of the kind activity, organ or part seen from the Domain Theory. However, when discussing organs, please notice the parallel with the German concept of *Maschinenelemente,* which are actually kinds of organs. Figure 9.6b shows the Moka Pot's part structure similar to the organ structure concerning the break down and relations. Note that we need to understand the mode of action of the pot for being able to identify organs, while the part structure is the composition of material entities.

Above we saw that an organ is identified and characterised by its ability to create an effect, but also parts are active and interact with other parts through their interface, i.e. assembly relations. We therefore choose also to relate functions to parts:

- A *part's function* (the part's *wirk function*) is based upon physical, chemical and biological phenomena. The part's interfaces with other parts and its surroundings, create the effects of the part.

Stimuli and effects may be material, energy, information or biological objects. One may ask if the distinction between organs and parts is actually only a question of resolution level. Our opinion is that there are fundamental differences between finding solutions for organs in an 'endless' amount of possibilities, and arranging parts in a limited number of roles and arrangements. The organ structure is setting the requirements for the part structure, but the part structure and parts' functionalities are also determined by the materialisation and assembly arrangement; this is unfortunately an area lacking in supporting theories and models, Andreasen and Howard [18].

Differing from the characteristics of activities and organs, we are able to classify the characteristics of parts as proposed by Hubka and Eder [7]:

- *Form* which can be modelled or interpreted as geometry.
- *Material* which determines properties like elasticity, strength, conductivity, etc.
- *Surface quality* which determines properties such as wear, smoothness and reflection.
- *Dimension* in the meaning size and measure; slightly unsystematically can tolerances be seen as dimensions also.

Hubka also adds 'state' to the list, reasoning from design situations where the specification of state is important: tension, magnetic, warm, etc. We disagree on this simple articulation of state and show in a section below the overall role of state changes for a product's functions and properties.

9.2.4 Product Synthesis

The Domain Theory offers evidently a vocabulary for analysing and describing three important aspects of a product; the question is how it can support the synthesis of new products. When a domain is modelled the model is only part of the product's realisation and is dependent upon the synthesis in other domains. An organ is not fully determined before its parts, and the part structure's final determination is not feasible without clarifying the activity and organ structures.

The three systems in their gradual determination may be seen as a supporting framework for the synthesis. One design strategy would be to follow what may be seen as a causal chain from user need → use activity result → determination of use activity → determination of the product's effects and functions → determination of organs and organ structure → determination of parts and part structure. The causality is principally correct but hardly a practical strategy; the mentioned activities have to overlap each other for proper fitting of activities, organs and parts. An important aspect is that the concept of domains enables the articulation of the degree's of freedom of a design's total solution space.

The three views may be seen as origin of three possible strategies for synthesis: Starting with ideas and concepts, either in the activity domain, the organ domain or the part domain. We may consider the following types of product development:

- *Designing a new product*: This seems like starting with an empty solution space. Typical approaches will be to create a new mode of use or start with creating mode of action for a central organ.
- *Incremental design*: where past solutions are utilised. Here rough models of the activity structure and organ structure may be very supportive for defining what shall be re-used and defining the cut in the interaction and interface between the re-used and new entities.
- *Platform-based design*: where a product family is created based upon past generations and expected new generations of products. We go into details with this topic below.

The navigation of the three views and the design strategy related hereto differs with the three types of development. Designing a new product may take its starting point anywhere, from a technical idea to a start in the need formulation and use activity, and progressing design from there. In incremental design, we may start by elaborated models of the systems in all three domains, so that we can be determine the parts of the structures we want to re-use and thereby defining the part to innovate. Platform-based innovation may be started from elaborated models of a company's central areas like production, sales and distribution, etc. The creation of alignment between these areas and the product's design is central for this type of development. Also, here is there a need for models of the three systems, for instance, articulation of a modular product family, see below.

Dependent upon the design task there may be a need for establishing overview of a solution space in one of the domains, for creating models focusing upon interactions between elements of the kind of activity, organ or part, for creating overviews of compositions of product family members, for creating detailed models of the total product or partial models, for determination of embodiment details and experiments or simulation of the design, etc. The challenge of the Domain Theory and its models (see below) is to cover these situations.

Product synthesis has to be based upon the understanding of the nature of the solutions and reasoning about functions and properties. These are the topics for the following sections.

9.3 The Link Model

The Domain Theory and the related concepts introduced above now create a vocabulary for the 'spelling of a product'; which means, how to articulate its structural composition. We have also introduced behavioural attributes like function and properties, but the questions remains: What relates function and properties to a user's perception of value and needs satisfaction? And: How to reason from functions and properties to the structural characteristics of the product and its use? Our capturing and modelling of these two phenomena, closely related to the Domain Model's concepts, are articulated in the so-called Link Model Fig. 9.7.

The model contains the following choices and reasoning, after Howard and Andreasen [17]:

- The need satisfaction is created by the 'Use Result' (the result of the use activity) producing the brewed coffee.
- The value perception is composed of by being the owner of the Moka Pot and able to brew coffee; the properties of the pot and its functions; the properties of the use activity and its functions, see the example above.
- The functions related to the product are called *wirk functions* because they relate to how the product works or more precisely the product's mode of action (German: *Wirkungsweise*) – in the case above the mode of action is how the arrangement of the brewing chamber functions 'to create the flow of liquids'.
- The functions related to the use activity are called *use functions* and represent what the operator intends to realise with the product,—in the example 'to brew coffee'.

The Link Model may be read in two different ways: The one is the analytical way or how the user experiences the need satisfaction and value (thin arrows). The other one is the direction of synthesis from need to determining the product and its use, by *function and property reasoning* (big arrow).

The notation of the use activity may be expanded to the totality of the product's life activities from manufacture to disposal. In each life activity,a stakeholder may ask for functions and appreciate certain properties, for instance may product

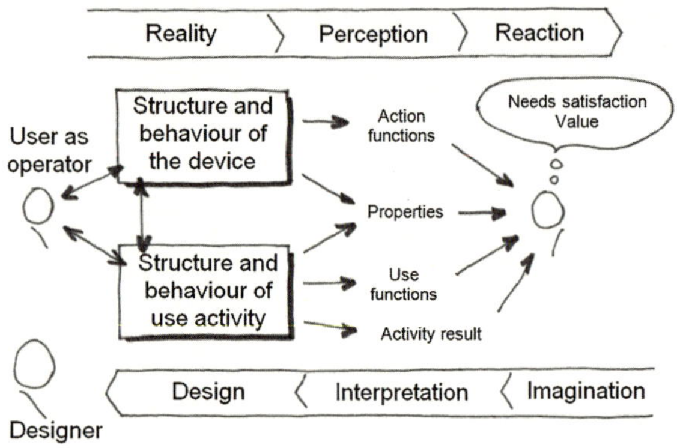

Fig. 9.7 The Link Model for the product's need satisfaction and creation of value, and for the designer's reasoning from need to structure of product and use activity [17]

functions be added for easing assembling and the product design may be supporting its ease of assembly; the installation operator may request certain functions for installing, adjusting and testing the product, and service delivered together with the product may call for functions, Tan [19].

The Link Model introduces the concept of 'function properties', which play an important role among the product's many properties, because the composition of functions in a product is closely related to reasoning about the performance and goodness of these functions. Any distinct type of function has a related set of properties which articulate the goodness of the function, see the example in Fig. 9.8 for functions related to a use activity, an organ and a part. The boiler organ is carrying more functions, and the function properties related to 'heat water' are shown. Also, the boiler cup part carries or contributes to more organs and therefore carries more functions. The function properties related to 'transmit heat' are shown.

9.4 The Role of State Changes

A precondition for designing is to understand how things behave and thereby realise the expected functions and carry the wanted properties. The simplification obtained by seeing a product's attributes as structural characteristics and behavioural properties (between those also functions) may lead to the statement that properties are indirectly determined by the designer's determination of characteristics. But behaviour only unfolds when the product is 'activated' and used, meaning that state changes occur. The properties depend upon behavioural aspects of the activity system, organ system and part system.

Any activity is as mentioned a transformation of certain operands (material, energy, information and biological objects [7], in which their states are changed; which means that values of their properties are changed. Any organ is based upon physical, chemical and/or biological effects, which are triggered by stimuli or input and influenced by the surroundings. An organ's mode of action is also a

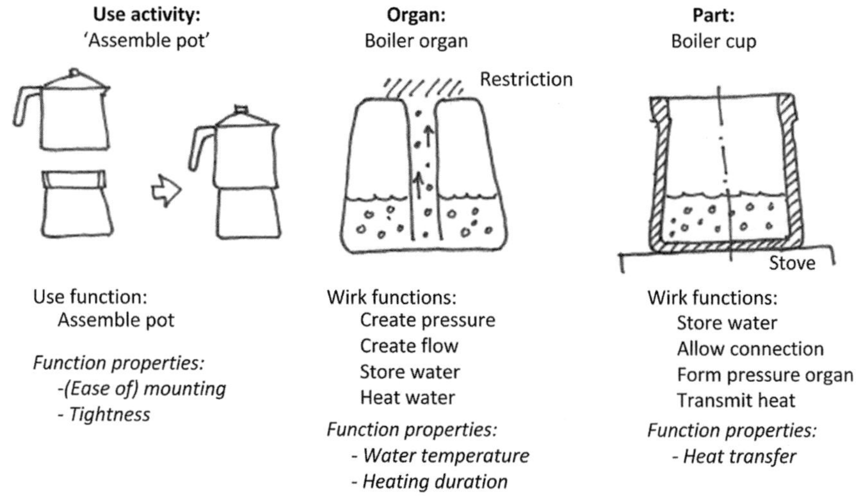

Fig. 9.8 Functions and function properties related to a use activity, an organ and a part

matter of state change; of special importance is the active effect created by the organ, seen as function. The parts, which are contributing to a certain organ, each have their role of delivering surfaces, support, heat transfer, conduct electricity, etc. It means that their roles are realised by state changes, even if some of the roles look static.

All three domains may be seen as systems and therefore have structure. What we have just described are the state changes in these structures' elements: activity, organ and part. But also the structure change. The composition of activities may change over time, certain organs may be passive in certain state transitions and the part structure may depend upon activities and operations made on this structure.

Example: Moka Pot. We have shown the use activities related to the pot in Fig. 9.5, explained its organ and part structures in Fig. 9.6, and used the pot to explain function properties. In the following, we will reason about state changes and properties of the product and its use.

The central organ in the product is the brewing chamber where filter plates keep the coffee powder in position and allow the water to pass through whilst sealing the lower chamber (over the water level) in order to create the pressure required for the brewing. We find here state changes as building up pressure, water flow and transfer of coffee oil from the powder to the water.

One of the parts is the boiling pot which has aluminium walls that transfer the heat from the hot plate to the water. But the pot also carries the thread allowing the assembly of the product and a rim, meeting the sealing, tightening the boiler and carrying the coffee container. There are several state changes 'experienced' by this part including its assembly and disassembly, transmitting heat from cold state to hot and back and carrying pressure from ambient state to brewing pressure and back.

The property 'ease of operation' is related to the activity structure by the efforts to dose, assemble (see Fig. 9.8), disassemble and clean, and the user's task to interrupt the heating. The organ structure asks for efforts to manage the brewing and especially the assembling organ, the simple thread connection, may trouble the user. The part structure is asking the user to understand the logic and operations, and the characteristics of the thread, influencing its 'gripping', also playing a role in the pot's ease of operation.

The property 'quality of the coffee' depends upon the technology choice for the central use activity: coffee brewing, but also of the user's proper dosing and timely stop of the heating. The organ structure's contribution is primarily the creation of pressure in the brewing chamber and proper filtering. The part structure influences through the choice of aluminium, chosen to not tarnish the coffee'.

The overall value or goodness of the Moka Pot is defined by the user or owner and may be composed by price, ease of operation, quality of coffee and pride of ownership. The original product from Bialetti 1933 is the aluminium product with hexagonal form, which may be the owner's preference and therefore adding to the pride of ownership. The hexagonal form is carried by the heating pot.

The role of state changes may be postulated in this way: A certain product property may be dependent upon the state changes of the use activity, the organ structure and the part structure. This means that each of the entities activity, organ and part system has time-dependant structure, characteristics and state, by which they realise the functions and properties. The implications for designing is that the view upon design as creation of seemingly passive parts in an assembly structure is very far from understanding what really matters: The active phenomena of the use activity, the organs and the parts. Understanding the dynamics is a precondition for proper concern on functionalities, properties and detailed quality questions related to production.

9.4.1 Function and Property Reasoning

Above we have explained the system's nature, the state change related to behaviour and the concept of function related to activities, organs and parts. It is easy to identify functions of existing products; the interesting part is to understand function reasoning, i.e. how to come from need, problem or task to the determination of how the solution shall work.

The wirk functions are achieved by the resulting effects of the organs which are normally predetermined by the designer and leave no alternatives for the user. However, the delivery of use functions seems to be an ambiguous phenomenon. The synthesis of the use activity and the product are therefore also different concerning function reasoning:

- The pattern of wirk functions is a question of (at least at the end) establishing a precise interacting pattern (German: *Logikstruktur*). Systematic approaches like formulating a 'function structure', a design matrix or a logical plan seem appropriate.
- The pattern of use functions is a question of interpretation of a need, of a problem or task, of the users' intended use and utilisation, and maybe even creating a need or other ways of using the product than intended. The synthesis call for approaches utilising creativity and playing with the language.

Example: Rescuing device. Imagine a fire situation in an airport building with many travellers who shall rescue themselves by leaving the building quickly. Some persons may be in wheelchair or have walking difficulties. Can we imagine a rescuing device like a sledge, operated by volunteers from the public, which can slide safely downstairs with the disabled persons? Early ideas may be based upon using words covering things which may be applicable or similar: a sledge, a tracked vehicle, a multi-wheeler, a multi-legged walker, people with straps at both end of the device, two persons folding hands for a 'throne chair', and we might start seeing sub functions like turning the device, fixation of the disabled, etc. And who shall bring the device up for the next transport? How? Identifying use functions is here related to both the users' and the helpers' roles which may happen through the gradual addition of precision in the language supported by sketches to explain the words and lead to settling of the use activity's and product's characteristics. It is noteworthy to see how our reasoning about functions is closely related to reasoning about properties, like whether it is easy to use, safe, clumsy, understandable, light, etc.

Many authors have commented on the multiple concepts of function we find in practice and literature Vermass [20]. It seems evident that practitioners can cope with this ambiguous and multiple concept of function, while many researchers struggle with creating 'the' definition. The following degrees of concretisation seem interesting; all of them might be called function or function modelling:

- *Purpose, goal, intention and task* as the starting points for designing or derived from whatever idea or design proposal there might be present.
- Verbally formulised statements in *daily language* for creative capture of what might be, what the product shall do and how it is to be used.
- Verbally formalised statements as *labels* for activities and organs, intentionally pointing to mode of use or mode of action, which are available and designable.
- *Black box* determination of organs, labelled by their functional identity, by identification of input and output and applied for instance to articulate a so-called 'function structure', but actually an organ structure (German: *Logikstruktur*).
- *Description of organs' mode of action* in models or descriptions of their parametric dependencies being physical laws.
- *Description of organs' embodiment* in the form of principal illustrations or technical drawings showing parts and their assembly structure.

The list is telling that what starts with considerations about the interaction among user, product and use activity (see the Link Model Fig. 9.6) gradually become the product's organ structure with functional relations, and later superimposed by the part structure and its assembly relations. What the list is not telling is how you actually determine the phenomenon suitable for an organ and give it a certain articulation in a mode of action. The Contact and Channel Approach, Albers and Wintergerst '[21], Chap. 8 in this book support analysis by pointing to the lines of state changes throughout an organ structure and thereby throughout the related parts, leading to a fundamental understanding of the functions.

Functional and property reasoning [17] and the closely related area 'property reasoning' is not only the question of requirements and creating organ structure, but a very delicate question of mastering the best knowledge about the relation between the product's characteristics and its properties as articulated by Weber [8], Chap. 16 in this book.

Function reasoning is one of the examples of the Domain Theory's nature. The idea is not to formulate normative methods but to give designers a language for 'spelling and reasoning' and supporting the creation of a designers mindset, i.e. understanding of design phenomena and thereby supporting the creation of strategies and models and the application of methods.

Domain Theory has strong similarities to Gero's Function-Behaviour-Structure framework and ontology, Gero and Kannengiesser [22], Chap. 13 in this book. Gero claims a generality of the framework and articulates it by identifying function, behaviour and structure of a wide palette of artefacts; The Domain Theory claims its validity for technical products and articulate function, behaviour and structure for three classes of structure, belonging to technical products. Both theories see the transformations or synthesis steps from function to behaviour and from behaviour to structure as essential. Other contributions to design science show similarities to the Domain Theory. Lindemann [9], Chap. 6 in this book presents a so-called 'Munich Model of Product Concretisation' which has certain similarities. Displayed in his model are working elements seen as abstraction of the components model, though it is not clear how these terms map on to our definitions of organs and parts as laid out in this chapter.

9.4.2 Product Modelling Based Upon the Domain Theory

If we shall create a Designer's Workbench (see the product modelling section) as an information tool which is able to support the design activity and composing the gradual product synthesis in the three domains,—then there is a need for product models. A product model is a description of the product's structure seen as an activity, organ and part structure, based upon these entities, their relations and characteristics.

Based upon the Domain Theory, Mortensen [23] proposed design languages for organs and parts, and Jensen [13] has studied conceptualisation based upon formal models of organ and part structures. Ideally seen, the models could carry fully detailed descriptions including function reasoning, designers' intent and the realised properties (Malmquist and Schachinger [24]) but the complexity is frightening. However, we have found the modelling formalism extremely well suited for modelling of product families sharing certain entities defined for instance as modules. Harlou [25] has developed generic organ diagrams for modelling structure and interaction, and a tool, the Product Family Master Plan, for modelling the commonality and variety of a family. Formally described entities, relations and characteristics allow the utilisation of object-oriented modelling and thereby application of standard software, Hvam et al. [26].

The basic idea of the mentioned authors' platform approach is the alignment between a company's products and their production, marketing, sales, distribution and the company's involvement in delivering service and actively participating to re-use, recovery and disposal. Alignment means fitting and optimising these structures to each other for overall high company performance; the fitting is partially supported by DFX methods and the basic pattern is explained in Olesen's Theory of Dispositions [27]. The three core aspects of alignment is illustrated in Fig. 9.9, where the variety and commonalities of the product family is shown in the engineering view formulated as an organ structure and aligned with customer views showing applications, and in the part view showing the part structure's relations to production.

Modularisation is based upon encapsulation leading to distinct functionality and rule-based interaction and interface of the modules. Pedersen [28] has expanded this idea to organ, part and activity encapsulation which allows identification of interaction between an organ, its related parts and their manufacturing processes, in order to create an optimised design with regards to variant creation, part commonality and optimisation of the manufacturing process. The Product Family Master Plan framework has been enlarged in work by Kvist [29] to also include the activities of the manufacturing process, for modelling interactions between parts and their manufacturing activities.

Organs are carrying a product's functions and properties, between these its performance, while parts are the physical embodiment of the organs and relate to production and assembly. Product modelling therefore contains the challenge to formally model the relations between the organ structure and part structure. Bruun et al. [30] have developed a design tool, Interface Diagram, for handling this relation. The tool supports encapsulation of organs and groups them by their functional identity becoming an organ structure, which is superimposed by the part structure and its assembly relations. The product model has been applied as a data model in a commercial PLM systems in order to identify the interacting information belonging to the organ and part domains.

Product models are not just the result of synthesis and product variety schemes, but also visual support for management decisions and configuration management. Visual product architecture models with multiple perspectives on products have

Theory, models and methods are interrelated. Several of the mentioned contributions have launched methods, for instance a framework and several methods for enhancing and designing a product's interface and operation for giving it a competitive edge, Markussen [36]. The role of the Domain Theory has been to supply a framework and basic concepts, allowing specific areas, like man/machine interaction to be articulated in specific models [36], for instance showing the characteristics of a product's interface. Application of the models for analysis and synthesis may result in methods.

The question related to this book's thematic is arising: What is the validity of such a theory, models and methods? Is there any rigour and scientific foundation? The Domain Theory contains mental constructs like function, behaviour and properties, which we try to give a meaning together with other concepts by definitions and integration into models. Our proposals shall be meaningful to practitioners, they shall be easy to integrate into their practice and support their reasoning; this is where the validity shall be found.

Design theories cannot be compared to engineering theories, which are based upon the laws of physics; rigour therefore may not be seen as 'accordance with physics', but in the efforts to link a theory to design practice. Design theories shall give understanding and support human design; the derived methods shall 'function in the pragmatics of professional practice', Sonalkar [37], Chap. 3 in this book, but we cannot expect that what humans choose to do in design situations shall be generally explainable. Design methods are 'soft'; they have only a certain probability for leading to results and often we cannot even postulate that the method was the reason for a certain result, Jensen and Andreasen [38].

We see two dimensions in a theory's goodness, namely its range and productivity. Range is the breadth of related phenomena that the theory is able to describe based upon a shared set of concepts. The sharing or fit of these concepts into ontology, may be seen as supportive to learning and mindset and respecting Occam's Razor and thereby to the design research area's consolidation [39].

An interesting signal about range is created by Storga et al. [40], who categorised key concepts and relations between them and brought them into an ontology. The basis is Mortensen's Genetic Design Model System [23], Hubka and Eder's Theory of Technical Systems and Theory of Properties (1988), Design Process Theories by Pahl and Beitz [42] and Hubka [42], and Olesen's Theory of Dispositions [27]. The feasibility of bringing these theories together we see as a sign of our theories' general range and comprehensiveness. Figure 9.10 shows the position of the ontology in the model of design research.

The productivity of a theory shall be found in its suitability for teaching its applicability for designers' practice and its utility for researchers to understand and analyse the phenomena of design. The applicability relates to building up creative and productive thinking, to lead to strong supportive models for both analysis and synthesis and for supporting visualisations for management and decision-making.

9.4.4 Summing Up

The aim of this chapter is the presentation of the Domain Theory as an example of a comprehension of theory, models, methods and ontology, illustrating the nature of these articulations. We view design theories as 'soft' in their origin and articulation, in their integration into designers' practice and in their application. Therefore, we judge that goodness of a theory may be articulated by its range and productivity, underlining that such a theory is model based and mainly a mental construct for practical designing. Hereby, we return to the aim of our research concerning the application of the Domain Theory, to be able 'to spell a product', and to perform function reasoning and property reasoning in the design route from need and value to the structuring or spelling of the product and its use.

Acknowledgments The authors would like to acknowledge the generations of engineering design researchers at the Technical University of Denmark that have contributed to and utilised the Domain Theory. We are in debt to Professor Emeritus Ken Wallace, Cambridge University, and to Professor Christian Weber, University of Ilmenau, for valuable discussions and critiques, and we gratefully acknowledge our colleague Anja Maier for her valuable comments and suggestions contributing to the development of this chapter.

References

1. Andreasen MM (1980) Machine design methods based on a systemic approach. Ph.D. Thesis, Lund University (in Danish)
2. Hansen CT, Andreasen MM (2002) Two approaches to synthesis based on the domain theory. In: Chakrabardi A (ed) Engineering design synthesis—understanding, approaches and tools. Springer-Verlag, London, pp 93–108
3. Chestnut H (1967) Systems engineering methods. Wiley, New York
4. Hall AD (1962) A methodology for systems engineering. Van Nostrand, Princeton
5. Andreasen MM (2007) How to spell a product? In: Unpublished lecture at TU, Ilmenau
6. Suh NP (1989) Principles of design. Oxford University Press, New York
7. Hubka V, Eder WE (1984) Theory of technical systems. Springer-Verlag, Berlin
8. Weber C (2013) Modeling products and product development based on characteristics and properties. Springer, Heidelberg (In this book)
9. Lindemann U (2013) Models of Design. Springer, Heidelberg (In this book)
10. Tjalve E (1979) A short course in industrial design. Newnes-Butterworth (Facsimile edition 2003 from Institute for Product Development, DTU). German edition: Systematische Formgebung für Industrieprodukte, VDI-Verlag, Düsseldorf 1978, Danish edition 1976)
11. Andreasen MM (1990) Designing on a designer's workbench. In: Unpublished notes from WDK Workshop, Rigi
12. Ferreirinha P, Grothe-Møller T, Hansen CT (1990) TEKLA, a language for developing knowledge based design systems. In: Hubka V (ed) Proceedings of ICED'90, Dubrovnik, Heurista, Zürich, pp 1058–1065, 1990
13. Jensen T (1999) Functional modeling in a design support system—contribution to a designer's workbench. Ph.D. Thesis, Technical University of Denmark

14. Eder WE, Hosnedl S (2008) Design engineering—a manual for enhanced creativity. Taylor & Francis, Boca Raton, London, New York
15. Smith J, Clarkson PJ (2005) Design concept modeling to improve reliability. J Eng Design 16(5):473–492
16. Houkes W, Vermaas PE (2010) Technical functions: on the use and design of artifacts. Springer, Dordrecht
17. Howard TJ, Andreasen MM (2013) Mindset of functional reasoning in engineering design. AIEDAM special issue on functional descriptions. Summer 27(3):233–240
18. Andreasen MM, Howard TJ (2011) Is engineering design disappearing from design research? In: Birkhofer H (ed) The Future of Design Methodology. Springer, Heidelberg, pp 21–34
19. Tan A, McAloone TC (2006) Understanding and developing innovative products and services: the essential elements. In: Marjanovic D (ed) Proceedings of the design 2006 9th international conference on design. Dubrovnik, Croatia, pp 647–654
20. Vermaas PE (2009) The flexible meaning of function in engineering. In: Proceedings of ICED'09, Stanford 2009
21. Albers A, Wintergerst E (2013) The contact and channel approach (C&C^2-A)—relating system's structure to its functionality. Springer, Heidelberg (In this book)
22. Gero J, Kannengiesser U (2013) The function-behaviour-structure ontology of design. Springer, Heidelberg (In this book)
23. Mortensen NH (1999) Design modelling in a designer's workbench. Ph.D. Thesis, Technical University of Denmark
24. Malmquist J, Schachinger P (1997) Towards an implementation of the chromosome model—focusing the design specification. In: Riitahuhta A (ed) Proceedings of ICED 97 Tampere 1997
25. Harlou U (2006) Developing product families based on product architectures—contributing to the theory of product families. Ph.D. Thesis, Technical University of Denmark
26. Hvam L, Mortensen NH, Riis J (2008) Product customization. Technical University of Denmark, Springer-Verlag
27. Olesen J (1992) Concurrent development in manufacturing—based upon dispositional mechanisms. Ph.D. Thesis, Technical University of Denmark
28. Pedersen R (2009) Product platform modeling. Ph.D. Thesis, Technical University of Denmark
29. Kvist M (2009) Product family assessment. Ph.D. Thesis, Technical University of Denmark
30. Bruun HPL, Mortensen NH, Harlou U (2012) Visual product architecture modelling for structuring data in a PLM system. In: Proceedings of PLM 12. Montreal, Canada
31. Hansen CL, Hvam L, Mortensen NH (2011) Proactive modeling of product and production architectures. In: Proceedings of ICED 11. Copenhagen, Denmark
32. Duffy AHB, Andreasen MM (1995) Enhancing the evolution of design science. In: Hubka V (ed) Proceedings of ICED'95, Praha 1995. Zürich : Heurista, pp 29–35
33. Culley SJ (2013) Re-visiting design as an information processing activity. Springer, Heidelberg (In this book)
34. Maier A, Wynn D, Howard TJ, Andreasen MM (2013) Perceiving design as modeling: a cybernetic systems perspective. Springer, Heidelberg (In this book)
35. Andreasen, MM (2011) 45 Years with design methodology. J Eng Des 22:293–332
36. Markussen TH (1995) A theoretical basis for creating interaction design (in Danish). Ph.D. Thesis, Technical University of Denmark
37. Sonalhar N, Jung M, Mabogunje A, Leifer L (2013) A structure for design theory. Springer: Heidelberg (In this book)
38. Jensen TE, Andreasen MM (2010) Design methods in practice: beyond the 'systematic' approach of Pahl & Beitz. In: Maejanovic D et al (eds) Proceedings of the 11th international design conference design 2010

39. Andreasen MM (2008) Consolidation of design research: symptoms, diagnosis, cures, actions?. Unpublished presentation at DS board meeting, Eltville
40. Storga M, Andreasen MM, Marjanovic D (2010) The design ontology: foundation for the design knowledge exchange and management. J Eng Des, 21(4):427–454
41. Pahl G, Beitz W (1988) Engineering design, a systematic approach. Springer-Verlag, London
42. Hubka V (1976) Theorie der Konstruktionsprozesse. Springer Verlag, Berlin

Chapter 10
Engineering Design: Role of Theory, Models, and Methods

W. Ernst Eder

10.1 Introduction

Some strict distinctions need to be made. "Design" in the English language has two usages. The noun, "the design" refers to that actual manifestation of a product, a tangible man-made object, an idea, a concept, a pattern, an artificial process, etc.—the way it looks, feels, and behaves, the result of a human intention. As a verb, "designing" refers to the mental and other processes that occur during this activity in order to establish "the design." In Design Research, the main interest lies in "designing," the verb, and in any underlying theory that can provide guidance for methods to enhance or enable designing. Design Practice at times looks for such guidance to overcome problems—when the design situation is nonroutine, when expertise and competence is lacking [8], for instance in enabling experienced engineering designers to explore the design space beyond their level of competence. Research for activities such as design engineering follows at least six parallel paths [18, 19]:

- The classical experimental, *empirical* way of independent observing, e.g., by protocol studies, including self-observation, and impartial observation of experimental subjects, etc., describing, abstracting, recognizing, perceiving, understanding, modeling, formulating hypotheses—observations capture a proportion of thinking, usually over short time spans;
- *Participative* observation, the observer also acts as a member of the design team and thus acts in the observed process [23]—which in consequence may be biased by the observer's participation;

W. E. Eder—Retired

W. E. Eder (✉)
Royal Military College of Canada, 107 Rideau Street, Kingston, ON K7K 7B2, Canada
e-mail: eder-e@kos.net

A. Chakrabarti and L. T. M. Blessing (eds.), *An Anthology of Theories and Models of Design*, DOI: 10.1007/978-1-4471-6338-1_10,

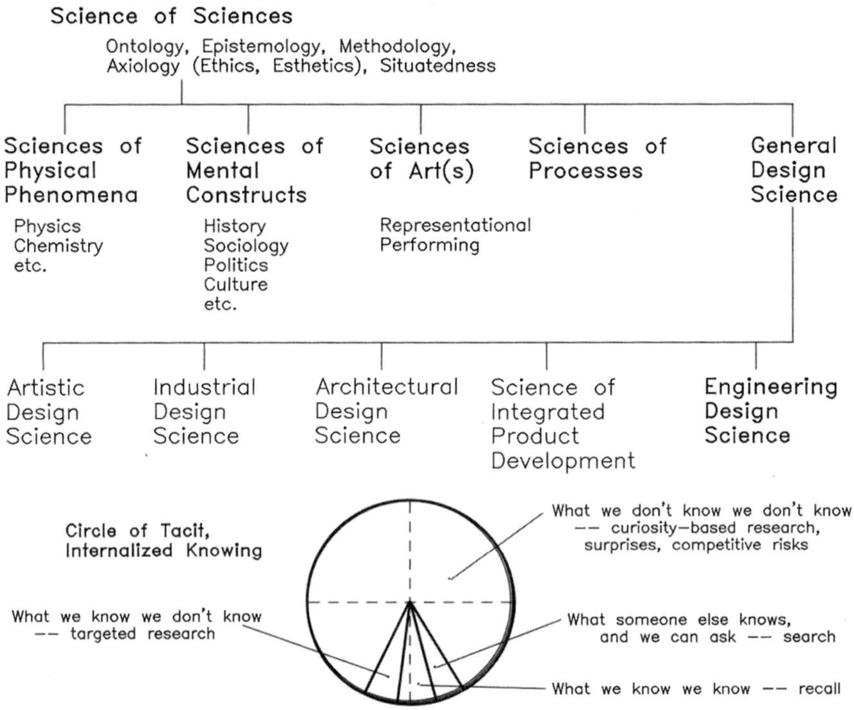

Fig. 10.1 Hierarchy of sciences ([18]; McMaster [38])

- A *reconstructive*, detective way of tracing past events and results by looking for clues in various places [40]—reconstructions never fully capture the original events, human memory is limited, and needs to be reconstituted for recall;
- Speculative, reflective, *philosophical* generating of hypotheses, and testing;
- Transfer between practical *experience* and the insights of knowledge; and
- *Development* of not-for-profit products [25].

These paths must be coordinated to attain internal consistency and plausibility. The theory and methods outlined here were developed from the fourth and fifth of these paths.

The purpose of design research is to clarify design processes, see Fig. 10.1, including designing in general, and particular forms of designing, e.g., design engineering, the focus of this chapter. In addition, design research should explore where assisting methods may be needed and useful—routine versus nonroutine design situations, and the needs of management versus practitioners [8].

A further necessary distinction is between a theory and a method. As formulated in cybernetics [34, 35], "*both theory and method emerge from the phenomenon of the subject*', see Fig. 10.2. A close relationship should exist among a *subject* (its nature as a concept or object), a basic *theory* (formal or informal, recorded or in a human mind), and a recommended *method*—the triad "subject—

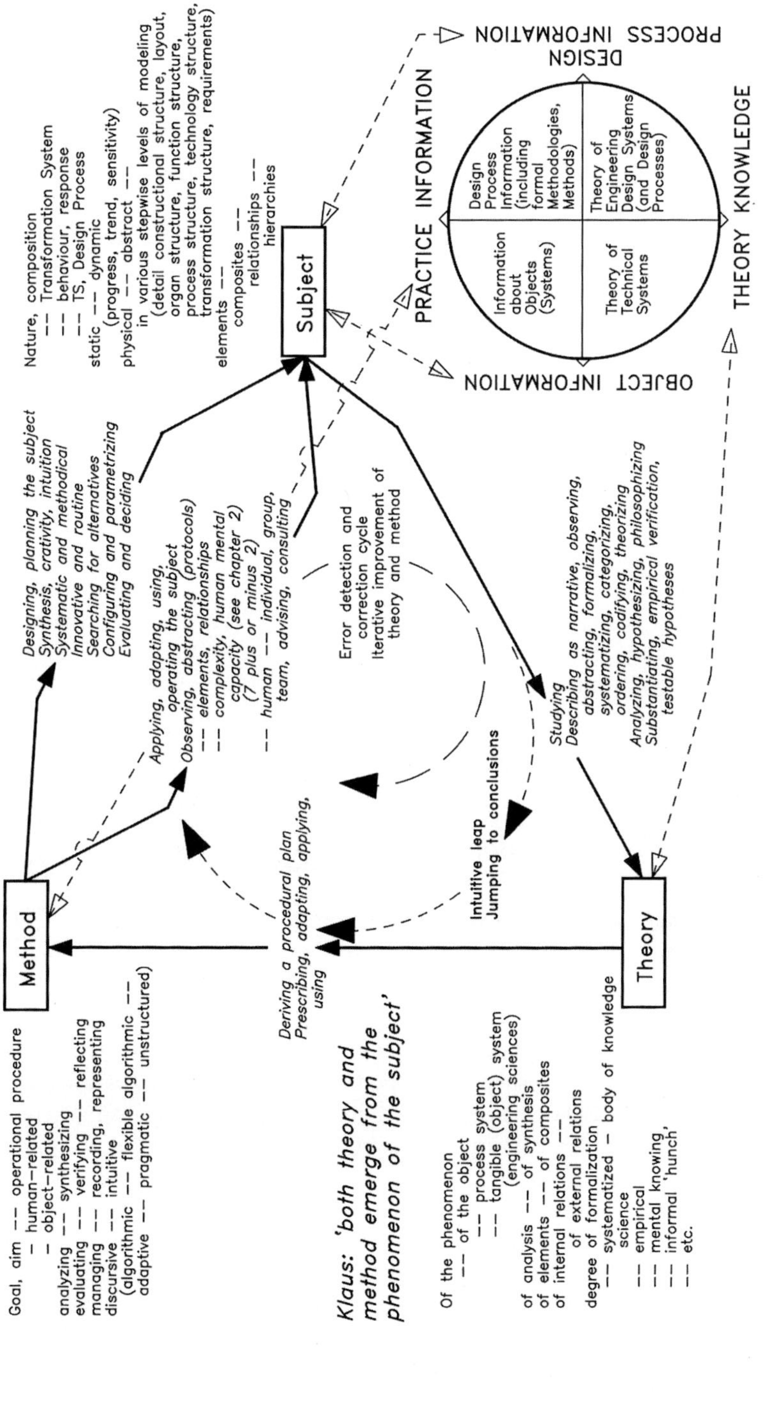

Fig. 10.2 Relationship among theory, subject, and method ([18, 19, 34, 35])

theory—method." The theory should describe and provide a foundation for explaining and predicting "the behavior of the concept or (natural or artificial, process or tangible) object," as subject. The theory should be as complete and logically consistent as possible, and refer to actual and existing phenomena. The (design) method, intended as a guide for more or less experienced engineering designers to assist them when their task reaches beyond their level of expertise, can then be derived, and consider available experience [19]. One aim is to separate declarative theory of the object being designed from both the theoretically logical method, and the suggested and voluntary method to be idiosyncratically applied.

In design engineering, the transformation process, TrfP(s), and/or the technical system involved in the TrfP, the TS(s)—see Sect. 10.4 of this chapter—are the subjects of the theory and the method. The *theory* should answer the questions of "why," "when," "where," "how" (with what means), "who" (for whom and by whom), with sufficient precision. The theory should support the utilized methods, i.e., "how" (procedure), "to what" (object), for the operating subject (the process or tangible object) or the subject being operated, and for planning, designing, manufacturing, marketing, distributing, operating, liquidating (etc.) the subject.

The *method* should be sufficiently adapted to the subject, its "what" (existence), and "for what" (anticipated and actual purpose). Graphic models that are part of the theory can be used as guides to design activity. Methods in application must remain voluntary, adaptable, and without guarantee of success. They must be learned, preferably on simplified examples, before they can be used (often from subconscious memory) on a problem of any substantial importance [8]. The phenomena of subject, theory, and method are of equal status.

10.2 Basic Considerations

Design engineering and the more artistic forms of designing (e.g., industrial design, architecture, graphic, and sculptural art) have much in common, with partly overlapping duties, but substantial differences, see Table 10.1—the descriptions show a contrast of extremes, rather than all aspects of designing.

If a product is intended to be visually attractive and user-friendly, its form (especially its observable shape) is important—a task for, e.g., *industrial designers*, and architects. Industrial design [20, 33, 44, 45], in the English interpretation, tends to be primary for consumer products and durables, emphasizes the *artistic elements*, appearance (size, shape, etc.), ergonomics, marketing, customer appeal, satisfaction, etc.—observable properties of a product. This includes color, line, shape, form, pattern, texture, proportion, juxtaposition, emotional reactions [21], etc.—these are mainly *observable properties* of a tangible product. The task given to or chosen by industrial designers is usually specified in rough terms. The mainly *intuitive* industrial design process emphasizes "creativity" and judgment, and is used in a studio setting in architecture, typographic design, fine art, etc., Industrial designers can introduce new fashion trends in their products.

Table 10.1 Scope of sorts of designing [18, 19]

Objectives, design conditions	Design engineering	Artistic-architectural-industrial design
The object to be designed, or the existing (designed) object	Transformation process and/or technical system; primary: functioning, performing a task	Tangible product; primary: appearance, functionality
Representation and analysis of the object as designed and its "captured design intent"	Preparing for TS(s) manufacture, assembly, distribution, etc., AI, CAD/CAM/CIM	Rendering for presentation and display, product range decisions
Design process (for the object), methodology, generating the "design intent"	Theories of designing, Engineering design science, formal design methodologies	Intuitive, collaborative, Interactive designing
Properties of the object as output of design	Mediating and elemental design properties, to establish observable properties	Observable properties to achieve customer satisfaction
Design phenomenology	Empirical, experimental, and implementation studies	Protocol studies
Responsibilities	Professional, ethics, reliability, safety, public, legal liability, enterprise, stakeholders	Organization, stakeholders (Architecture adds organizational and contract responsibility)
Location	Design/drawing office	Studio

For industrial designers, "conceptualizing" for a future tangible product consists of preliminary sketches of observable possibilities (even if somewhat abstract)—a direct entry into observable hardware (the constructional structure) and its representation. The sketches are progressively refined, and eventually "*rendered*" (drawn and colored, or modeled by computer or in tangible materials—*maquettes* or full-size clay models) into visually assessable presentation material, full artistic views of the proposed artifact, to provide a "final" presentation, for management approval. Considerations of engineering may need to take place, depending on circumstances, e.g., stability and self-strength of a sculpture. Industrial designers usually work "outside inwards", defining the observable envelope, thus constraining any envisaged internal constituents and actions.

In contrast, if a tangible product should work and fulfill a purpose by helping to perform a transformation process, TrfP, e.g., by mechanical, electrical, chemical, electronic, etc., means, its *functioning and operating* (verb form) are important—a task for engineering designers. Anticipating and analyzing this *capability for operation* is a role of the engineering sciences. *Engineering* intends to create what does not yet exist, that is likely to work, even if the way it works (its mode of action) is only partially understood by scientific means. Engineering needs designers to be aware of a wide range of existing information (e.g., scientific and experience-based heuristic) and its complex interactions, and to consider and accommodate all relevant influences (e.g., scientific, technical, economic, societal, political) to achieve a technically and economically successful and optimal product. The outcome of design engineering is a set of *manufacturing instructions* (detail and assembly drawings to scale, including tolerances and raw material specifications [1]—these, more recently, are likely to be computer-resident) for each constructional part, including instructions for assembly, adjustment, testing, use, spare parts, etc., see Fig. 10.3. These were traditionally produced manually in a design/drawing office, using drafting machines. Computer "seats" have more recently taken over some duties. In addition, documented analytical verification of anticipated performance in all life-cycle phases must be delivered by a qualified professional engineer. The resulting tangible product is a *technical system* (TS).

Design engineering is more constrained than industrial design, because:

- a design specification is usually prescribed by a customer or a marketing department, and is often the basis of a legally binding contract for delivery of a desired *performance* (a transformation process, TrfP),
- the relevant engineering sciences must be applied,
- societal norms and regulations (including laws) must be satisfied,
- risks and hazards must be controlled, the existing information must be respected, and
- economic considerations apply, e.g., survival and profitability.

Design engineering has available a theory of technical systems [28], see Sect. 10.3, and its associated engineering design science [31], which suggests several abstract models and representations of structures for existing transformation processes, TrfP-Str, and technical systems, TS-Str. These structures can be used stepwise during designing as tools for establishing requirements, and for verbal, graphical, cognitive, and conceptual modeling of novel or redesigned products (tangible and process), see Sects. 10.4 and 10.5—mathematical modeling is well established in the engineering sciences.

In fact, design engineering must consider a wide spectrum of information, and fit into the various cultural schemes applicable to different regions and countries, see Fig. 10.4. This is one of the many challenges facing engineering. Conversely, design engineering influences many of the cultural, social, political, and other environments. The process of implementing any technology (process or tangible object, old or new) almost invariably begins with design engineering.

Is a car an engineering product? The steering mechanism, suspension, motor and drive train, instruments, and a range of other items internal to the car (often hidden from view) are certainly engineering products, to which industrial-artistic designers can have little input. Mostly, these items cannot normally be observable for the driver, passenger, or casual observer, they are described by the *mediating* and *elemental design properties* of a technical system. Some of these intermediate products are OEM or COTS parts (original equipment manufacturers supplies, commercial off-the-shelf engineering products) manufactured by other organizations, e.g., springs, starter motors, alternators, computers, etc. Even the interior of

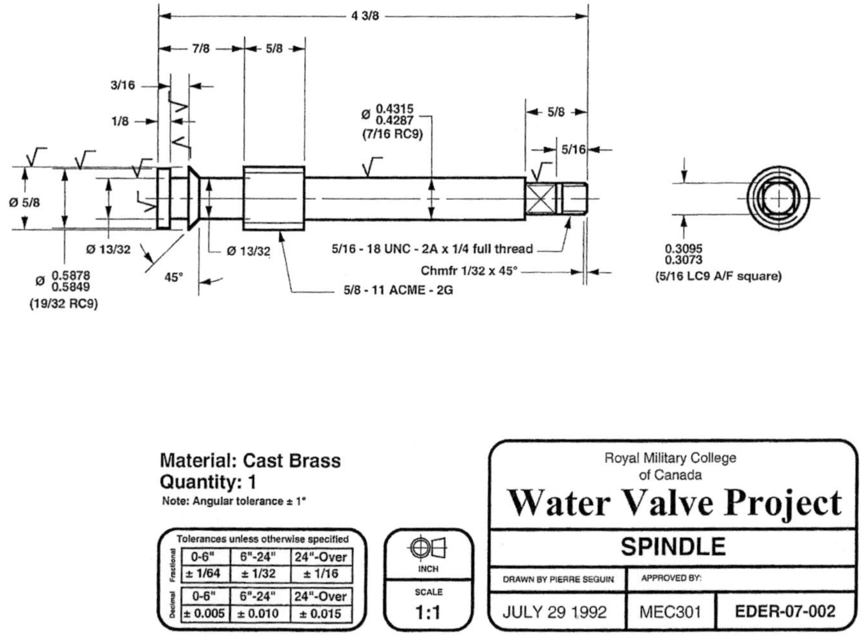

Fig. 10.3 Engineering detail drawing with typical geometric features [15]

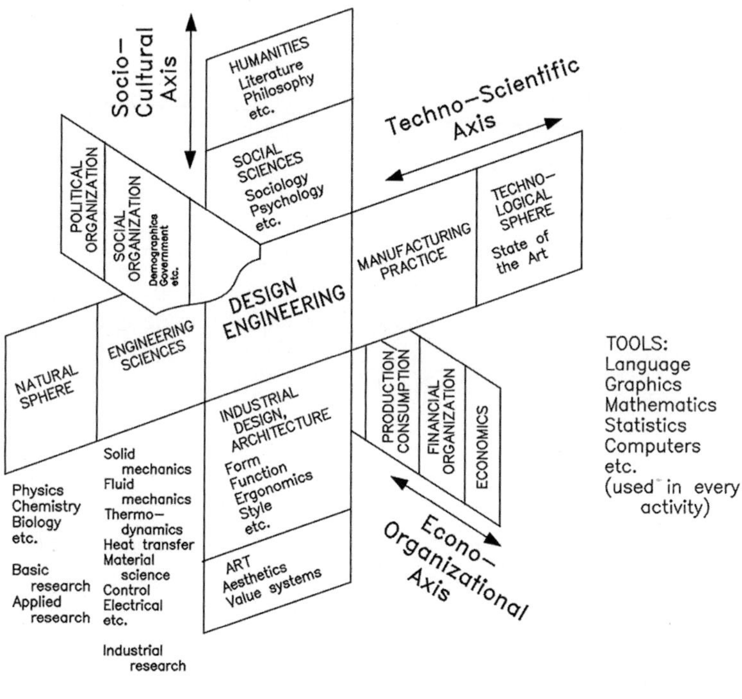

Fig. 10.4 Dimensions of design engineering in technology and society [16]

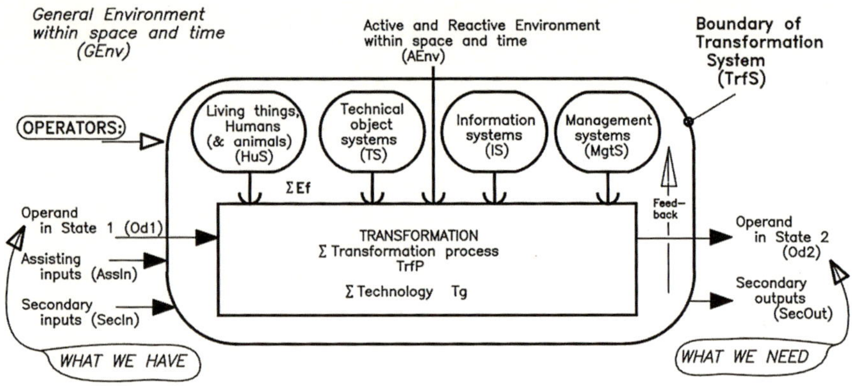

Fig. 10.5 General model of a transformation system [18, 19]

doors and other body parts (structural members, stiffeners, window mechanisms, etc.) are much more engineering than artistic. The exterior of the body parts (including the enclosed volume of the passenger compartment) is certainly more industrial-artistic. But the arrangement and division of individual body panels and their strength and structural integrity are engineered for durability, manufacturability, etc.—an engineering responsibility. In fact, a car is definitely an engineering product—without the engineering you only have an essentially decorative monument. Without the industrial design, the appearance and appeal of the car may be unsatisfactory, reference the "U.S. Army General Purpose Vehicle (GP)" of the 1940s, the original Jeep. Is this is a reason why the industrial designer often gets named, but the engineering designers are not ever mentioned, and credit for the engineering items is often given to "(applied) science?" In contrast, an electrical power transformer (500 MVA, 110 kV) hardly needs industrial design.

This comparison of artistic versus engineering designers is, of course, extreme and exaggerated, the truth is somewhere in between, many technical systems also need industrial design, and cooperation is often essential. The comparison is based on the author's personal experience in industry and life—10 years in industry (1951–1961) "on the drawing board" for electrical power transformers and switchgear, vehicles for alpine forestry, and other nonconsumer engineering products [11].

10.3 Outline of Technical Systems

This section presents a brief outline of the Theory of Technical Systems and its supporting graphical models, under development since about 1965—its latest version is Eder and Hosnedl [19]. Figure 10.5 shows the basic model on which the theory and method are based, the transformation system, TrfS, which declares:

- *An operand (materials, energy, information, and/or living things—M, E, I, L) in state Od1 is transformed into state Od2, using the active and reactive effects (in the form of materials, energy and/or information—M, E, I) exerted continuously, intermittently, or instantaneously by the operators (human systems, technical systems, active and reactive environment, information systems, and management systems, as outputs from their internal processes), by applying a suitable technologyTg (which mediates the exchange of M, E, I between effects and operand), whereby assisting inputs are needed, and secondary inputs and outputs can occur for the operand and for the operators.*

This model, initially proposed in 1974, is now recognized as the prototype for a Product-Service-System, PSS, recently the focus of research in product development [37]. Hubka's theory (and consequently the recommended design methodology, see Sects. 10.4 and 10.5 of this chapter) also includes many other considerations. The operators can be active or reactive in their interaction with each other and in their technology interaction with the operand. A hand power tool is reactive to its human operator, but active toward the operand. An automotive automatic transmission is mainly active.

The operators of a TrfS can in most cases be regarded as full transformation systems in their own right. For instance, the management system (MgtS) performs its management process, driven by human managers, management technical systems, a management environment, a management information system, and an upper-level management system.

Both the general environment (regional, national, and global) and the active and reactive environment cover physical, chemical, societal, economic, cultural, political, ideological, geographic, ecological, and all other influences directly or indirectly acting on or reacting to the transformation system, its process, and its operators.

The transformation process, TrfP, that is the main purpose of the transformation system, TrfS (and therefore is the task of the technical system as its operator), has a structure of operations and their arrangement or sequencing. The transformation process, TrfP, can take place if (and only if): (a) all operators of the transformation system, TrfS, are in a state of being *operational*, they (especially the TS) should be able to operate or be operated, if appropriate inputs are delivered to the operator; (b) an operand in state Od1 is available; and (c) both are brought together in a suitable way, with an appropriate technology. The TrfP must therefore be totally external to the operators.

A typical life cycle of a technical system, TS(s), is defined as a sequence of TrfS: *LC1*—Planning of TS(s) and TrfS(s), product planning, *LC2*—Designing of the TrfP(s) and the TS(s), *LC3*—Technological and organizational preparation for subsequent life cycle processes, *LC4*—Manufacture of the TS(s), *LC5*—Distribution of the TS(s), *LC6*—Operational usage to perform the TrfP(s), plus (*LC6A*) TS(s) maintenance, repair, refurbishing, etc., and *LC7*—Liquidation, re-engineering, recycling, etc. Life cycle stages *LC6* and *LC6A* are often referred to as the Product-Service-System, PSS [37].

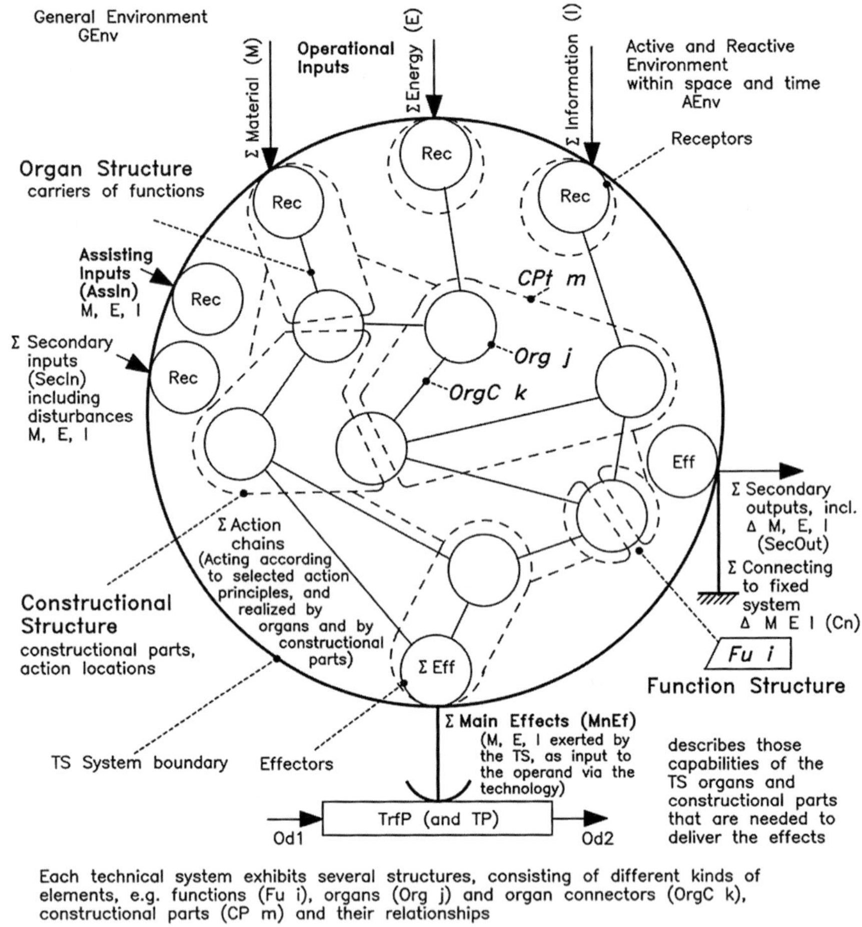

Fig. 10.6 Model of a technical system—structures [8, 18]

Various useful structures can be recognized, see Fig. 10.6: (a) transformation process, TrfP(s), and its structure of operations (see above), (b) technology, Tg, (c) TS-function structure, FuStr, a structure of TS-internal and cross-boundary capabilities of operation—also adopted in [41], (d) organ structure, OrgStr, action locations on constructional parts interacting—[41] replaces this with "physics", (e) constructional structure, CStr, the acting constructional parts—the main emphasis of [41]—for engineering design this structure is represented (usually graphically [1]) in (e1) preliminary layout, (e2) definitive/dimensional layout, and (e3) detail, assembly, parts-list, etc. These structures are, of course, closely interrelated, but almost never in a 1:1 relationship. Such structures are recognizable in technical system, but usually not in artistic-designed objects (products).

The TrfP and the TS exhibit properties. These are arranged in classes appropriate to each constituent of the TrfS derived from Fig. 10.6, and the classes are arranged in major groupings of *observable*, *mediating*, and *elemental design* properties. The mediating properties are separated into the *intrinsic* (heuristic and experience information) and *general* (engineering sciences) properties.

The *states* of TS-properties exist and change among the different states of existence, e.g., various life-cycle phases of a TS(s), and under various operating states, the "duty cycle" of an operational TS(s): (a) at rest, no operation; (b) during start-up; (c) during normal operation—idling, full-power and part-load, overload, etc., for self-acting operation (automatic), or running and ready to be operated by another operator, e.g., human or another TS; (d) during shut-down, ending an operational state and returning to "at rest" conditions; (e) in fault conditions—(e1) internal faults—overload, safe trip-out, breakage or equivalent, and (e2) external faults—damage, wrecking, etc.; (f) during maintenance, repair, testing, etc.; (g) at "life ended"; (h) any other states. The TS(s) can thus be operational, and even operating, in the absence of the operand of the TrfP.

Further considerations include mode of action, development in time, and other items of interest for engineering design processes.

The models of Hubka's theory are closely interconnected, and have been extended into considerations of engineering education [18], engineering management [18, 19], the design process itself [8, 18], and others. These models allow explanation of a usual operation of a manufacturing organization, the need for a distribution and servicing network for the product, the need for a supply chain to the manufacturing organization, and several other societal and economic factors.

10.4 Outline of Engineering Design Methodology

Using the models in Figs. 10.5 and 10.6 as bases, the stages and steps for a novel design process [18, 19] use the various structures in a sequence from abstract to more concrete, and considering all the aspects described by the theory of technical system, see Sect. 10.3. Constituting life cycle stage *LC2*, they are summarized as:

- *task defining*:

(P1) establish a design specification for the required system, a list of requirements; partly clarified also in [41];

(P2) establish a plan and timeline for design engineering;

- *conceptualizing*:

(P3a) from the desirable and required output (operand in state Od2), establish a suitable transformation process, TrfP(s), with alternatives;

(P3.1.1) if needed, establish the appropriate input (operand in state Od1);

(P3.1.2) decide which of the operations in the TrfP(s) will be performed by technical systems, TS, alone or in mutual cooperation with other operators; and which TS(s) (or parts of them) need to be designed;

(P3.1.3) establish a technology, Tg, (structure, with alternatives) for that transformation operation, and therefore the effects (as outputs) needed from the technical system, TS(s);

(P3b) establish what the technical system needs to be able to do (its internal and cross-boundary functions, with alternatives);

(P4) establish what organs (function carriers in principle and their structure, with alternatives) can perform these functions. These organs can be found mainly in prior art, especially the machine elements, in a revised arrangement as proposed in [4, 5, 51];

- *embodying/laying out and detailing*:

(P5a) establish what constructional parts and their arrangement are needed, in sketch outline, in rough layout, with alternatives;

(P5b) establish what constructional parts are needed, in dimensional-definitive layout, with alternatives;

(P6) establish what constructional parts are needed, in detail and assembly drawings, with alternatives.

The suffix "(s)" indicates that this TrfP(s) and/or TS(s) is the subject of design interest. Adaptation for redesign problems (probably about 95 % of all design engineering tasks) proceeds through stages (P1) and (P2) above, then analyzes from (P6) or (P5b) to (P4), and/or to (P3b) to reverse engineer these structures, modify them according to the new requirements, and use the stages in the usual order to complete the redesign.

The classes of properties of existing TrfP(s) and the TS(s), and the classes of properties related to the life cycle phases *LC1–LC3* (the manufacturing organization), lead directly to the list of primary and secondary classes of *requirements* that are the basis for step (P1), establishing a design specification [19].

Only those parts of this engineering design process, possibly coupled with other design methods, that are thought to be useful are employed. Such an "idealized" procedure cannot be accomplished in a linear fashion—iterations, and recursions are essential, using analysis and synthesis [6], see also Sect. 10.5.

The Hubka engineering design methodology allows and encourages the engineering designers to generate a wider range of solution proposals at various levels of abstraction from which to select—one of the hallmarks of creativity is a wide range of proposed solutions. Designers should also use serendipity, opportunism, spontaneity, pragmatic, and "industry best practice" methods, etc. The apparent linearity of this procedure is only a broad approximation [8, 39], parts of the TrfP(s) and/or TS(s) will inevitably be at different stages of concretization, and of different difficulty (routine to safety [39]), and will force iterative working—repeating a part of the design process with enhanced information to improve the solution proposals, within a stage or step of the engineering design process, and between stages—and recursive working—breaking the larger problem into smaller ones, subproblems and/or subsystems, to recursively solve (e.g., using the same systematic design methodology) and recombine. In this design process, the perceived or assumed TS-boundary is frequently redefined to restrict and focus, or expand, the designer's "window" of observation [40]—using the hierarchical nature of TrfS.

CAD—computer-aided design—can effectively be used in stages (P5a), (P5b), and (P6)—in earlier stages the representations are often too abstract for computer graphic processing (including semantics and implications), but mathematical analysis and simulation in earlier stages are often useful—CAE, computer-aided engineering.

Stage (P3b), development of a TS-function structure, reveals a special position. For instance, the TS-cross-boundary functions can include such nonobvious functions as "present a pleasing appearance to the TS(s)" or "allow easy and ergonomic operation by a human"—a direct connection to the need for involvement of industrial design. Also, the TS-internal functions can include "adjust" or "regulate and control" with respect to some TS-properties—this can be solved mechanically, electrically, fluidically, electronically (plus software), etc., and can provide a direct connection to mechatronics and other disciplines.

10.5 Problem Solving

Superimposed on the systematic approach to design engineering is a subprocess cycle of problem solving, frequently and repeatedly applied in every one of the design stage listed in Sect. 10.4, see Fig. 10.7.

Iterative working is related to TrfP/TS properties, requirements, and both heuristic and analytical use of the mediating properties, the engineering sciences, and the problem solving cycle [6, 7, 49, 50], see Figs. 10.7, 10.8 and10.9. Observable and mediating properties of future "existing" TrfP(s)/TS(s) can be analytically determined from the established elemental design properties, giving a reproducible result. The inversion of this procedure, synthesis, is indeterminate; each required observable property is influenced by many different elemental design properties that therefore need to be iteratively established to approach the desired state of the observable property. Analysis is in essence a one-to-one transformation, convergence to one solution. Synthesis goes far beyond a reversal of analysis; it is almost always a transformation that deals with alternative means and arrangements, involving divergence as well as convergence, a one-to-many (or few-to-many) transformation. Synthesizing, as part of Op-H3.2 "Search for Solutions," is the more difficult kind of action [6, 7]. Figure 10.8 constitutes proof that iterative procedure is a theoretical necessity in engineering design science, and a practical necessity in design engineering.

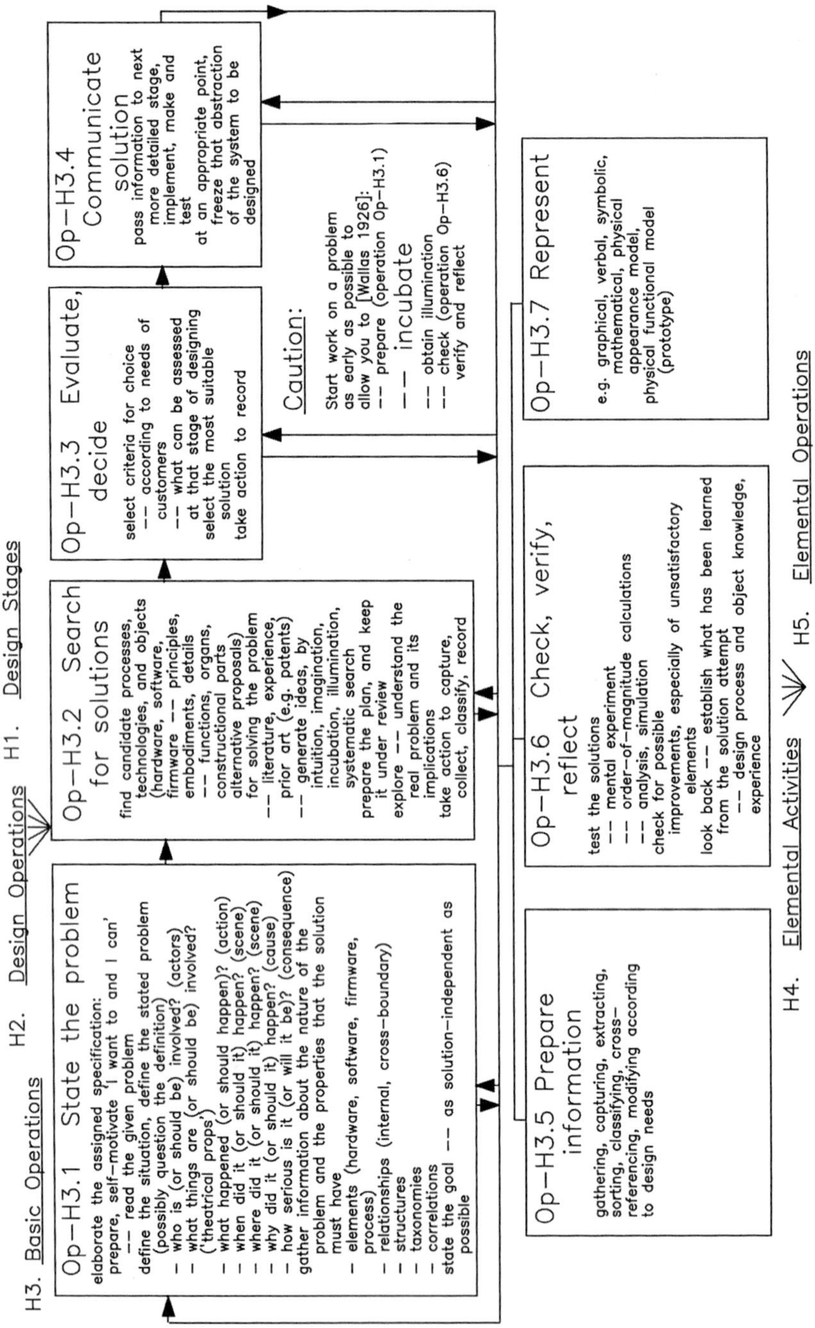

Fig. 10.7 Basic operations – problem solving in the engineering design process [18, 19, 22, 31, 36, 42, 43, 46–48]

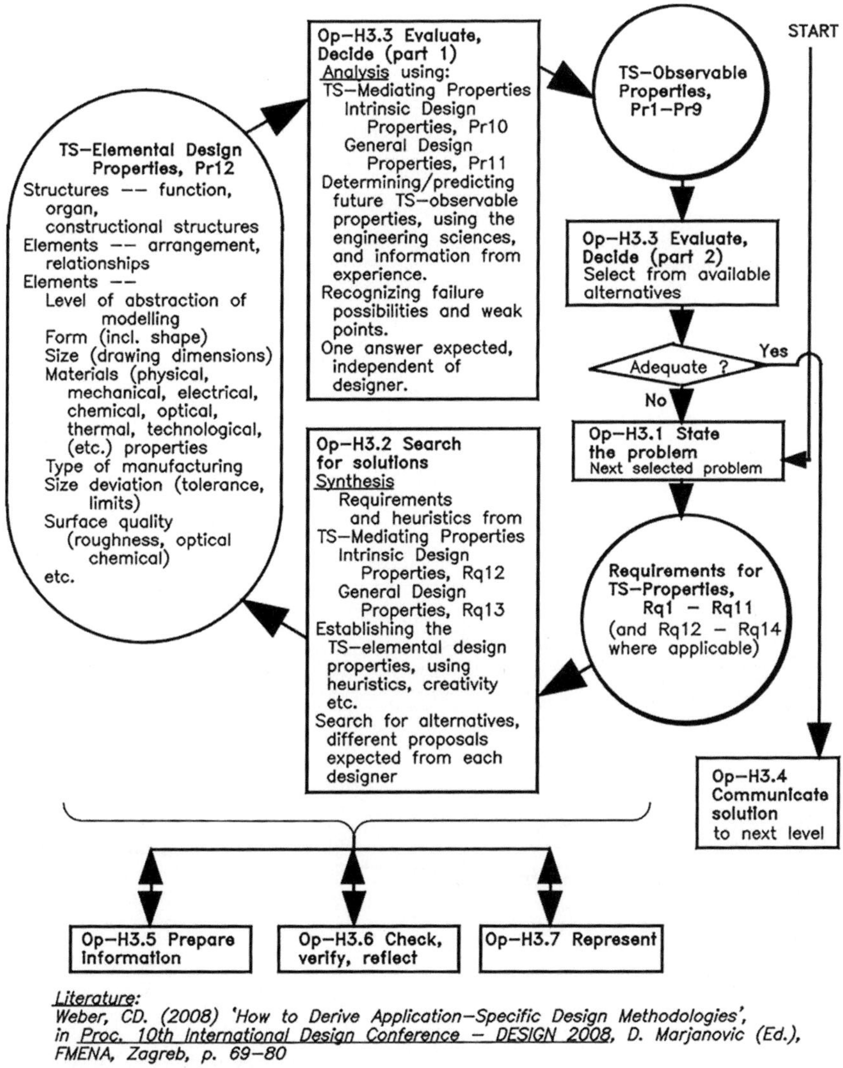

Fig. 10.8 Main relationships between problem solving, and mediating elemental design and observable properties (adapted from [19, 50])

10.6 Clarification and Verification

As shown in [8], such a fully systematic procedure is only necessary in limited situations, when an engineering designer is faced with an unfamiliar and non-routine situation [8, 39]. Systematic design engineering as a procedure is the heuristic-strategic use of a theory to guide the design process—Engineering Design Science [31] is recommended as guiding theory. Methodical design

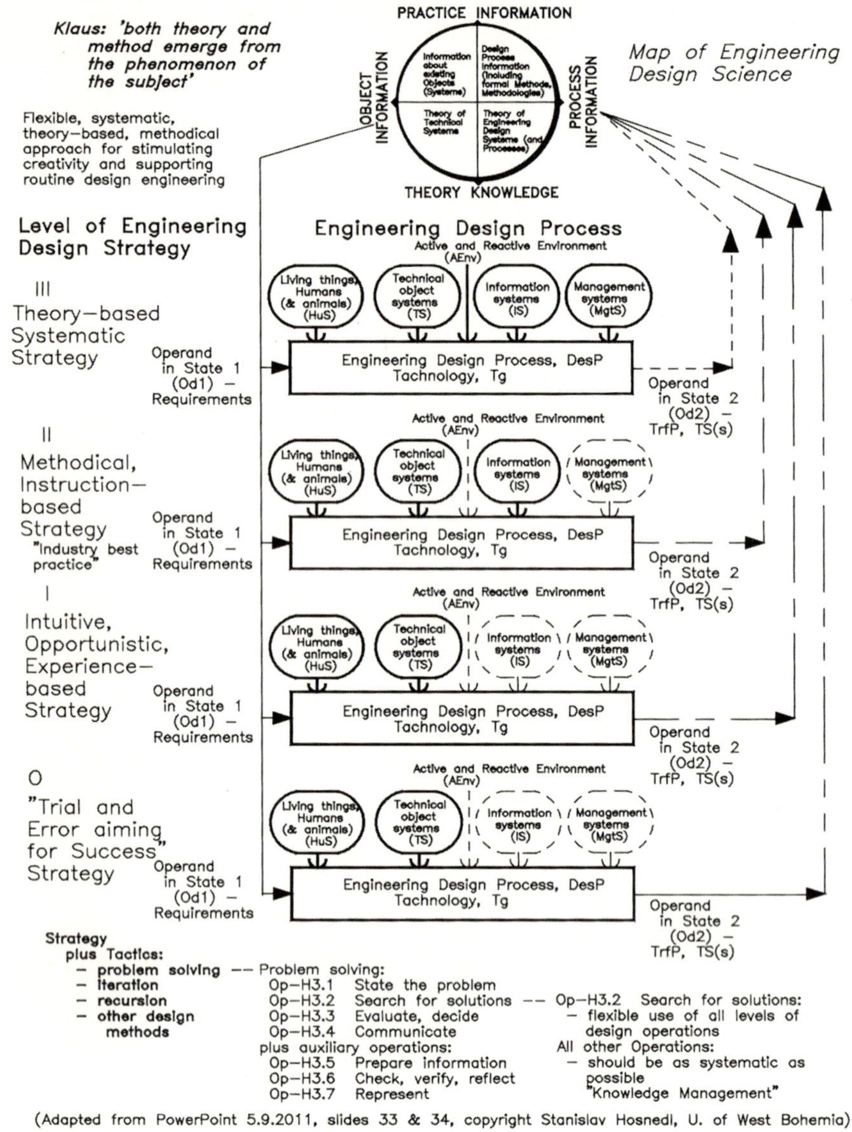

Fig. 10.9 Strategies for design engineering and problem solving

engineering as a procedure is the heuristic use of newly developed and established methods within the engineering design process, including theory-based, pragmatic, and "industry best practice," strategic and tactical, formalized and intuitive methods. Systematic and methodical procedures have a substantial overlap, but are not coincident. The engineering and other sciences can provide some assistance, especially for heuristic "what-if" investigations, and for analyzing expected behaviors. Engineering designers can then apply their intuition, trial-and-error procedures, and other methods, coordinated by systematic design methods, to the specific project, and still be aware of systematic project management, see Fig. 10.9.

Figure 10.9 indicated that whatever basic strategy is used during engineering design (level 0—trial and error, level 1—intuitive, level 2—methodical), it is probably advantageous to bring the results into the full theory-based systematic strategy of level 3, which provides the best documentation of the design process. Such documentation may be needed in case of litigation.

Creativity [3] is usually characterized by a wide search for solutions, especially those that are innovative. This search can be supported by the recommended systematic and methodical approach. All generated alternatives should be kept on record, to allow re-tracing and recovery from subsequent detection of a better alternative. Each step in the overall procedure should be concluded by selecting the most appropriate (one or two) solutions for further processing, in order to control a tendency toward "combinatorial complexity."

Various comparisons have been made between over 100 proposals for engineering design theories and methods [14, 31]. Some of these have caused clarifications, almost all have been found to enhance aspects of the Hubka theory and/or methodology.

The primary purpose of the case examples based on the Hubka design methodology is to present teaching examples for procedural application of the recommended engineering design method, especially for the conceptualizing phases of the engineering design process. Students and practitioners can follow and study them to help learn the scope of the method and its models, and show that the systematic method can be made to work. This purpose has been applied in courses at the EidgenössischeTechnischeHochschule (ETH) by Dr. Vladimir Hubka (1976–2000), at The Royal Military College of Canada (1981–2006), and at the University of West Bohemia (1990-present)—for all levels of education and for industry consultations. A secondary purpose was to verify and validate the theory, models, and methods, check for correctness, illustrate and document the theories, procedures, methods, and models that can be used within systematic design engineering, and to show up deficiencies—which were corrected in the theories, models, and methods. The emphasis in all case studies was on the engineering design procedure and use of the models. Care should be exercised when reading these case examples, they were not intended to show a plausible optimal proposed TS(s), and some of these cases are doubtful in that respect, the chosen technical systems were not necessarily optimal.

The systematic procedure must be adapted to the problem. The cases demonstrate that an engineering designer can idiosyncratically interpret the models to suit the problem, and develop information in consultation with a sponsor. Opinions will vary about whether, e.g., a requirement should be stated in a particular class of properties, or would be appropriate in a different class.

Hubka's engineering design methodology is demonstrated by the scope and variety of our case examples. As far as the author is aware, no other engineering design methodology is accompanied by such case examples. The initials after the case title show the originator—(VH) = Vladimir Hubka, (MMA) = Mogens Myrup Andreasen, (WEE) = W. Ernst Eder, and (SH) = StanislavHosnedl.

The first case study, systematic according to the state of the theory and method at that time, appeared in [26]—a machine vice (VH). Hubka and Eder [29] included the second case study—a welding positioner (VH). An English edition of case studies was published in [32], and included a riveting fixture (VH), a milling jig (VH), and a powder-coating machine (MMA), a P–V-T-experiment (WEE), a hand winding machine for tapes (VH), a tea brewing machine (MMA), a wave-powered bilge pump for small boats (MMA), and an oil drain valve (VH)—the (MMA) cases took a more industrial-artistic design approach, and only loosely followed the systematic method. Three further case studies were published in [18]—the tea machine revised to current systematic procedures showing enhanced engineering information (WEE); re-design of a water valve (WEE—first demonstration of systematic re-design); and an electro-static smoke gas dust precipitator, with rapper for dust removal (WEE—first demonstration of treatment for sub-problems, and the hierarchical nature of TS) [9]. The most recent book in this sequence [19] contains three new case studies, a portable frame for static trapeze display demonstrations (WEE) [10] which was built and used, re-design of an automotive oil pump (WEE—second demonstration of re-design) [17], and a hospital intensive care bed (SH—second demonstration of treatment for sub-problems)—the latter shows cooperation between industrial design and design engineering [24], and is one of many projects operated in cooperation with Czech industry. Hosnedl has also introduced the Hubka theories and methods into industrial use. Two new cases were presented at the International Conference DESIGN 2012 (WEE) [12, 13], both subsystems from the Caravan Stage Barge [2] which has been in operation in Canadian and U.S.A. coastal waters, and now in the Mediterranean, since 1995. The Canadian Engineering Education Association 3rd Annual Conference 2012 received two further case examples (WEE), a subsystem of the Caravan Stage Barge [15], and an auxiliary subsystem for a wind tunnel balance [16].

10.7 Closure

Depending on the nature of the (tangible or process) product, it is obvious that both engineering designers and artistic-industrial designers must in many cases work together. Their duties are partially overlapping. The Theory of Technical Systems [18, 19, 28, 31] is partially applicable to architecture and to industrial design (as demonstrated in [37]—of the five cases presented in this booklet from the Technical University of Denmark, only one refers to an engineering product, but exclusively with the external observable properties).

Nevertheless, engineering design is distinct from other forms of designing, and this needs to be acknowledged.

References

1. Booker PJ (1979) History of engineering drawing. Northgate, London
2. Caravan Stage Barge (2010) http://www.caravanstage.org
3. Eder WE (ed) (1996) WDK 24—EDC—engineering design and creativity—proceedings of the workshop EDC, Pilsen, Czech Republic, November 1995, Zürich, Heurista
4. Eder WE (2004) Machine elements û integration of some proposals. In: Proceedings of AEDS 2004 workshop, the design society—AEDS-SIG, Pilsen, Czech Republic, on CD-ROM, 11–12 Nov 2004 http://www.kks.zcu.cz/aeds
5. Eder WE (2005) Machine elements—revision and outlook for design education. In: Proceedings of second CDEN international conference, University of Calgary, Alberta, 18–19 July 2005 at Kananaskis Resort, paper 10006 on CD-ROM
6. Eder WE (2008) Aspects of analysis and synthesis in design engineering. In Proceedings of CDEN 08, Halifax, N.S., 27–29 July 2008, on CD-ROM
7. Eder WE (2009a) Analysis, synthesis and problem solving in design engineering. In: Proceedings of international conference on engineering design, ICED 09, Stanford University, Stanford, California, USA, 24–27 Aug 2009, paper 2–23, session W3 SIG-AEDS
8. Eder WE (2009b) Why systematic design engineering? In: Proceedings of 6th symposium on international design and design education, DEC 6, ASME, San Diego, California, USA, New York, 30 Aug–2 Sept 2009, paper number DETC2009-86067
9. Eder WE (2009c) Case study in systematic design engineering—smoke gas dust precipitation. In: Proceedings of 6th symposium on international design and design education, DEC 6, San Diego, California, USA, 30 Aug–2 Sept 2009, paper ASME DETC2009-86069
10. Eder WE (2010) Case study in systematic design engineering—trapeze demonstration rig. In: Proceedings of 7th symposium on international design and design education, DEC 7, Montreal, Quebec, Canada, 15–18 Aug 2010, paper ASME DETC2010-28065
11. Eder WE (2011) Engineering design science and theory of technical systems—legacy of Vladimir Hubka. Jnl Eng Design 22(5):361–385. Online 19 Nov 2010, InformaworldiFirst. 10.1080/09544828.2010.522558
12. Eder WE (2012a) Case example in systematic design engineering—leeboard mounting. In: Marjanovic D (ed) Proceedings of international design conference – DESIGN 2012, FMENA, Zagreb, Dubrovnik, Croatia, 21–24 May 2012
13. Eder WE (2012b) Case example in systematic design engineering—propeller shaft bearing arrangement. Marjanovic D (ed) In: Proceedings of international design conference—DESIGN 2012, FMENA, Zagreb, Dubrovnik, Croatia, 21–24 May
14. Eder WE (2012c) Comparison of Several design theories and methods with the legacy of Vladimir Hubka. private publication (74 pages) available from eder-e@kos.net, submitted for web-site of The Design Society, www.designsociety.org
15. Eder WE (2012d) Case study in systematic design engineering—bow thruster covers. In: Proceedings fo Canadian engineering education association CEEA 2012 conference, University of Manitoba, Winnipeg, MB, 17–20 June 2012, paper number 8
16. Eder WE (2012e) Case study in systematic design engineering—wind tunnel balance model support. In: Proceedings of Canadian engineering education association CEEA 2012 conferencem, University of Manitoba, Winnipeg, MB, 17–20 June 2012, paper number 9
17. Eder WE, Heffernan PJ (2009) A case study in systematic and methodical design engineering. In: Proceedings of CDEN/C2C2 conference 2009, McMaster University, Hamilton, ON, 27–29 July 2009
18. Eder WE, Hosnedl S (2008) Design engineering: a manual for enhanced creativity. CRC-Press, Boca Raton
19. Eder WE, Hosnedl S (2010) Introduction to design engineering—systematic creativity and management. CRC Press/Balkema, Leiden

20. Flurscheim CH (1983) Industrial Design in Engineering: a marriage of techniques. The Design Council, Springer, London, Berlin/Heidelberg
21. Green WS, Jordan PW (2002) Pleasure with products: beyond usability. CRC Press, Boca Raton
22. Gregory SA (1966) Design Science. In: Gregory SA (ed) The design method. Butterworths, London, pp 323–330
23. Hales C (1991) Analysis of the engineering design process in an industrial context 2nd edn, 1st edn 1987, Gants Hill Publ, Winetka
24. Hosnedl S, Srp Z, Dvorak J (2008) Cooperation of engineering and industrial designers on industrial projects. In: Marjanovic D (ed) Proceedings of 10th international design conference—DESIGN 2008, FMENA, Zagreb, pp 1227–1234
25. Howard T (2011) (Assistant Professor, DTU Management Engineering, Engineering Design and Product Development, Technical University of Denmark, Lyngby, Denmark), *The Design Society Newsletter*, e-mail 12 Sept 2011, supplement
26. Hubka V (1976) Theorie der Konstruktionsprozesse (Theory of Design Processes). Springer, Berlin
27. Hubka V (1982) Principles of engineering design. Butterworth Scientific, London, translated and edited by W.E. Eder from Hubka, V. (1980) WDK 1—AllgemeinesVorgehensmodell des Konstruierens (General Procedural Model of Designing), Zürich, Heurista (reprint by Zürich: Heurista, 1987)
28. Hubka V, Eder WE (1988) Theory of Technical systems: a total concept theory for engineering design. Springer, New York
29. Hubka V, Eder WE (1992a) Engineering design, 2nd edn, Heurista, Zürich of [27]
30. Hubka V, Eder WE (1992) Einführung in die Konstruktionswissenschaft (Introduction to Design Science). Springer, Berlin
31. Hubka V, Eder WE (1996) Design science: introduction to the needs, scope and organization of engineering design knowledge, Springer, London http://deseng.ryerson.ca/DesignScience/; completely revised edition of [30]
32. Hubka V, Andreasen MM, Eder WE (1988) Practical Studies in Systematic Design. Butterworths, London
33. Julier G (2000) The culture of design. Sage, London
34. Klaus G (1965) Kybernetik in philosophischerSicht (Cybernetics in Philosophical View), 4th edn. Dietz Verlag, Berlin
35. Klaus G (1969) Wörterbuch der Kybernetik (Dictionary of Cybernetics). Fischer, Frankfurt
36. Koen BV (2003) Discussion of the method: conducting the engineer's approach to problem solving. Oxford University Press, New York
37. McAloone T, Bey N (2010) Environmental improvement through product development. DTU Management Engineering, Engineering Design and Product Development, Technical University of Denmark, Lyngby
38. McMasters JH (2004) The biomechanics of flight: many possible solutions looking for problems. Int J Eng Educ 20(3):398–404
39. Müller J (1990) Arbeitsmethoden der Technikwissenschaften B Systematik, Heuristik, Kreativität (Working Methods of Engineering Sciences B systematics, heuristics, creativity). Springer, Berlin
40. Nevala K (2005) Content-based design engineering thinking. Academic Dissertation, University of Jyväskalä, Finland, Jyväskalä: University Printing House, http://cc.oulu.fi/~nevala
41. Pahl G, Beitz W, Feldhusen J, Grote H-K (2007) Engineering design, 3rd edn. Springer, London (1 edn. 1984) (Edited and translated by K. Wallace and L. Blessing), translated from 2003-5th edn. of Pahl G, Beitz W, Feldhusen J, Grote H-K Konstruktionslehre, methoden und anwendungen, 7th edn. Springer Berlin/Heidelberg (1 edn. 1977)
42. Schön DA (1983) The reflective practitioner: how professionals think in action. Basic Books, New York

43. Schön DA (1987) Educating the reflective practitioner: towards a new design for Teaching and learning in the professions. Jossey-Bass, San Francisco
44. Tjalve E (1979) A short course in industrial design. Newnes-Butterworths, London
45. Tjalve E, Andreasen MM, Schmidt FF (1979) Engineering graphic modelling. Butterworths, London
46. Wales CE, Nardi AH, Stager RA (1986a) Professional decision-making. Center for Guided Design (West Virginia University), Morgantown
47. Wales C, Nardi A, Stager R (1986b) Thinking skills: making a choice, Center for Guided Design (West Virginia University), Morgantown
48. Wallas G (1926) The art of thought. London, Cape, (reprint 1931) B pp 79–96 reprinted in Vernon PE (ed), Creativity. Harmondsworth, Penguin, 1970, pp 91–97
49. Weber C (2005) CPM/PDD – an Extended theoretical approach to modelling products and product development processes. In: Proceedings of PhD 2005, Srni, Czech Republic, pp 11–28, 7–9 Nov 2005
50. Weber C (2008) How to derive application-specific design methodologies. In: Marjanovic D (ed) Proceedings of 10th international design conference - DESIGN 2008, FMENA, Zagreb pp 69–80
51. Weber C, Vajna S (1997) A new approach to design elements (Machine elements). In: Riitahuhta A (ed) WDK 25 – proceedings of ICED 97 Tampere, vol 3. Tampere University, Tempere, pp 685–690

Chapter 11
10 Years of C–K Theory: A Survey on the Academic and Industrial Impacts of a Design Theory

Marine Agogué and Akin Kazakçi

11.1 Introduction: Bibliometrics on C–K Theory

Over the last 10 years, Concept-Knowledge theory (or C–K theory) has gained a growing academic and industrial interest. In 2002, Hatchuel and Weil [32] presented their first French conference paper exposing the main principles of C–K theory. This theory is based on the distinction between two expandable spaces: a space of concepts, the C-space (concepts are defined as undecidable propositions), and a space of knowledge K. The process of design is thus defined as the co-evolution of C and K through four types of independent operators (C–C, C–K, K–C, K–K). Exploration in the K-space encompasses the mapping of the knowledge base necessary for the understanding and the success of design path. The K-space is formed by the different pockets of activated knowledge that is used for the generative process of C-space. A concept is defined as a proposition without a logical status in the K-Space, i.e. an undecidable proposition: it is impossible to say if a concept is true or false. The C-space is a tree of undecidable propositions, each node of the tree corresponding to a partition (in the mathematical sense) in several sub-concepts of the mother concept. The C-space is tree-structured and describes the progressive and stepwise generation of alternatives, which are generally undecidable propositions before a conjunction can be interpreted as a solution. In other words, the designer who created the concept cannot tell whether such an object is possible or not before a design process is undertaken. The designer can than elaborate the initial concept by partitioning it—that is, by adding further properties to C. Current writings about C–K theory distinguish two kinds of partitioning. *Restrictive partitions* add to a concept a usual property of the object being designed *and expansive partitions* add to a concept novel and unprecedented properties. For more details, the reader is referred to [31–33] and the references in the Sect. 11.3.

M. Agogué (✉) · A. Kazakçi
Centre de gestion scientifique, Mines ParisTech, 60 boulevard Saint-Michel, 75006 Paris, France
e-mail: marine.agogue@mines-paristech.fr

A. Chakrabarti and L. T. M. Blessing (eds.), *An Anthology of Theories and Models of Design*, DOI: 10.1007/978-1-4471-6338-1_11,

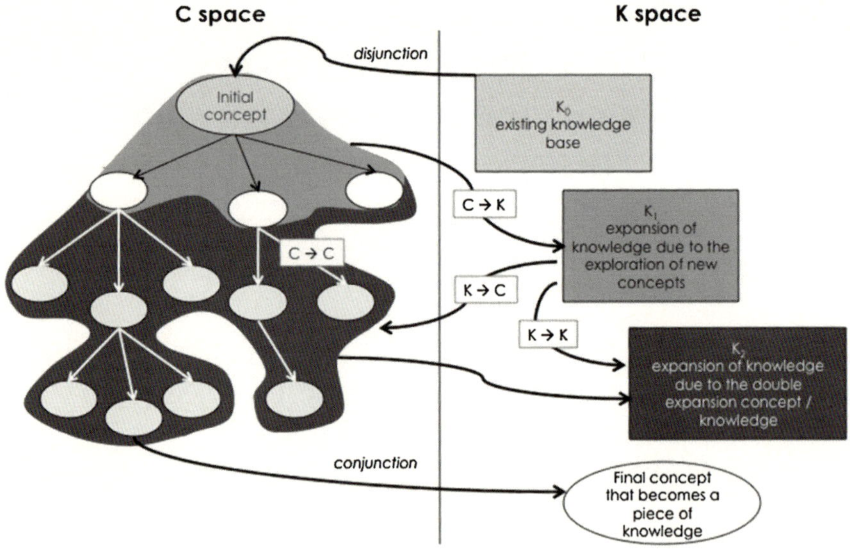

Fig. 11.1 Features of C–K theory: the expansion of two spaces

The Fig. 11.1 below summarises the basic features of C–K theory.

Since the seminal English-written paper from 2003 [33], the features of C–K theory have been recognised as being unique for describing creative reasoning and process in engineering design, as stated by Ullah, Rashid and Tamaki [68]. Specifically, these scholars highlight the fact that one of the most noticeable features of C–K theory is its foundation on the notion of a creative concept—a concept being an undecidable proposition with respect to the existing knowledge at the time it emerges. The impact of C–K theory is not limited to the design community: for instance, in 2012, a paper was presented in the French International Management Conference on the impact of C–K theory in management science over the last 10 years [9]. And today, this work echoes strongly in the industrial field: in 2010, the French company Thales, which designs systems and services for the aerospace, published a book on its design process and advocated C–K theory as a way to organise innovative design activities [19]. Since 2003, the RATP, the public transport operator for the city of Paris operating the subway, has deployed C–K driven tools [40]: they indeed use regularly the KCP approach, a method for collective creative design, on subjects such as ‘Bus Rapid Transit’, ‘Twenty-first century Metro’, ‘Local bus services’, ‘Walking’ or ‘Night bus stations’.

The aim of this chapter is to grasp at the variety of impacts fostered by C–K theory in academics as well as in empirical contexts. We therefore present a survey on the impacts of the work on C–K theory in the design community as well as in other academic fields, by studying the literature on C–K theory in English and in French. We looked at all the work done on C–K theory since its first premises in 1999: we gathered all the publications in blind peer-reviewed journals, as well as

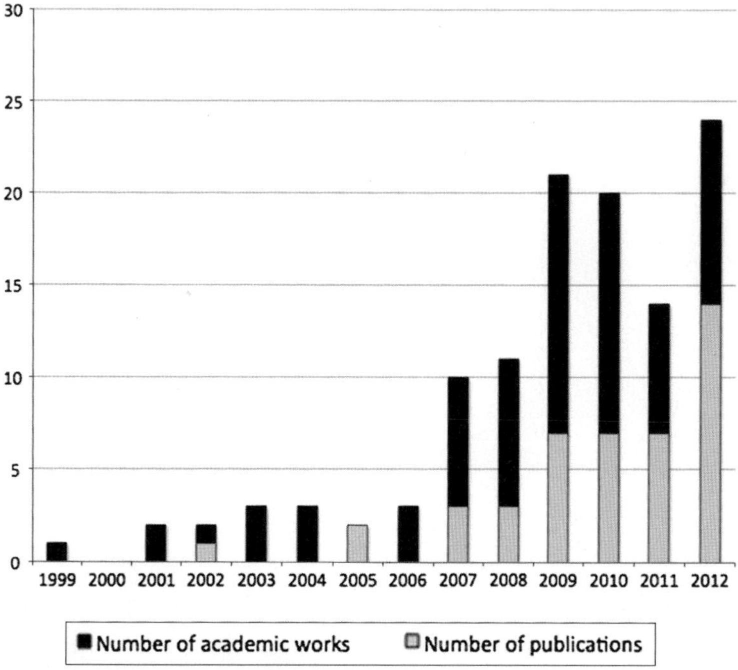

Fig. 11.2 Evolution of the publications on C–K theory (publications = in peer-review journals, works = thesis, book, conference paper, book chapter)

books, thesis, book chapters, conference papers with peer-reviews on abstracts and/or full papers. We analysed this material regarding different aspects: the discipline of the journal/authors, the purpose of mobilising C–K theory (as a theoretical framework or as an object of discussion), and the type of contributions in either a theoretical perspective or a more practical one. We completed our data collection with interviews and feedbacks from students and practitioners who applied C–K methodologies and tools.

Since the seminal paper of Hatchuel and Weil [32] on the foundation of a unified design theory modelling creative reasoning, over 40 papers have been published on the matter, more than 50 conference papers have been presented, 11 thesis and 5 books have built on C–K theory (Fig. 11.2).

On the 44 peer-reviewed publications around C–K theory, 13 are French-written, 12 discuss the theoretical aspects of the theory, 27 use it as a theoretical framework and 5 propose methods and tools derived from the theoretical foundations of C–K theory (Fig. 11.3).

If this stream of research initiated within the Design Theory and Methods for Innovation team at Mines ParisTech, the interest for C–K theory has now widely spread over the globe: over 40 % of the academic work has been conducted outside the team of Mines ParisTech (50 academic productions out of a reported 116 in total) (Fig. 11.4).

Language	Type of article	2002	2005	2007	2008	2009	2010	2011	2012	Total
English	On theoretical aspects of C-K theory		1	2		1		1	5	**10**
	Presenting C-K tools and methods					1	1		1	**3**
	Using C-K theory as a theoretical framework	1		1	2	3	2	3	6	**18**
French	On theoretical aspects of C-K theory		1			1				**2**
	Presenting C-K tools and methods						1		1	**2**
	Using C-K theory as a theoretical framework				1	1	3	3	1	**9**
Total		**1**	**2**	**3**	**3**	**7**	**7**	**7**	**14**	**44**

Fig. 11.3 Typology of peer-reviewed publication around C–K theory

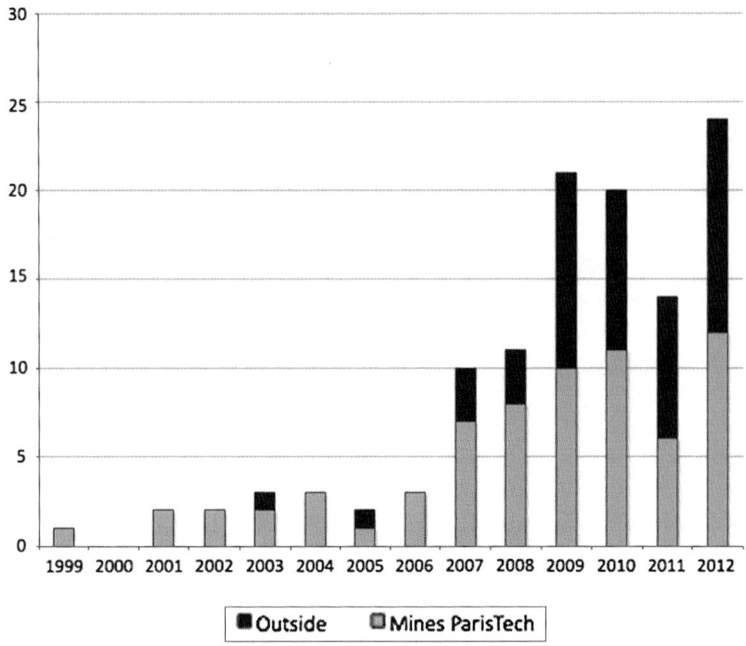

Fig. 11.4 Expansion of the work of C–K theory outside the design theory and methods for innovation team at mines ParisTech

11.2 C–K Theory and Engineering Design Theories: Revitalising a Lineage of Theories

Mathematical approaches to design have been developed in the 1960s, notably by Hiroyuki Yoshikawa. They tended to model the dynamic co-evolution between design solutions and requirements. There is today a revival of work on design theory. Since the development of Yoshikawa's General Design Theory, Suh's Axiomatic Design has appeared as well as Reich and Shai' Infused Design, Braha and Reich's Coupled Design Process and Hatchuel and Weil's C–K theory. The Design Society, the main scientific academy in the field, announced in August 2007 the creation of a Special Interest Group on Design Theory.

Within the field of engineering design, C–K theory opens new modelling directions that explore connections with basic issues in logic and mathematics, which differs from the classic use of scientific models in design. And today, C–K theory is debated by linking its formalism with other theories. C–K theory is discussed in [36] regarding Coupled Design Process [11] as C–K theory is used to interpret Braha and Reich's topological structures for design.

In Shai et al. [67] the authors present the Infused Design method to generate new concepts and methods in the classic discipline of statics, in addition to its prior use in the generation of a number of creative designs. The use of the Infused Design methodology in the creative scientific discovery process is modelled with C–K theory, leading to a deeper understanding of both Infused Design and C–K theory.

In another paper published in *Research in Engineering Design*, Reich et al. [61] modelled a creativity method, ASIT [46], with the help of C–K theory. They showed that using the modelling power of C–K as a theory of creative design reasoning improved their understanding of existing creativity methods. Specifically, they argued: '*ASIT is a specific, yet paradoxical creativity method, based on 'stay in the box' principle that seems contradictory with standard views about creativity. Modelling ASIT as a special instance of the C–K theory resolved the paradox and showed that ASIT is well adapted to a class of design situations*'.

As stated by Hatchuel et al. [39], '*the evolution of design theories can be interpreted as an attempt to increase their generative power without endangering their robustness*'. Indeed, the Special Interest group on Design theory grounded its approach on the following idea: having a diversity of design theories allows to compare the different perspectives, as the newer theory shed new light on some theoretical assumptions of the previous ones: hence the generativeness of design theories.

11.3 C–K Theory and Foundational Aspects of Logic: Unveiling Design Reasoning

The claim of C–K theory is that the above conceptive reasoning process is the defining essential characteristic of design and it is fundamentally different from the usual processes prevalent in formal sciences (i.e. deductive or inductive processes). Strangely, although design is a ubiquitous activity, design reasoning has not been considered *per se* as an object of scientific interest in traditional fields related studying reasoning processes, such as mathematics or logic. Moreover, when the need to better understand design and study associated processes formally emerged, the temptation has been to reduce or to apprehend design reasoning in terms of readily available formal constructs such as the classical logic, whose reasoning process relies on deductive operations, and machine learning, whose underlying paradigm is most of the time decision or search. Instead of using these formal constructs and their underlying paradigm, formal research on C–K theory focused on finding or building a set of design specific models and operators describing design reasoning [38].

As a starting point, a suitable source of inspiration was advanced set theory [31, 32]. From a general point of view set theory can be seen as a domain where particular objects (sets) and their properties are studied by formal means. The intuition that a concept (in C–K theory) must be a different kind of set since, during design, it is not possible to be certain whether it contains any elements led to the introduction of the idea of rejecting the *axiom of choice* (AC) in set theory, for any modelling attempt of C-space based on set theory: in standard set theory ZF (based on the axiomatic proposed by Zermelo–Fraenkel), the AC states that for every set there is a choice function. In C–K theory, this has been interpreted as the existence of elements are warranted for any set—which is contradictory with the intuition behind C–K theory according to which objects cannot be stated to exist during the process (or, else there is no need for design, but only for decision).

In Hatchuel and Weil [32], it was suggested that the axiomatic behind C-space was ZF without the axiom of choice while K-space was considered to be a CAT, the set of categories in category theory. In Hatchuel and Weil [33], the latter idea was abandoned and the knowledge space was defined as a space of propositions that have a logical status for some designer. The role and the implications of rejecting the axiom of choice for modelling design reasoning are explained in depth in that paper. Salustri [62] suggested the use of action logic ALX3d, to build a formal descriptive version of C–K theory. ALX3 considers a significantly different ontology than the original formal basis of C–K that is able to represent aspects such as the actions, preferences, beliefs and knowledge of collaborating, imperfect agents. This allows a more specific and granular structure of knowledge. Let us note that, contrary to C–K theory, Salustri [62, p. 9] sees no convincing reason to exclude axiom of choice. Instead, he states it can be used to increase flexibility of the proposed formal system.

In 2007, Hatchuel and Weil [34] presented a new result on the theoretical foundations of C–K. They proposed a correspondence between their theory and *Forcing*, a method for the invention of new sets in set theory. Forcing is a technique invented by Paul Cohen in order to prove the independence of the axiom of choice from set theory. This is a two-step process. First, a generic filter G is created using elements of a ground model M of ZF. The particularity of this step is that, although G is built using the existing sets in M, but it is not contained in M: it is a completely new object. Then, a naming process takes place where the sets in the old universe are renamed and reordered to accommodate the generic filter. The end result is a new model N of ZF, containing the old model M and the new object G. As an example, the authors use the concept of '*It exists a class of tires without rubber*'. Such proposition is undecidable within present standard knowledge. Existing tiyes are all made with rubber; yet, there is no established truth that forbids the existence of such no rubber tyres. When design begins, available knowledge cannot warrant the existence of tyres without rubber. In forcing language, it means that there is no generic set of conditions (i.e. generic filter) that extends the model of tyres (tyres with no rubber). Therefore, knowledge expansion becomes a crucial tool as it *generates new potential partitions*, i.e. new potential forcing conditions. For instance, introducing new materials or new knowledge about tyre shapes, cars or clients may enable such a generic filter and the design of tyres *without* rubber. At that point, our understanding and activities related to tyres will change (the authors use the terme *reordering*) since rubber is no longer seen as an essential attribute of tyres: tyre is seen now as an object independent of rubber (just as ZF is independent of the Axiom of Choice). Hatchuel and Weil argue that the correspondence with Forcing warrants and justify the properties of a design process as described by C–K theory, providing thus a foundational basis for C–K. In their interpretation, a C–K type reasoning process can be seen as a generalised Forcing operation on richer knowledge structures (than that of the set theory) creating at least one significantly new object while reorganising the old objects to preserve the meaning.

Hatchuel [30] gives further details on the relationship between set theory, Forcing and C–K theory. The driving theme of the presentation is the theoretical obstacles in defining a *thing*, which is a common issue in design as well as mathematics. From this perspective, techniques such as Forcing is interesting for foundations in design theory, since they provide means for creating fundamentally new objects (with respect to what is known) in a formal way.

Kazakci et al. [52] suggests a new formalisation of C–K theory based on *Wang Algebras* [70, 71]. The logical structure underlying Wang's system is a term logic with frequency and confidence values on weak inheritance relations. In 2005, Kazakci ([51], 2007) makes use of this flexible and expressive language to build a design assistant based on his interpretation of C–K type reasoning. In Kazakci et al. [52], the language is preserved, stripped of the memory system proposed by Wang, to propose an interpretation of the operators for C–K theory. In this work, they also introduce a notion of *models of K*. suggesting that, based on the formalism used to represent knowledge, C–K type reasoning may take different forms. Moreover, using different knowledge structures for K-space allows testing to what extent their underlying formalism allows a C–K type reasoning process. Kazakci [47] uses as a model of K the *intuitionist logic*, which is different from classical first-order logic in the interpretation of logical connectives and by the reject of the *law of the excluded middle* (LEM) which is a consequence of AC. Kazakci defines the concept space as a tree of formulae containing free variables and knowledge space corresponds to an incomplete theory in the logical sense. A set of operations is defined to model the progressive elaboration of the concept space, the expansion of the knowledge and the interaction of concepts and knowledge. Said in other terms, a formal interpretation of C–K theory's operators is provided the first time. The use of intuitionist logic opens the path to the investigation of the constructivist mathematics and philosophy as a basis for modelling design.

Kazakci and Hatchuel [49] studied the intuitionist mathematics pioneered by Brouwer [12, 13]. Intuitionist mathematics is one of the major constructivist approach in mathematics, challenging the dominant axiomatic approach fostered especially by Hilbert and several fundamental concepts in mathematics such as LEM, transfinite and actual infinity. Brouwer describes mathematics as a mental activity constructing progressively mental mathematical objects by the free choices of the mathematician—that Brouwer calls a *creative subject* [49]. The free choices provide a unique mechanism inside mathematics to continue the definition (or, construction) of partially determined mathematical objects in novel and unprecedented ways, breaking away thus with any static and algorithmic descriptions. The paper opens a unique debate about the parallels between design and mathematical activity, as described by Brouwer.

Hendriks and Kazakci [42] investigated the logical implications of the *dual expansion of concepts and knowledge*. This property is claimed to be the main engine through which a design reasoning progress. The contribution provided is a step towards better understanding the theoretical roots of C–K theory and design reasoning in general. They suggested using an *extended Kripke structure* by considering a partial order on what they call *design stages*. A design stage is a tuple $< K, C >$ where K is a partial theory representing knowledge and C is a (single) concept. A design stage might be extended by a *design move* either by expanding the concept, or by expanding the knowledge. As they discuss, this captures formally the principle of conjoint expansions of concepts and knowledge.

Going a step further, Kazakci et al. [50] presented an investigation of C–K's formal foundations based on a simulation study. They used a graph as a minimalist knowledge structure where vertices and edges represent objects. They gave an interpretation of C–K process based on this simple formalism. The presented formal model is used to conduct simulation of two contrasted design learning strategies; a concept driven strategy that is based on the concept being developed and a knowledge driven strategy that is based on learning based on the missing knowledge. They argued that the concept-strategy is much faster while the knowledge driven strategy provides a robust and well-connected knowledge space. Hendriks and Kazakci [43] built on [42, 47] to consider design as an aspect of rational agency hardly even mentioned in traditional logical theory. As an engineering discipline, design involves reasoning but seems to depend much more on a mix of factual knowledge, experimenting and imagination. They provided algorithmic descriptions of operators allowing these interactions within their logical interpretation of C–K theory.

Hendriks and Kazakci presents a framework demonstrating how C–K type design reasoning can be applied *within logic*. They extend and generalise the well-known method of Semantic Tableaux, invented by Beth for logical theorem-proving, to Design Tableaux—a general, formal procedure allowing to implement expansive reasoning within the formalism of logic. Their contribution is twofold. First, they give a formal, verifiable procedure that explicit and apply the ill-defined operators of C–K theory. Second, they contribute to the notion that design science can be useful to other fields and theories (in that case, logic) by proposing a mode of creative reasoning within a logical framework stemming directly from a theory of design. In Hendriks and Kazakci [44], they suggest a system architecture for a design assistant based on the design tableaux.

Kazakci [48] investigates forms of constructivism in design. He argues that the notion of constructivism should be integrated as a foundational element into the design research. Present forms of constructivism considered in design research, such as interactive [65, 66] or social constructivism [14, 15] lack the explicit consideration of creativity as a central issue of design. To explore how creative and constructivist aspects of design can be taken into account conjointly, he considers the roots of constructivism in mathematics, namely, the Intuitionist Mathematics and the process of mental mathematical constructions realised by a creative subject over time. One of the most original features of Intuitionist Constructivism is the introduction of *incomplete objects* into the heart of mathematics by means of lawless sequences and free choices. This allows the possibility to formulate undecided propositions and the consideration of creative acts within a formal constructive process. Based on his analysis, the author suggests a third form called imaginative constructivism. He uses this in interpreting design processes with the example of Manhattan project.

Hatchuel et al. [41] present new propositions about the ontology of design and a clarification of its position in the general context of rationality and knowledge. Their ontology is derived from a comparison between formal design theories developed in two different scientific fields: Engineering and Set theory. First, they review the evolution of design theories and their relationship with C–K theory. Then, the Forcing technique in reviewed and interpreted as a general design theory since it offers a process of controlled invention of new sets. Studying similarities and differences between C–K theory and Forcing, they highlight a series of common notions like "d-ontologies", "generic expansion", "object revision", "preservation of meaning" and "K-reordering". It is suggested by the authors that these notions form altogether an "ontology of design" which is consistent with unique aspects of design.

11.4 C–K Theory and Other Disciplines: Design as a Form of Creative Rationality

Today, the impact of C–K theory goes beyond the sole domain of engineering design or mathematics (cf. Fig. 11.5). It is a very strong sign of the generativeness of C–K theory, underlining the possible to use C–K theory as a framework to model very diverse issues.

11.4.1 Design Theory in Management Studies

In the management field, the formalisms of C–K theory have been used in different settings to highlight, model, and sometimes solve managerial challenges. In his conference paper retracing the influence of C–K on management research,

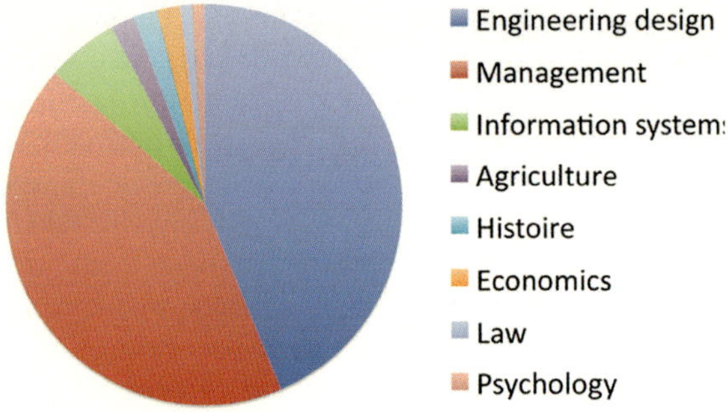

Fig. 11.5 Publications on C–K theory in diverse academic fields

Benguigui [9] stated that C–K theory is an excellent theoretical framework to explain the process of early phases of innovation, to interpret the misunderstandings (or *quiproquos*) in management context, to develop managerial tools and to relate the history of inventions.

C–K framework has indeed been used to frame understandings of historical projects, such as the Tenerife accident [69], the Manhattan project [57], the Swatch history [24], or the emergence of carbon markets [16]. In these different case studies, the historical events leading to the achievement of the design of a new product, a new process or a new system are explained by stating the different expertises (in the K-space) mobilised to explore different paths of innovation (in the C-space).

Some scholars have proposed to build tools for radical innovation management [45, 53], for building a common purpose [25, 26, 28], for handling patent issues [22, 23], or for organising collective action in innovation activities [7, 8, 20, 27]. Furthermore, the need to understand contemporary industrial dynamics led to mobilise C–K theory-driven diagnosis of innovative fields [1, 5].

11.4.2 Links Between Cognition and Design Theory

In the domain of cognition, Hatchuel et al. [37] have shown how C–K theory can help overcome fixation effect, i.e. being fixed on a small number of solutions, binding creativity. They stated that the outcomes of C–K theory-based design curriculum can be measured, being a possible catalyst while teaching creative thinking to students with the ability of creative thinking. Building on the notion of fixation effect, Agogué et al. [3, 4] claimed that there are two types of examples that C–K theory helps to characterise: (1) restrictive examples that do not change the definition or the attributes of the object and (2) expansive examples that modify its identity by adding unexpected attributes. Using an experimental protocol, they showed in the field of cognitive psychology that the solutions proposed by the group exposed to restrictive example are less original than those given by groups exposed to expansive examples.

Deepening this first set of experiments, Agogué and Cassotti [2] have used C–K theory modelling to understand more precisely the link between some activated knowledge and the solutions that are consequently explored, in order to model the fixation that occurs during design reasoning. The authors showed how different populations can be fixed in different ways and how their C–K based framework allowed making sense of this variety of fixation in design processes. They concluded by proposing three capabilities that are required to both understand fixation and overcome it: restrictive heuristics development, inhibitory control and expansion.

11.4.3 Expanding to Ecology

In ecology, a stream of research focuses on identifying and exploring effective solutions for integrating development of agriculture and conservation of biodiversity at a landscape scale. Berthet et al. [10] presented a case study on an intensively farmed French cereal plain, where the reintroduction of grasslands has been proposed to protect the Little Bustard, a threatened European bird species. They analysed the design reasoning that fostered this idea in order to highlight the innovative paths that were opened. They used C–K theory to do so, and revealed the links between the production of scientific knowledge and the generation of various solutions. It allowed them to state that specifying the ecological functions of grasslands facilitates their management.

11.4.4 Interactions with Other Fields

The implications of C–K theory have disseminated as well in other fields, such as creativity research (Le Masson et al. 2011) [37], data mining and knowledge management [59, 60, 29], history of engineering design [54, 55], psychology and cognition [3, 39], ecology [10, 11], philosophy [63, 64] and economics [18, 58]. There is today an impact of C–K theory in a branch of philosophy, called contemporary epistemology. Traditional epistemology discusses the truth or proof of truth of sciences. Contemporary epistemology is interested more in how science can create new techniques and control processes through ethics and democratic principles. Interestingly, researchers in this field have found in C–K theory an operational framework to describe processes and principles for generic epistemologies [64].

These different links between C–K theory and research questions in other fields underline how a design theory, by being a model of creative rationality, can be a framework for disciplines outside of design, whenever there is a need to model and understand creative reasoning.

11.5 Industrial Applications and Education: Increasing Needs for Stimulating and Teaching Innovative Design

Many applications of methods and tools derived from C–K theory have been deployed and tested in diverse industrial contexts [28, 40, 24, 44]. Typically, Hooge et al. [45] presented 14 industrial case studies conducted from 2009 to 2011, through an action-research methodology. Stemming from the diverse objectives that cover the innovation process, they propose practical guidelines to use C–K theory-driven tools. They suggest that such practical guidelines enable managers and designers to manage an important amount of knowledge and to structure the potential design paths of innovation projects.

Some industrial partnerships have led practitioners to present and publish vulgarisation work using C–K theory-driven methodologies and tools. In a white paper published in 2010, the experts of the International Technological Roadmap for Semiconductors[1] proposed to used C–K methodologies to explore the potentiality of innovating beyond the 'More Law' paradigm [6]. In 2011, the French company Thales, which designs systems and services for the aerospace, published a book on its design process and advocated C–K theory as a way to organise innovative design activities [19]. In its internal journal on Research and Innovation, the SNCF, the French railway company, exposed their approach using KCP, the C–K theory-based method for collective creative design.

Moreover, C–K theory formalisms are taught today in various contexts (engineering schools, management schools, business schools, design curricula, entrepreneurship schools, universities) and in different countries (France, Sweden, US, Israel, Tunisia). Over the last 5 years, the team from Ecole des Mines de Paris has supervised closely 41 master students doing internships using C–K theory in French institutions and firms (big firms, medium size firms and start-ups). They worked in sectors such as transports, energy, food, NTIC, health, nanotechnologies and urbanism.

Observations through empirical investigations (interviews with consultants specialised on C–K methodologies, industrial partners, students) show that today, the diffusion and adoption of C–K theory through teaching and companionship leads to the emergence of practices outside of the scope of the Design Theory and Methods for Innovation team at Mines ParisTech. Those practices are indeed adapted very finely to the technological, social and organisational contexts of their applications. Interviews with practitioners underlined that the feature the most useful for them was the notion of expansion in both C and K-spaces. Indeed, interviews show that during design activities, the expansible characteristic of the

[1] A group of semiconductor industry experts representative of the sponsoring organisations based in the US, Europe, Japan, South Korea and Taiwan who intend for technology assessment only and without regard to any commercial considerations pertaining to individual products or equipment.

C-space (i.e. the ability to add unusual or unexpected attributes to a concept) leads to surprises that lie at the heart of any innovative process. Deviating from the known identity of objects is therefore said to require design-based methodologies and tools, and designers mobilising C–K tools acknowledge the ability of C–K approach to structure design reasoning and to support the elaboration of original, expansive design paths.

11.6 Discussion: Features of a Design Theory

Studying the effects of C–K theory over the last 10 years confirms the great contemporary expectations towards a design theory. First, it aims at revitalising the knowledge accumulated in engineering design. Then, deepening the formal aspects of a design theory helps to both unveil and explain the surprises, the paradoxes, the oddness of design reasoning that goes beyond classic rationality and logics. Moreover, a design theory being a model of creative rationality, it can circulate and become a framework for disciplines outside of design, where there is a need for innovation and for building understanding on creative reasoning. A design theory therefore can become a reference model to do so. And last but not least, it is certainly not a coincidence that the rapid diffusion of C–K at least in global companies corresponds precisely to a decade where innovation has been considered as the most important competitive asset. All our interviews with sponsors showed that the most important impact of C–K has been on the development of new radically or disruptive systems.

Like a product reveals the market, C–K theory revealed the latent need and potentials of a design theory. Yet, there is a paradox for the non-specialists: it is a difficult theory to understand but easy to practice. Of all design theories, it is the most abstract one, and understanding its foundations requires very specific high-skilled background, typically in mathematics. As for today, there is no easy software to implement C–K theory and its development rests on the work of consultants and researchers. But in spite of these difficulties, the development of C–K theory shows three important conditions that any design theory that wants to address contemporary issues to fulfil: generality (i.e. the capacity to propose a common language on the design reasoning and design processes), generativity (i.e. the capacity to model creative reasoning and to relate to innovative engineering in all its aspects) and relatedness to contemporary knowledge and science (i.e. the capacity to relate to advances in all fields even when they seem far from the design community, such as mathematics or cognitive psychology : a design theory enables a dialogue that either benefits from or contributes to other disciplines).

How can we test this model of three parameters of a design theory? For sure, this model would be more convincing if it was tested for instance on Nam Suh's Axiomatic Design theory, which has also a large impact. It would be interesting to compare our data with a similar approach on the developments of Axiomatic Design theory. From the perspectives of generality and generativity, we can assume that both theories have very strong features. However, it seems, but needs more confirmation, that the links between Axiomatic Design and other disciplines are narrower than those with C–K theory.

These conditions (generality, generativity and relatedness to other disciplines) are quite demanding, but they seem to be essential to ensure that a design theory has an impact, both in academia and in empirical contexts. It is impossible to say what will be the next generations of design theories but it is sure that they should progress on these directions.

Acknowledgments The research presented in this chapter has benefited from a research grant given by the French government (department of Defense) and by the sponsors of the chair *Theory and Methods for Innovative Design* at Mines ParisTech (Dassault, RATP, Renault, SNCF, STMicroelectronics, Thalès and Vallourec). We would like to thank our sponsors and industrial partners for their feedbacks on their practice with C–K theory. We would like also to thank Armand Hatchuel, Benoit Weil and Pascal Le Masson for their insights on the evolution and the dissemination of their work on C–K theory.

References

1. Agogué M (2013) L'innovation orpheline. Presses de Mines, Paris, p 246
2. Agogué M, Cassotti M (2013) Understanding fixation effects in creativity: a design-theory approach. In: ICED conference 2013, Séoul, 2013
3. Agogué M, Kazakçı AO, Weil B, Cassotti M (2011) The impacts of examples on creative design: explaining fixation and stimulation effects, In: International conference on engineering design, Copenhague, 2011
4. Agogué M, Kazakçi AO, Hatchuel A, Le Masson P, Weil B, Poirel N, Cassotti M (2013) The impacts of examples on originality: explaining fixation and stimulation effects. J Creat Behav (in press)
5. Agogué M, Le Masson P, Robinson DKR (2012) Orphan innovation, or when path-creation goes stale: a design framework to characterize path-dependence in real time. Technol Anal Strateg Manag 24(6):603–616
6. Arden W, Brillouet M, Cogez P, Graef M, Huizing B, Mahnkopf R (2011) "More-than-Moore" white paper, In: International technological roadmap for semiconductors, Annual report
7. Arnoux F, Béjean M, Hatchuel A (2010) Strategies for radical innovation: exploring the role of collaborative creative design methods. In: IPDM Conference, Murcia
8. Béjean M (2008) Le management des entreprises à prestations artistiques. Activités de conception, régimes de signification et potentiel de croissance. Thèse de doctorat en sciences de gestion. MINES ParisTech, Paris
9. Benguigui J-M (2012) Les 10 ans de la théorie C-K: revue de littérature, XXIth international french conference on strategic management (AIMS). Lilles, France
10. Berthet ET, Bretagnolle V, Segrestin B (2012) Analyzing the design process of farming practices ensuring little bustard conservation: lessons for collective landscape management. J Sustain Agric 36(3):319–336
11. Braha D, Reich Y (2003) Topological structures for modelling engineering design processes. Res Eng Design 14(4):185–199
12. Brouwer LEJ (1907) Over de Grondslagen der wiskunde (On the foundations of mathematics), p 183

13. Brouwer LEJ (1908) De Onbetrouwbaarheid der logische principes. Tijdschrift voor Wijsbegeerte 2:152–158
14. Bucciarelli L (1988) An ethnographic perspective on engineering design. Des Stud 9(3):159–168
15. Bucciarelli L (1994) Designing the engineers. MIT Press, Cambridge
16. Cartel M, Aggeri F, Agogué M (2012) Enabling perfomativity in 'skunk labs': the untold story of carbon markets design. In: EGOS conference 2012, Helsinki, 2012
17. Cohen JP (1964) The independance of the continuum hypothesis II. In: Proceedings of the national academy of science, USA
18. Colasse S, Nakhla M (2011) Les démarches de contractualisation comme processus de conception : l'émergence du contrôle de gestion médicalisé à l'hôpital, Revue Politiques et Management Public, 28/3, juillet-septembre, pp 311–331
19. Defour M, Delaveau C, Dupas A (2010) Avionique. Des technologies innovantes au services des plus belles réussites aéronautiques. Gallimard Loisirs, Paris
20. Elmquist M, Segrestin B (2007) Towards a new logic for front-end management: from drug discovery to drug design in pharmaceutical companies. J Creat Innov Manage 16(2):106–120
21. Elmquist M, Segrestin B (2009) Sustainable development through innovative design: lessons from the KCP method experimented with an automotive firm. Int J Automot Technol Manage 9(2):229–244
22. Felk Y (2011) Evaluation et pilotage des activités de recherche pour la rupture dans la R&D centrale de STMicroelectronics": réviser les classiques du management de la recherche industrielle. Thèse de doctorat en sciences de gestion. MINES ParisTech, Paris
23. Felk Y, Le Masson P, Weil B, Cogez P, Hatchuel A (2011) Designing patent portfolio for disruptive innovation—a new methodology based on C–K theory. In: International conference on engineering design, Copenhague, 2011
24. Garel G, Mock E (2012) La fabrique de l'innovation. Dunod, Paris
25. Gillier T (2010) Comprendre la génération des objets de coopération interentreprises par une théorie des co-raisonnements de conception. Thèse de doctorat en Génie des Systèmes Industriels. INPL, Grenoble
26. Gillier T, Piat G (2011) Exploring over the presumed identity of emerging technology. Creat Innov Manage 20:238–252
27. Gillier Piat, Roussel Truchot (2010) Managing innovation fields in a cross-industry exploratory partnership with C–K design theory. J Prod Innov Manage 27(6):883–896
28. Gillier T, Kazakci AO, Piat, G (2012) The generation of common purpose in innovation partnerships: a design perspective. Eur J Innovation Manage 15(3):372–392
29. Goria S (2010) Proposition d'une méthode d'expression d'idées et de problèmes d'innovation. Revue ESSACHESS 5
30. Hatchuel A (2008) Mathématiques et conception, In: Hatchuel A, Weil B (eds) Les nouveaux régimes de la conception: Langages, théories, métiers. Edition Vuibert, Paris, pp 115–131
31. Hatchuel A, Weil B (1999) Pour une théorie unifiée de la conception, Axiomatiques et processus collectifs. CGS Ecole des Mines, GIS cognition-CNRS
32. Hatchuel A, Weil B (2002) C–K Theory: notions and applications of a unified design theory. In: Herbert Simon international conference on "Design Science", Lyon, 15–16 March 2002
33. Hatchuel A, Weil B (2003) A new approach of innovative design: an introduction to C–K theory. In: XIVth international conference on engineering design, Stockholm, Sweden, 19–21 Aug 2003
34. Hatchuel A, Weil B (2007) Design as forcing: deepening the foundations of C–K theory. In: XVIth international conference on engineering design, Paris, 28–31 Aug, 2007
35. Hatchuel A, Weil B (2008) Entre concepts et connaissances: éléments d'une théorie de la conception. In: Hatchuel A, Weil B (eds) Les nouveaux régimes de la conception: langages, théories, métiers, sous la direction d', Vuibert, Cerisy, Paris, 115–131
36. Hatchuel A, Weil B (2009) C–K design theory: an advanced formulation. Res Eng Design 19:181–192

37. Hatchuel A, Le Masson P, Weil B (2011) Teaching innovative design reasoning: how C–K theory can help to overcome fixation effect. Artif Intell Eng Des Anal Manuf 25(1):77–92
38. Hatchuel A, Weil B, Le Masson P (2013) Towards an ontology of design: lessons from C–K design theory and forcing. Research in engineering design, 1–17
39. Hatchuel A, Reich Y, Le Masson P, Weil B (2011) A systematic approach of design theories using generativeness and robustness. In: International conference on engineering design, Copenhague, 2011
40. Hatchuel A., Le Masson P, Weil B (2004) CK theory in practice: lessons from industrial applications. Design Society Conference, Dubrovnik, May 2004
41. Hatchuel A, Weil B, Le Masson P (2013) Towards an ontology of design: lessons from C–K design theory and forcing. Res Eng Design 24(2):147–163
42. Hendriks L, Kazakci A (2010) A formal account of the dual expansion of concepts and knowledge in C–K theory. In: International design conference—design 2010, Dubrovnik, May 2010
43. Hendriks L, Kazakçi A (2011a) Design as imagining future knowledge, a formal account. In: Grossi D, Minica S, Rodenhauser B, Smets S (eds) Logic and interactive rationality, Yearbook 2010, p 111–126
44. Hendriks L, Kazakci A (2012) A design assistant architecture based on design tableaux, conference of the design society, Dubrovnik, May 2012
45. Hooge S, Agogué M, Gillier T (2012) A new methodology for advanced engineering design: lessons from experimenting C–K theory-driven tools. Design Society Conference 2012, Dubrovnik
46. Horowitz R (1999) Creative problem solving in engineering design (Doctoral dissertation, Tel-Aviv University)
47. Kazakci A (2009) A formalisation of CK design theory based on intuitionist logic. In: Chakrabarti A (ed) International conference on research into design (ICORD09), Research Publising Services, Bangalore, pp 499–507
48. Kazakci A (2013) On the imaginative constructivist nature of design: a theoretical approach. Res Eng Design 24(2):127–145
49. Kazakci A, Hatchuel A (2009) Is "creative subject" of Brouwer a designer?—an analysis of intuitionistic mathematics from the viewpoint of C–K design theory, International conference on engineering design (ICED'09). Stanford, CA
50. Kazakci A, Hatchuel A, Le Masson P, Weil B (2010) Simulation of design reasoning based on C–K theory: a model and an example application. International Design Conference 2010, Dubrovnik
51. Kazakçi AO, Tsoukias A (2005) Extending the C–K design theory: a theoretical background for personal design assistants. J Eng Des 16(4):399
52. Kazakci A, Hatchuel A, Weil B (2008) A model of C–K design theory based on a term logic: a formal background for a class of design assistants. International Design Conference 2008, Dubrovnik
53. Kokshagina O, Le Masson P, Weil B, Cogez P (2012) Risk management strategies in a highly uncertain environment: understanding the role of common unknown. International Product Development Management Conference, Manchester
54. Le Masson P, Weil B (2010) La conception innovante comme mode d'extension et de régénération de la conception réglée: les expériences oubliées aux origines des Bureaux d'études. Entreprises et Histoire 58(1):51–73
55. Le Masson P, Weil B (2010) Aux sources de la R&D: genèse des théories de la conception réglée en Allemagne (1840–1960). Entreprises et histoire 58(1):11–50
56. Le Masson P, Weil B, Hatchuel A (2010) Strategic management of design and innovation. Cambridge University Press, Cambridge
57. Lenfle S (2012) Exploration, project evaluation and design theory: a rereading of the Manhattan case. Int J Manag Proj Bus 5(3):486–507
58. Nahkla M, Colasse S (2011) Les démarches de contractualisation comme processus de conception. Rev Polit Manag Public 28(3):311–331

59. Ondrus J, Pigneur Y (2009) C–K design theory for information systems research. In: 4th International conference on design science research in information systems and technology, New York
60. Poelmans J, Elzinga P, Viaene S, Dedene G (2009) A case of using formal concept analysis in combination with emergent self organizing maps for detecting domestic violence, advances in data mining. Appl Theoret Aspects 5633(2009):247–260
61. Reich Y, Hatchuel A, Shai O, Subrahamanian E (2012) A theoretical analysis of creativity methods in engineering design: casting and improving ASIT within C–K theory. J Eng Design, 23(2):137–158
62. Salustri FA (2005) Representing CK theory with an action logic [online]. In: Samuel A, Lewis W (eds). ICED 05: 15th international conference on engineering design
63. Schmid AF (2009) Les sciences de l'ingénieur entre modélisation et conception, Primeras Jornadas interdisciplinarias del Universidad del Nord-Este
64. Schmid AF, Mambrini-Doudet M, Hatchuel A (2011) Une nouvelle logique de l'interdisciplinarité. Nouvelles perspectives en sciences sociales: Revue internationale de systémique complexe et d'études relationnelles, 7(1):105–136
65. Schön DA (1983) The reflective practitioner. Basic Books, New York
66. Schön DA, Wiggins G (1992) Kind of seeing and their functions in designing. Des Stud 13(2):135–156
67. Shai O, Reich Y, Hatchuel A, Subrahamanian E (2013) Creativity and scientific discovery with infused design and its analysis with C–K theory. Res Eng Design, 24(2):201–214
68. Sharif Ullah AMM, Rashid MM, Tamaki J (2012) On some unique features of c-k theory of design. CIRP J Manufact Sci Technol 5(1):55–66
69. Szpirglas M (2006) Gestion des risques et quiproquos. Revue française de gestion 161(2):67–88
70. Wang P (1995) Non-axiomatic reasoning system: exploring the essence of intelligence. PhD Thesis, Indiana University
71. Wang P (1998) Grounding the meaning of symbols on the system's experience. In: Working notes of the AAAI workshop on the grounding of word meaning: data and models. Madison, Wisconsin

Chapter 12
Designing the Inventive Way in the Innovation Era

Denis Cavallucci

12.1 Introduction: From the Age of Quality to the Age of Innovation

Since its very beginnings, our industry has regularly experienced important upheavals throughout its history. Each of these changes signals the birth of a new industrial 'age' where new rules must be respected. Moving from one age to the next undeniably implies changes to adapt to these new rules, because each age is weighed down with a novel set of problems considered as obstacles when encountered by companies in the specific context of perpetual evolution. Historical descriptions of industrial ages are diverse. Kao puts forward that four ages characterising the industrial phases (Agriculture, Industry, Information and Creation) [1]. For their part, Freeman & Louca also speak of four ages but different ones (mechanisation through using water, mechanisation through using steam, electrification, and motorisation and wars) and suggest that we are currently entering a fifth age: the total computerisation of the economy [2]. Finally, Judit Kapás defines three ages (the advent of the factory, the setting-up of multi-sector organisations, the information and knowledge industry) [3]. Although these analyses are often inherited from economic and management sciences, they are nonetheless interesting since they are based on cross-disciplinary observations (society, politics and business). Yet we should note that aspects linked to R&D methods and theories are often absent from these analyses, which means it is not easy to make a descriptive use of them and link such contributions to the essence of this chapter.

As a basis for the analyses below, we therefore put forward a representation specific to the links between theories, methods, design tools and society. This description stems from the observation that an industrial upheaval obliging all businesses to undergo deep restructuring is a rare event in the space of a century. The observation of demographic, geopolitical and historical facts combined with

D. Cavallucci (✉)
INSA, Strasbourg, France
e-mail: denis.cavallucci@insa-strasbourg.fr

A. Chakrabarti and L. T. M. Blessing (eds.), *An Anthology of Theories and Models of Design*, DOI: 10.1007/978-1-4471-6338-1_12,

social and theoretical causes leading to significant changes in practices, points to three ages: production, quality and innovation [4]. Each of these ages has its set of problems and it is these problems that incite scientists in engineering design to develop new theories, which, in scientific organisations first and then within the business environment, will cross through a certain number of phases before being accepted as common practice.

To analyse the current situation and put forward some hypotheses as to what is likely to happen over the next few decades, let us return to the age of quality and what it can teach us. At the beginning of the 1950s (for the highly industrialised countries), the rare companies able to impose their products on customers were quickly caught up by ever-increasing competition from their rivals. The social context changed, the buying power of families increased and a new paradigm entered into play. It notably imposed the need on companies to stand out from competition by adopting approaches aimed at optimising output and minimising losses of all kinds. Very rapidly, industrialists became deeply preoccupied with the need for their practices to evolve in order to cope with this new situation. Larry Miles and others then put forward the notion of Value, and while, theoretically, the aim was to reduce the new evils of industry, it had to be transcribed in the form of methods, and then enhanced by tools to have an effective impact on companies. Initially, methods such as Value Analysis, Kanban, MRP and more recently 6sigma came to the fore as potential means for companies to attain progress in terms of their practices. In a second phase, partly overlapping with the first phase, methods were enhanced by tools aiding optimisation. CAD tools and computer-assisted calculation (structure, simulation, and optimisation) surfaced and helped businesses attain optimisation, reduce unnecessary expenses and therefore maximise the industrial aim in terms of its value.

Today, we are once again witnessing a change in paradigm, whose roots can be traced back to the 1980s. Here again, a socio-political evolution, resulting in a higher buying power for families and lower costs for high-tech products (and therefore an increased access to novelties), has generated greater stress among businesses with the need to renew their products and systems more frequently in order to meet the impatient demands of customers who became avid for new hight-tech products. Companies have to make their mark in an ever-more competitive environment where scientific discoveries abound leading to broader potential evolution paths for artefacts.

What are the fundamental incidences underlying innovation's arrival on the scene?

One of the major incidences for industries when a new age occurs is the need to adapt their internal practices both in terms of human and methodological competencies. Among others, this evolution leads to a redefinition of the limits of knowledge, methods and tools designed to help industrial stakeholders in their everyday tasks.

From the lessons of the past, how can we anticipate all the difficulties our industry will be confronted with? How can we list these difficulties and position them on a timescale? In what directions should we focus research in engineering design taking all these observations into account?

In scientific and industrial environments, there is a major difference between innovation and invention [5]. Several works have highlighted these differences, but they can be summarised in the field of engineering as follows.

The status of innovation (associated with an object, or more broadly what man has built: the artefact) is acquired when society adopts this, apparently new, artefact within a community of consumers that boasts a certain legitimacy.

On the other hand, the relationship is confused between creativity, often perceived and envisaged from the angle of a process or a phase within the process sometimes called 'ideation' and the creative skills of the designer which is rather translated by a special capacity to solve certain specific problems.

Creativity, which increasingly seems to be an acknowledged process within the more general activity of design, is at the heart of many reflections. As evidence of this, we can take the important increase in contributions over the last few years focusing on the creative phases in the field of information sciences, and their attempts to explain that it is high time their efficiency was increased. Let us add to this the emergence of theoretical and practical propositions to characterise them, such as TRIZ [6], and the trend is confirmed by the desire of several researchers in the scientific fields to take them into consideration [7]. It would seem that this new fact is a logical development of a practice studied in the human and social sciences, where little computerisation would seem to be possible. We therefore postulate that understanding the creative act up to the synthesis of its ontology must inevitably happen before computerisation can occur. In the presented work, creativity is considered as a human capacity, therefore beyond the scope of what this chapter intends to present.

This chapter proposes a possible frame for a re-design of early stages of design activities in the innovation age. This proposal is the result of 15 years of research in the domain of Design science, invention and innovation. It has been synthetised in a methodology for our hypothesis to be potentially tested in a real-life situation (mostly R&D departments). While being at very first simple TRIZ-based applications on case studies in the first years, it became a complete frame for inventive design activities at the early stages of any complex project aiming at provoking an innovation as a result.

After the above introduction, the present chapter is divided into four parts. The first part is dedicated to summarise the research that brought us from TRIZ limitations to a global framework namely IDM (Inventive Design Methodology). IDM will also be briefly described through it deployment algorithm. The second part is focused on the central notion of both TRIZ and IDM approaches: the notion of contradiction. The third part consists in developing a case study conducted using IDM to illustrate and discuss the benefits but also the limitations of our current statement regarding IDM. A fourth part will conclude and discuss each current axis of research we are involved in for the overall IDM's enhancement and improvement.

12.2 How TRIZ Can Contribute to Design Theories

A theory or model of Design is supposed to provide designers with answers to their everyday professional difficulties. Along each tasks assumed by designers, a relevant Theory of Design should provide first theoretical roots, scientifically proven, then a methodological declination of it for appropriate use and practice. It should describe the world and its realities through a prism from which, when observed through, designers could envision useful insights as regarding their designing tasks. These useful insights could be provoked by an original description, a clear definition and allow designers to anticipate with artefacts design processes with some kind of robustness.

The notion of robustness can only be reached if what the theory proposes matches with temporal realities. As explained in introduction section, nowadays realities are facing with the problematic of Innovation. They fundamentally differ from the problematic disclosed during quality era. Therefore, one would be mistaken to think that with a theory of Design built during quality era, we could cope with the difficulties brought by innovation era. It is also an error in my mind to think that just an evolution of existing Design theories could solve this problem. We are in need of fundamentally different theories of Design that will engender fundamentally different methods and tools for Design activity's framing. But if we rapidly look at what most of existing methodologies proposes, they all focus on the very same notions as during quality era: customers listening (satisfying) and value criteria are the only targets. Some 15 years ago, we decided to investigate what TRIZ theory proposes and how this new approach differs from existing ones, this section is dedicated to further detail how we switched from TRIZ theoretical groundings to a new approach in support to design activity in the innovation era.

For nearly two decades, TRIZ has appeared as a set of methodological tools useful for supporting inventive aims in industry. This theory represents a significant breakthrough in driving problem statement and solving in a direction that is expressed through the idea that technical systems are driven by objective laws. But the difficulties to fully benefit from TRIZ in companies are strongly felt as TRIZ itself has some incomplete concepts and incoherencies. Upstream phases of the design process are often associated with market feedback [8], documentation research [9], a state-of-the-art review and idea generation [10], followed by a sorting of these ideas to select those to be used in the downstream development phase. A good illustration of this subject is the stage gate process [11], currently quite popular in companies. Yet these approaches provide little assistance, in either the multidisciplinary formalisation of knowledge [12] required to understand an initial situation that is unsatisfactory or the design of an innovative system it is intended to improve, or in opening up new knowledge streams to resolve the key issues of its underlying initial situation.

IDM method is the fruit of our recent research that uses a structured process upstream of innovation projects [13]. IDM takes the place of the standard or routine design process in the early stages and seeks to rapidly arrive at a reasonable number of inventive Solution Concepts to evolve a complex initial situation that is currently unsatisfactory. IDM therefore aims to be part of the so-called innovation process in companies. In other terms, IDM is intended to be implemented in a company predisposed to assuming risk after experiencing failed solutions and requiring significant R&D phases to arrive at adequate resolutions. The idea here is not so much that a company state that it operates as an innovation-driven entity—a nearly universal claim made nowadays—but rather that it indicates that it is prepared to accept the risk of investigating knowledge that it does not yet fully control.

IDM finds its roots into TRIZ methodology. Yet if one can perceived TRIZ as a theory, we are there mainly focused on the methodological aspects of TRIZ and its limitations. In other terms, why TRIZ was not sufficient and for which situations TRIZ was in need of an enhancement based on other theories from different theoretical but complementary background?

These limitations are of five orders. Here is a brief explanation of each of them:

- About initial and exhaustive investigations: TRIZ is not designed to investigate initial situations (gathering thoroughly all knowledge necessary and known to qualify the problem) [14]. There are no means in TRIZ to efficiently start with a complex initial situation and engineers in R&D departments are rarely confident with the classical TRIZ way of dealing with such task. Often there is a minimum of time allowed to this task and it seems that the TRIZ expert is randomly (or based on intuition) considering the problem at a given systemic level. As a result, the level of confidence that the overall problematic has truly been taken into consideration is low. Of course, when a nice solution appear, the level of satisfaction is increased, but the confidence that we really addresses the problem at the right level and right way is not assumed in TRIZ with evidences.
- About contradiction's quantity... and choice: TRIZ is designed for solving a single contradiction. How to choose the most appropriate one since contradictions quantity increase exponentially with any system's complexity? In real life engineering situations, rarely a problem, even the simplest ones, only shows a set of three parameters involved in the problematic. The parameters quantity involved to fully describe all influenced characteristics of a given problem is increasing as the problematic gets complex. Assuming that a set of three parameters are already defining a contradiction, how to fully formulate a full set of contradictions related to a problem? We believe that in really complex situations, this set of contradiction can be of several hundred [15]. TRIZ is not proposing anything in such cases. It seems that the contradiction intuitively sticks-out of the inventor's mind among a set of potentially involved other parameters. Then, how can we ensure that the one disclosed is the one we should solve in priority?
- About a formal methodology to disclose a contradiction: There are no accurate ways to disclose appropriately a contradiction. The definition of a contradiction (at different levels of formulation) is given in almost all TRIZ books. But these descriptions are rather simple and it seems obvious in simple situations. Only

OTSM framework made an interesting proposal to separate parameters typology in two kinds (control and evaluating) [16]. We believe that if a contradiction is not defined up to its physical level, then its level of definition is incomplete or doubtful [17]. As a consequence, we shall not consider technical contradictions (with no physical oppositions) as 'true' contradictions but only as 'prototypes' of contradiction as their physical roots are incomplete. With technical contradictions, we know which parameters are in opposition but not why this opposition exists.

- Solution Concepts that emerge from TRIZ technique usage are often inventive (since TRIZ refuses compromise). Therefore, as creativity of people led by TRIZ techniques is efficiently put into practice, the quantity of solutions might exceed a single one. In our past experience of using TRIZ, in most cases, the population of solutions was between 5 and 20. Now how to choose among several solutions the best one that addresses, not only a single contradiction, but a set of contradiction behind a set of problems? In classical TRIZ, only intuition is leading to the choice of a solution, then how can we be convinced that among a set of solutions there is no necessity to further find an additional one? To be confident with such a choice, we believe there must be mathematical means, based on statistical expert's votes on potential influences. Such an approach leads us not only to highlight the solution concept, but also to rank all solutions ordered by their capacity to solve sub-sets of contradictions disclosed in the overall problematic [18].
- About TRIZ corpus (or body of knowledge) consistency: We are not aware of any 'glossary' related to TRIZ. Therefore, there are no logical links/coherence between TRIZ components as each TRIZ component appears at a given level of abstraction, being more or less categorised or related to another term, but a completed description of all TRIZ components with their definition and interrelations does not exist. This is what is commonly called in Artificial Intelligence an ontology. Such ontology work starts to appear within several publications [19–21] but is only a proposal of a team, here we need a commonly agreed work on TRIZ ontology from all the TRIZ community to go further with scientific tools to investigate its potential and make it progress.

To summarise, these five limitations were at the basis of the challenges behind IDM framework of research. It has been conducted by both scientists from different fields (Artificial Intelligence, Computer science, Engineering science, Social and Educational science).

As shown in Fig. 12.1, IDM breaks down into four stages briefly described here.

Step 1: Initial situation analysis: This phase consists of investigating all knowledges having to do with the initial unsatisfactory situation and transposing this tacit and explicit knowledge [22], which may exist either in textual documents or in the minds of experts, into an exploitable mathematical model in order to determine by which means to enter into a more detailed or parameterised description of the problem. The objective here is to build a 'problems graph'

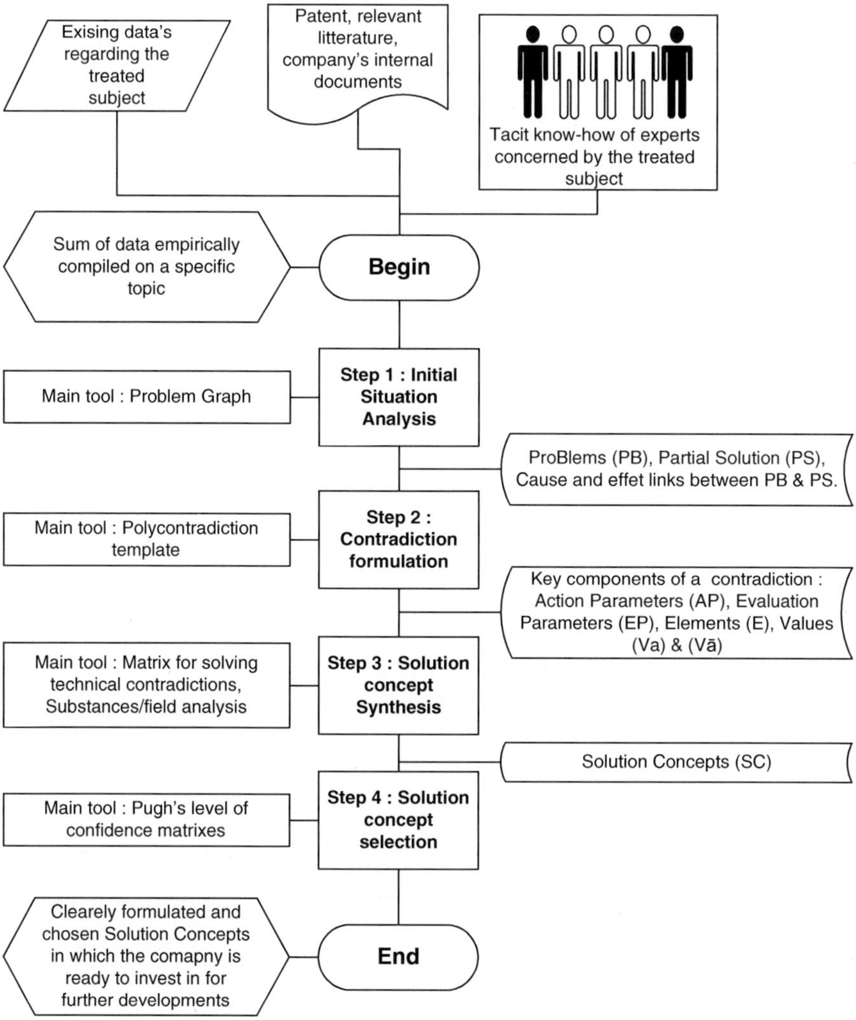

Fig. 12.1 Macrostructure of the IDM framework

resulting from this transposition in order to develop from a situation that is fuzzy, often resulting from an empirical heaping of studies experiences, toward an exploitable graphic model that uses rules and algorithms of the Graph theory [23]. The problems graph is made up of elements of a simple symbolic graph, which rapidly facilitates clarification and recording of tacit knowledge bits gleaned from questioning. It is proposing a graphic model in which two knowledge categories coexist: knowledge representing problems as yet unresolved from the initial situation and knowledge representing known partial solutions in the same domain.

Step 2: Formulating contradictions: With Stage 1 decision-making conventions in place, what appears as a key issue in the study is subsequently used as a departure point for producing a detailed formalisation of a range of polycontradictions, from which the contradictions of the area being examined will be extracted and ranked. These contradictions are the technical and physical issues to resolve in order to have an impact on root problems that render the initial situation unsatisfactory. The following phases are found within this stage:

- Formulating polycontradictions;
- Extracting contradictions;
- Creating a priority-based hierarchy of each contradiction depending on a given scenario.

Step 3: Generation of Solutions Concepts: Each contradiction stated as a priority in the previous stage then becomes an entry point for implementing TRIZ techniques and tools to achieve a resolution without trade-offs. The problem resolution processes are used for each of the priority contradictions and may be successive or iterative. They exploit the technical contradictions resolution matrix related to inventive principles, the Substances-Field modelling related to the Inventive Standards system and the ARIZ-85C algorithm [24]. This stage produces a limited number of solution concepts that are pertinent to the initial situation and exhibit full traceability.

Step 4: Selection of solution concepts: In this stage, the hypothetical impact of each solution concept is weighed against the problems graph created in Stage 1. The purpose of this is to evaluate the impact of each of the solution concepts on the initial unsatisfactory situation and to choose which one or ones among them to develop more in detail. These stages were detailed in a previous publication [25].

The closest set of procedure, methods and tools to IDM is, as said in introduction, OTSM. Nevertheless, even if we found many advanced definitions in OTSM that were appropriately moving in the right direction, we could not clearly define a coherent corpus of components useful for fully operating OTSM both in education and research. In Table 12.1, we summarise the major components of OTSM and how have they been either replaced or reconstructed in order to fit with the overall ontology of IDM.

12.3 Focusing on the Central Notion of IDM: Contradiction

While most well-known design approaches rely on customers voice listening (or anticipating), TRIZ-related methods rely on its main axiom: Laws of engineering Systems evolution as formulated by Altshuller. They are the drivers of any artefact's evolution. In 'classical' TRIZ, laws were simply expressed and illustrated with many examples. IDM framework has more accurately defined the mechanisms underlying behind the laws and connected them to contradictions (TRIZ's second axiom). This places the considerations for a Design process inspired by

Table 12.1 TRIZ, OTSM and IDM major differences

TRIZ [1946–1985]	OTSM [1985–2009]	IDM [starts in 2006]
Intuitive human expertise estimates the problematic situation	Notion of network of problems	Problem graph Problem + accuracy of the
	Notion of problem	syntax
	Notion of partial solution	Eligibility of a partial solution
	Notion of core problem (intuitive)	Core problem automatic—graph theory
	Notion of network of contradictions	Automatic derivation of a problem graph Into a set of contradictions
Human skills-based theory (unachieved)	Expert-based theory	Ontology-based theory
Laws are described	Laws exists	Laws are connected to contradictions
Validation of the solution if no compromise is made	No solution concept impact measurements	Solution concepts are ranked according to their Capacity to shrink the graph
No differences between parameters	Control parameter	Action parameter (ontology consistency)
Dedicated to solve engineering problems	Dedicated for developing thinking skills	Dedicated for becoming an industrial practice

TRIZ on another logic: the transition between the present and the future of the lifecycle of a given system consists in overcoming contradictions. This section is dedicated to further detail the notion of contradictions.

Dialectics is a philosophical school having roots in the old Greek philosophy, represented by Heraclitus and Aristotle. In particular, Aristotle, in his 'Metaphysics', speaks about an 'ontological non contradiction principle' which establishes that an object cannot simultaneously have and not have a given property P.

The philosophy grows out of the Hegelian discussion about the relation (or contradiction) between ideas and reality. Thus, the key concept of dialectics is 'contradiction'. A contradiction consists of two aspects, which are mutually dependent and opposed to each other at the same time. All complex phenomena consist of several contradictions, one of them dominating the others and characterising the phenomenon. This one is the principal contradiction. Furthermore, all processes consist of a movement of contradictions from the beginning to the end. Through time, the principal contradiction may change. According to dialectics, the causes of change and evolution are:

- the changing relation between two aspects of each contradiction and
- the changing relation among the contradictions of a certain phenomenon [26].

In short, the main characteristics of dialectics are duality, opposition, influence and dynamicity. The analysis of a situation in this context consists in collecting what changes and in what way changes affect other changes, i.e. by detecting influences of changes on other changes that can reveal a contradiction.

The author worked in the past on the definition of the way to represents complex problems as networks of problems in the framework of OTSM-TRIZ [27]. Today, this author think that the notions introduced at that time were not formal enough. Consequently, after an updated and more accurate definition of terms and notions, we have incorporated them in the ontology presented in [25, 28]. Based on this ontology, we have developed a computational system that helps to solve inventive design problems using IDM. The concepts of 'problem' and 'partial solution' have been precisely defined and used to set the problem and analyse it and the 'parameters' derived from them are used to constitute the contradictions.

Contradiction and laws of engineering systems evolution are two pillars whereon TRIZ lies. Only the first of these two notions will be considered here, how IDM cope with the evolution laws is described in a previous publication [29].

There are three types of contradictions, administrative, technical and physical. The last two are more concrete and interconnected; they will be described in more detail.

An administrative contradiction does not reveal any contradictory aspect, it often describes a desire to improve a characteristic of a system without having an emerging direction of resolution. Its syntax is:

> I would like to [describe the desired characteristic of the studied system], but I don't know how.

For example, in the case of a clothespin, an administrative contradiction might be: I would like to [prevent the clothespin from marking the laundry] but I do not know how. An analogy can be made at this stage between the part of the sentence regarding willingness to get an improvement and the notion of function as in functional analysis.

A technical contradiction describes the state of a system where there is an action having a useful effect but causing simultaneously an undesirable effect. In our example of the clothespin, it could be said that trying not to mark the laundry generates an undesired effect: the laundry will be loosely fixed on the wire.

The resolution of a technical contradiction is mainly performed using reasoning by analogy, facilitated by the use of two TRIZ components: the inventive

[1] G. Altshuller (the creator of TRIZ) screened patents in order to find out what kind of contradictions were resolved or dissolved by the invention and the way this had been achieved. From these works, he developed a set of 40 inventive principles and later a 'matrix of contradictions'. Rows of the matrix indicate the 39 system features that typically would need improvement, such as speed, weight, accuracy of measurement and so on. Columns refer to typical undesired results. Each matrix cell points to principles that have been most frequently used in patents in order to solve the contradiction.

principles and the matrix of contradictions.[1] This analogical reasoning consists in matching the two opposed features to the generic parameters used in the matrix. As an output, the matrix summarises the inventive principles that have statistically been often used in similar situations by inventors. At this stage, the resolution is not finished because the retained inventive principles have to be interpreted by the user. This interpretation is not easy, because of the high abstraction level of these principles (See to recall the TRIZ solving process).

A technical contradiction requires the definition of several parameters associated to the technical system or of any of its elements. When acting on a parameter to satisfy a requirement of the problem; it may appear that it must simultaneously have at the same time a value a and $\overline{a}$ (the opposite value of a). In the example of the clothespin, it can be said that the 'the stiffness of the spring' must be 'hard' to fix the laundry and 'soft' not to mark it.

The physical contradiction addresses the part of the technical contradiction centred on that parameter that must have at the same time two opposite values.

In the following, the notions of TP (for Technical or Physical) contradiction or simply contradiction will be used to point out a configuration of parameters that can be interpreted as a technical contradiction and where a subpart of it is a physical contradiction.

12.3.1 First Definition of the Tp Contradiction

The TP contradiction (the most precisely defined in classical TRIZ) is characterised by a set of three parameters and where one of the parameters can take two possible opposite values Va and $\overline{Va}$ It is important here to distinguish two types of parameters: 'action parameters' and 'evaluation parameters'.

Indeed, an action parameter ('the stiffness of the spring' in the case of the clothespin) is characterised by the fact that it has a positive effect on another parameter when its value tends to Va and that it has a negative effect on another parameter when its value tends to $\overline{Va}$ (that is, in the opposite direction). For example, the value of the parameter 'stiffness of the spring' tending to 'soft' satisfies the parameter 'depth of the mark left on the laundry'. When the value of 'stiffness of the spring' tends to 'hard', it will satisfy the parameter 'keeping of the laundry on the wire'.

The fact that a parameter may tend towards two opposite values and may have an impact on one or more other parameters makes it an action parameter (AP). For such parameters, the designers have the possibility of modifying them (the designer can decide if the 'stiffness of the spring' is going to be hard or not).

An evaluation parameter (EP) can evolve under the influence of one or more action parameters. It allows measuring the degree of satisfaction or dissatisfaction; it is characterised by a value considered to be 'desirable'. The main characteristic

Fig. 12.2 Possible representation model of contractions (physical and technical)

$$AP \begin{matrix} Va \\ \overline{Va} \end{matrix} \overset{\begin{matrix} EP_1 & EP_2 \end{matrix}}{\begin{pmatrix} -1 & 1 \\ 1 & -1 \end{pmatrix}}$$

of an evaluation parameter is its capacity to evaluate the positive aspect of a choice made by the designer.

Thus, in the case of the clothespin, the evolution of the parameter ‘depth of mark left on the laundry’ is only interesting for the designer if it tends to a minimum, the ‘maximum depth mark’ having no meaning or interest. Similarly, the parameter ‘keeping of the laundry on the wire’ should tend towards a maximum, the direction towards ‘clothes slipping and falling down from the wire’ being undesirable.

The contradiction has to express that each of both directions of the variation of the AP has two opposite influences. If AP tends towards *Va* then EP_1 tends towards a value that is satisfactory and EP_2 tends towards an unsatisfactory value. And if AP tends towards $\overline{Va}$ then EP_1 tends to an unsatisfactory value and EP_2 tends towards a satisfactory value.

The fundamental aspect of a contradiction resides in the opposition of the values of the considered action parameter. The opposite values have to be explicit and the opposition must be based on the fact that, if in a certain state, a value *Va* implies a positive feeling, then $\overline{Va}$, must imply a negative feeling. So it is indispensable to investigate $\overline{Va}$, the opposite of $\overline{Va}$ to make the contradictory aspects of the analysis appear clearly.

So, for each contradiction, it is necessary to ensure that the change of the AP in each of the opposite directions causes opposite effects on the evaluation parameters. If this is not the case, there is no contradiction in the sense of TRIZ and IDM.

Thus in the case of the clothespin, it is necessary to check that a ‘soft stiffness’ implies really a dissatisfaction about the parameter ‘keeping of the laundry on the wire’ and that a ‘hard stiffness’ leads to ‘a very deep mark left on the laundry’ which is unsatisfactory. Figure 12.2 depicts a possible representation model of technical and physical contradictions, where AP is the action parameter and EP_1 and EP_2 are the two evaluation parameters. The values of the EPs in the matrix are the reason of the physical contradiction: ‘AP cannot have both values *Va* and $\overline{Va}$’.

Note that in other chapters [30] and in the case study presented in paragraph 4, the values 1 and -1 are represented respectively as smileys (☺, ☺). These values have been chosen in order to allow weighting of the EPs. Indeed, it is interesting to differentiate their importance that can be relatively different following the preferred scenario of the designer.

The description made until now merits further consideration: the meaning of ‘competing values’ (for *Va* and $\overline{Va}$) and ‘satisfactory’ (the *1* value in the matrix) and ‘unsatisfactory’ (the *-1* value in the matrix) values has to be clarified in order to build a formal model of contradiction. This formal model is then used as the basis of problem characterisation in either simple cases (like the clothespin case) but can also, together with other tools from both TRIZ and IDM, be applied in complex situations. In the next section, a typical complex study is presented.

12.4 Idm Framework on Complex Studies: The Case of Enamelling of Steels

During the last 5 years, enamelled steel and ceramic or glass coatings came out as a potential way aiming at finding new solutions for different applications in Appliance, Construction and even Metal Processing. When a problematic such as long term corrosion protection, temperature resistance, abrasion, chemical resistance, recyclability, sustainability, long-term durability and clean ability were discussed. Porcelain enamels were found to show several interesting properties regarding the requirements [31]. Enamelling is a niche, a very specific coating process today known only by a few specialists. It is defined as a 'post-treatment', that is to say, it is engaged as a final stage, after forming and welding by end customers. But the current existing post-treatment techniques are perceived as over-dimensioned yet not cost-effective.

Obviously, a pre-enamelled steel—understood as steel coated by a glass layer—could not be the solution as this association is based on two antagonist products: steel is formable, light, robust and used mostly for these properties, while enamel (glass) is brittle and cannot be deformed. Key actors on the scene of steels (steelmakers) propose a precursor of the final vitrified coating, a metal sheet coated with one or more formable layers, precursors of the final coating.

A porcelain enamel coating is thick: this property is not a must. This characteristic is the result of two facts. First, until now, enamel is applied after forming on complex shapes, operation that leads to a wide dispersion. Second, the link between steel and enamel obtained at high temperature is a rather thick interface of about 40 μm. This interface needs to be fully covered, as it does not have the required properties.

Our industrial partner already tried in the past to introduce enamels in powder form in a 'precursor' of the final vitrified layer. It was impossible to reduce thicknesses of the layers. This project, known internally as 'pre-enamelled steel' was patented and closed. Even if the main substrate used today is Cold Rolled Steel, enamel is also applied on other metals & alloys such as aluminium, copper, stainless steel, cast iron and NiAl. That is why it was decided to launch an IDM study on enamelling in order to identify trends, priorities and maybe proposals of further action plans. As shortly mentioned earlier, the main driving forces for this breakthrough are:

1. from thick to thin ceramic (glass) coating technologies and products;
2. from post-treatment to pre-treatment to keep on knowledge and added value;
3. working on steel or multi-metal systems;

4. allowing also for the evolution of the processes (forming, firing, welding, coating…) as compared to the state of the art.

12.4.1 Methodology and Planning for Operating IDM Process

Based on IDM and TRIZ, a software tool namely STEPS (Systematic Tool for Efficient Problem Solving) was built. Its structure totally matches IDM process as illustrated Fig. 12.1. Typically, to go through the software steps, ten sessions are necessary according to the planning pictured in Fig. 12.1.

Each session corresponding to one working day, consecutive sessions can be separated by a few weeks during which the study can be carried on internally; this is especially advised for the three sessions making up the first step, because it allows for further maturation of the network. This step consists in identifying first the main, central problem of the study. Then based on the objective explained in the §3 a problem graph was built. The description also includes, when there are any, the partial solutions known in the art to the problems thus identified. For instance, in the case of enamelling, the starting point was the fact that enamelled products are expensive. The reasons for this were found to be:

- Steel for enamelling, itself, is expensive;
- The enamelling process is complex and not continuous;
- The enamelling process involves high temperatures (energy cost);
- The enamel layer is oversized (product cost).

This part of the study required the contribution of 6 people from diverse and complementary backgrounds.

- a specialist in ceramics and enamelling (formulations, processes, properties, applications);
- a specialist in metallurgy (chemistry, treatments and processes, properties, applications);
- a scientist with a large scientific knowledge in materials and processes and in Intellectual Property issues;
- an engineer having knowledge in metallurgy (processes, products, customers approach) and pilot project for enamelling steel since around 1995;
- a scientist having scientific knowledge in materials, particularly in ceramics;
- an animator, expert in IDM, STEPS and particularly in TRIZ.

The resulting set of links PB → PS; PS → PB; PB → PB represented a problem graph made up of green and yellow boxes (see Fig. 12.3). The green boxes correspond to the problems and the yellow ones to the partial solutions. These last can be processes already used in industry or ideas found in chapters or patents, but they must have been experienced and found to—at least partly—

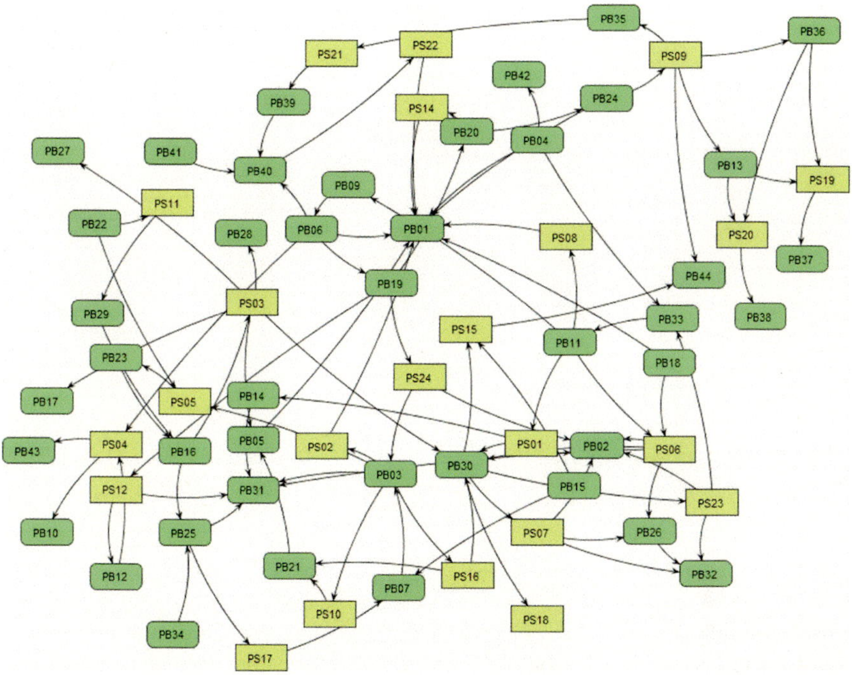

Fig. 12.3 Problem graph associated to enamelling case study

succeed in solving the given problem. However they often give rise, in turn, to new problems: this is why they are called 'partial solutions'. As a consequence, a yellow box can never be found at the end of a sequence (if so, the software highlights them in red colour). Partial solutions are therefore always followed by a green box (a problem), otherwise it would mean that the global solution has been found. Figure 12.3 displays the problem graph after roughly 25 h of work, they emphasise how complex a problem can get, with 44 problems (green boxes) and 25 partial solutions (yellow boxes). Important is to note that, once completed, the network not only constitutes the essential basis for the rest of the exercise, but also provides a schematic representation of the state of the art which:

- must be agreed by all the participants;
- can be re-used;
- can be added, in the future, with new findings.

As such, the network of problems has been considered as a very interesting starting point for new projects by the team members.

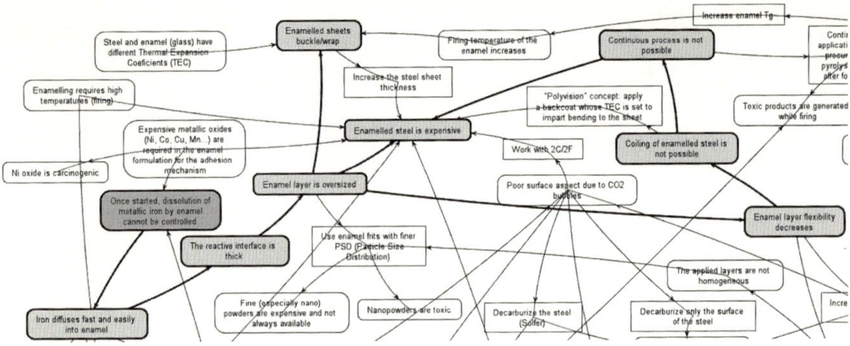

Fig. 12.4 STEPS software interface and tools

12.4.2 Manipulating the Problem Graph to Extract Useful Information

As the overall problematic is now having a graph-like structure, we can easily build algorithmic rules and extract useful information out of it. As an example of problem graph use, Fig. 12.4 illustrates the result of a simple algorithm that points out the longest chain of problems, not disturbed by the presence of a partial solution. Such problem can be interpreted as a core problem (or a problem with which it can be very beneficial to start the solving problem).

12.4.3 Identification of the Parameters

Each problem or partial solution is then associated to one or several parameter(s) and their units characterised (as an example if length is an *AP*, its unit could be mm). The problems are also granted a weight according to their importance in the whole network, from 0 (not very important) to 5 (critical). These parameters allow qualifying, qualitatively or quantitatively, the evolution of the problems (PB), like a kind of 'tracer'. For a given problem, the associated parameter(s) can be identified by asking the question: 'which parameter(s) can be used as a tracer for this problem?' In some cases, the answer is obvious: for example, the problem 'Enamelled steel is expensive' can be followed by the parameter 'Total cost' expressed in €/m2. Sometimes it is not so evident: the problem 'Hydrogen-traps concentration (cementite) decreases' has to be evaluated by measuring the permeation time, given in min.

The parameter(s) derived from each Partial Solution (PS) can be identified by asking the question:'which parameter(s) of the system was (were) worked on to operate this solution?' Let us take the example of the PS 'Increase the roughness' to the PB 'Adhesion decreases' acts on the parameter "roughness" measured in μm. Once identified, the parameters are sorted into two categories: Evaluation Parameters (EP) and Action Parameters (AP).

According to what has been exposed §3, EPs are parameters whose evolution is only wished in one direction; there is absolutely no interest to see it move in the other one. For example, the problem 'Enamelled steel is expensive' was characterised by the parameter 'Total cost' expressed in €/m2: this parameter is not desired to increase, but only to decrease. *EPs* can also be described as parameters that we cannot directly control but are subjected to. Once defined, they have to be granted a weight, called 'importance' that represents how critical it is. This weight is not introducing subjectiveness since the idea is just to commonly agreed in a group, that EPs are of a certain level of importance as regarding their role in the accomplishment of a given objective (a scenario). So this ranking is based on facts and harmonised experiences between the experts in the group.

In contrast with the case of EPs, the evolution of Action Parameters (APs) is interesting in both directions. They can be seen as a tuner on which it is possible to act. Alike EPs, they are given a coefficient representative of the level of influence.

Most of the time, the parameters associated to problems and to partial solutions are EPs and APs, respectively, but they have to be studied individually to confirm this distribution. Besides, in some cases several problems can be associated to a single parameter; and the other way around: one single problem may be followed by several parameters. This is why the numbers of EPs and APs do not have to equal those of problems and partial solutions, respectively; for instance in our case, 36 EPs and 20 APs were identified.

At this point of the study, the problem is already extensively developed. However to achieve a complete description, further examinations are required. This is the purpose of the next step.

12.4.4 System Analysis

Towards a complete description of the problem in link with the basics of TRIZ, STEPS software enables us to fully define the system's structure, contradictions and hypothesis of evolution. To achieve all these steps, series of tools coming from TRIZ and often rearranged and enhanced are proposed:

- system completeness,
- multi-screen,
- system maturity,
- evolution laws,
- DTC operators.

The goal is to approach the problem from points of view that are not usually considered.

12.4.5 Focus on Multi-screen

Also known as '9 windows', this tool represents the system:

- In a temporal environment ('x axis'): the '-1' (on the left hand-side), '0' (in the middle) and '+1' windows (on the right) correspond to the past, present and future, respectively;
- In a systemic environment ('y axis'): the studied system (middle windows) belongs to a larger one, called 'super-system' (top windows), and includes a smaller one, called 'micro-system' (bottom windows);

In the present case, the super-system consists of the steel, the enamel, the thermal treatment and the application process, while the micro-system is made of the chemical elements Fe, C and O, the iron oxides, the inter-metallic layer, the bonding oxides (referred to as 'catalysts' in Fig. 12.3) and the roughness of the steel. It will be noticed that the super-system includes the tool and the object of the system; as for the micro-system, it is made of some of the parts of the tool (engine, transmission, controls). The purpose of these nine windows is to describe how the studied system and its super- and micro-systems have been evolving in time. By translating this evolution into parameters that have been either improved or damaged, it can allow for the identification of other parameters than those derived from the problems and partial solutions of the problem graph. Looking at the past is also the opportunity to search for the last technical jump that revolutionised the studied field. In our case, it was pointed out that generalisation of pickling and nickeling in the 1980s induced a major change in enamelling; indeed the nickel thus brought to the surface of the steel highly promoted the enamel adhesion.

12.4.6 Focus on Evolution laws

This tool presents the 9 evolution laws defined in TRIZ methodology, i.e.:

(1) System completeness: any technical system is made of several well-defined parts properly connected to each other, at least one of them has to be controllable to make possible the control of the whole system;
(2) Energy conductibility: none of the constitutive parts of the system must slow down the energy flow that makes it work;
(3) Harmonisation: tendency of a system to make all its constitutive elements evolve homogeneously;
(4) Ideality: tendency of a system to have all its desired and un-desired properties maximised and minimised, respectively;
(5) Irregular evolution of the parts: tendency of a system to solve contradictions resulting from an heterogeneous evolution of its different parts;
(6) Super-system transition: tendency of a system to disappear as such to the benefit of the super-system it belongs to;

(7) Micro-level transition: tendency of a system to see its element 'work' evolve towards the micro-level;
(8) Dynamisation: tendency of a system to have a flexible structure, able to quickly change and adapt;
(9) Substances-fields interaction: tendency of a system to see new connexions appear within its structure, increasing the ways to control it.

It will be noticed that the laws 'system completeness', 'super-system transition' and 'micro-level transition' are extensively exploited in the first two points of the system analysis. For each of the nine laws, the software gives the question to ask in order to determine whether the law is relevant in the present case or not; if it is, it suggests how to modify the system to make it conform to the given law. To illustrate this part of the study, let us take the example of the fifth law, dealing with the evolution of the different parts of the system. Here the question asked is: 'are all the system's components at the optimum of their development for maximizing the main useful function or not?'; if the answer is positive (all parts optimised), this law can be passed over, otherwise it is suggested to imagine how to fix this particular component's problem to unlock the evolution of the whole system. This is what was done, giving rise to a situation where the adhesion mechanism would be of a different kind, e.g. velcro-mechanical, thus requiring no redox reaction anymore. These laws can eventually help to define the evolution desired for the system and its super- and micro-systems, to go on fulfilling the right-side windows of the multi-screen. This examination enables to bring new elements to the problem graph; it is only once the system analysis has been completed that the problem graph is considered as finalised.

12.4.7 Contradiction Synthesis and Analysis

When the problem has been properly settled, lots of contradictions appear. Indeed, some parameters are wished to evolve in different directions depending on the cases. One of the strongest points of IDM methodology consists in taking into account all these contradictions to analyse them using computing calculation, a task that would not be feasible in a reasonable time by hand. To allow for this automatic analysis, all the contradictions arising from the confrontation between the parameters have to be listed. To do so, the *APs* are first classified according to the element of the system they refer to. Then, each *EP* is successively placed in front of all the *APs* to determine whether the latter can impact the former. There is a contradiction as soon as a given *AP* positively influences different *EPs* when it evolves in opposite directions. For example, the *AP* 'Firing time' can be either long or short:

- if short, it positively impacts the Evaluation Parameters:
  - 'Reactive layer thickness' (since it has no time to grow);

  - 'Total cost' (since the energy consumption decreases);
  - 'Enamelled sheet bending' (since wrapping is less likely to occur);

- if long, it positively impacts the Evaluation Parameters:

  - 'Enamel adhesion' (since the adhesion mechanism have more time to take place);
  - 'Surface quality of the enamel' (since the glass has more time to spread).

The contradictions are gathered into groups called poly-contradictions, according to the *AP* they correspond to; as a result, there are as many poly-contradictions as *APs* (here, 20). A poly-contradiction is thus defined by a set of data:

- an Action Parameter (in the example above: 'Firing time');
- the element it refers to (Enamel);
- the two opposite values it can take ('long' and 'short');
- the list of the Evaluation Parameters impacted ('Reactive layer thickness', 'Total cost', 'Enamelled sheet bending', 'Enamel adhesion' and 'Surface quality of the enamel').

This 'confrontation operation' applied to every *EPs* and *APs* resulted in our case in a total of 127 contradictions. To each poly-contradiction the software associates a weight, calculated from the importance of the *EPs* and the coefficient of the *AP* involved, as well as a balance between the two possible values of the *AP*.

12.4.8 Suggested Contradictions: Bubble Graph

STEPS software splits the poly-contradictions into 'mono' contradictions. Contrary to the poly-contradictions, the contradictions as explained in Sect. 12.3, only involve two Evaluation Parameters from those of the poly-contradictions, impacted by the same Action Parameter but in opposite ways. For example, it has been chosen to rank contradictions in the Fig. 12.5, based on the scenario of 'Enamel flexibility'. Each contradiction is characterised by:

- X axis: its weight,
- Y axis: its universality, i.e. the number of other contradictions involving the same *EPs*,
- Z axis (each bube diameter) the quantity of *EP* each *AP* is impacting.

The contradictions are then represented on a graph like in Fig. 12.5 where the three properties appear through the position and the diameter and colour of the bubbles. Each one corresponds to a contradiction. Some sets of contradictions (represented by the same colour) stem from the same poly-contradiction; they thus share the same *AP* and as a consequence they have the same size; however they can be bigger or smaller to make it possible to distinguish superposed bubbles from each other.

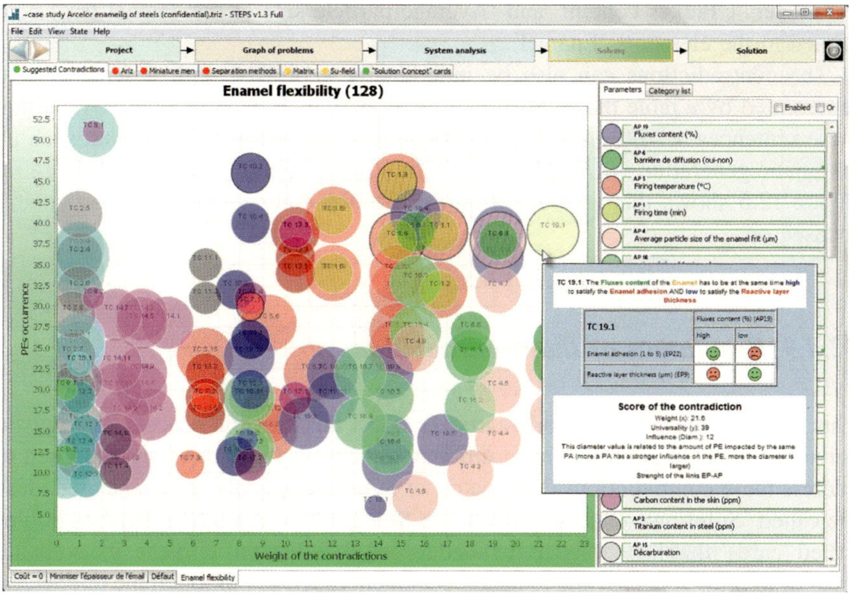

Fig. 12.5 Bubble graph diagram of all gathered contradictions and scoring

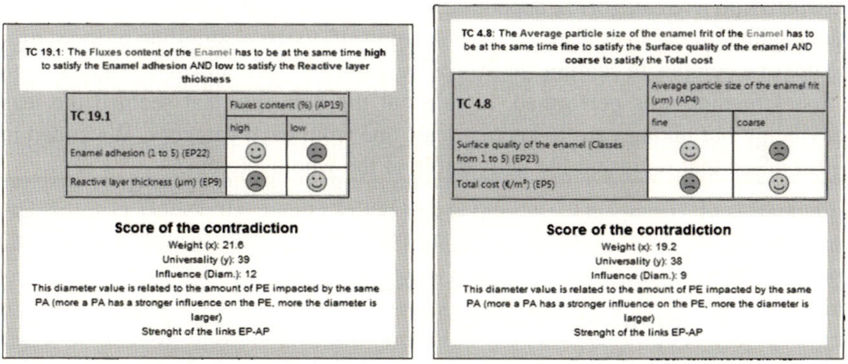

TC 19.1: The Fluxes content of the Enamel has to be at the same time high to satisfy the Enamel adhesion AND low to satisfy the Reactive layer thickness

TC 19.1	Fluxes content (%) (AP19) high	Fluxes content (%) (AP19) low
Enamel adhesion (1 to 5) (EP22)	☺	☹
Reactive layer thickness (µm) (EP9)	☹	☺

Score of the contradiction
Weight (x): 21.6
Universality (y): 39
Influence (Diam.): 12
This diameter value is related to the amount of PE impacted by the same PA (more a PA has a stronger influence on the PE, more the diameter is larger)
Strenght of the links EP-AP

TC 4.8: The Average particle size of the enamel frit of the Enamel has to be at the same time fine to satisfy the Surface quality of the enamel AND coarse to satisfy the Total cost

TC 4.8	Average particle size of the enamel frit (µm) (AP4) fine	Average particle size of the enamel frit (µm) (AP4) coarse
Surface quality of the enamel (Classes from 1 to 5) (EP23)	☺	☹
Total cost (€/m²) (EP5)	☹	☺

Score of the contradiction
Weight (x): 19.2
Universality (y): 38
Influence (Diam.): 9
This diameter value is related to the amount of PE impacted by the same PA (more a PA has a stronger influence on the PE, more the diameter is larger)
Strenght of the links EP-AP

Fig. 12.6 Contradiction no 19.1 and 4.8 represented as tables

The purpose of such a diagram (Fig. 12.5) is to suggest the contradictions that are the most relevant to start the solving phase with: the ones with the highest weight, universality and diameter. Practically speaking, the contradictions represented by the biggest bubbles near the top right-hand corner shall be treated first. Furthermore, it is advised to work on different Action Parameters (i.e. different colours) related to different elements. In our case we focused on two contradictions, one is pointed out by the arrow in Fig. 12.6, although both Action

12.4.12 Evaluation Results Analysis

The 15 Solution Concepts evaluated are represented by the green boxes on the right-hand side, and the problems by those at the bottom. The blue ellipses figure the contradictions lifted when the SCs are processed, leading to the solving of the problems. It is also possible to check how many problems are solved by each SC, as shown on the graph Fig. 12.7 (bottom): those directly solved are highlighted in orange, and the sub-problems they resulted in (which are also solved in a domino effect) are highlighted in green.

The Solution Concepts are classified by decreased number of solved problems. Given that the results can dramatically change if the opinion of one expert highly differs from the other ones, simulations were made, successively ruling out one expert at a time. The number of Solution Concepts able to solve problems was found to vary (from 10 to 6) as well as their ordering. However, whatever the combination of four experts taken into account, two SCs always appeared among the first three ones: SC no 1.1.2 and SC no 1.1.5. These SCs are currently in experimentation phase and provided to the industrial partner, reliable and robust R&D programs with high impact on an initial problematic that, at first, was not showing any relevant direction on where to search and place R&D efforts.

12.5 Conclusions and Future Work

Through this chapter, we intended to present our recent achievements in terms of applied research based on TRIZ, namely Inventive Design Method. We first introduced why the paradigm of quality era is arriving to an end and is unable to cope with the difficulties brought by innovation era. Despite the fact that innovation is always cited in strategic speeches (by managers) the reality regarding its formalism to be efficiently deployed in R&D department is far to be acquired. As a potential methodology fulfilling this requirement, TRIZ has been increasingly used by large-scale companies in the last two decades. Nevertheless, highly industrialised countries discovered TRIZ in the early nighties with its lack of accurate formalism. We have thus investigated through IDM framework, the possibility to enhance TRIZ with research activities. It started with the analysis of its limitations and the planning of several research programs to reach a sufficient level of formalisation. IDM research framework was thus constituted; it coincided with the willingness of large-scale companies to cope with TRIZ limitations in building new methods and tools derived from it. After 6 years of investigation, analysis and case studies, we were able to propose a complete framework in support to inventive projects aiming at provoking innovation in R&D departments. Upon these research results, it has been possible to build a software to support the deployment of IDM and speed-up the process without losing the accuracy of the methodology, namely STEPS. In order for IDM to be comprehensive and transferable to industry, we beforehand had to build a complete ontology of its major concepts, so as to publish on its major tools to explain other researchers and industrialists our beliefs and their roots together with examples. The central notion of our approach lies in the roots of TRIZ: the contradiction notion. This chapter also provides an enhanced version of the notion of contradiction in comparison with its original version in Classical TRIZ. We detailed how they are disclosed out of a problematic situation and how they constitute a point of entrance into the solving phase.

Our upcoming work follows three directions: The first one is to establish appropriate measurement means (indicators) of inventive practices in R&D in order to scientifically prove the added value of practicing IDM when in innovation contexts. The second is to investigate the possible links between inventive Design and routine Design tasks like calculation (or optimisation) in order to be more confident with the Solution Concept true robustness and thus financial investments in developing them. The third one is going in the direction of patent mining as we believe invention can be provoke due to the smart use of available knowledge (as distant as possible from the original field). In this direction, patent represents more than 80 % of engineering knowledge available but definitely underused in context of inventive design.

Today, we are thankful to the TRIZ Consortium to allow us to publicly present the result of our research, hoping that it will serve TRIZ world and become for all of its members, researchers, industrialists, educators and newcomers, a way to efficiently put Inventive Design into practice and as a result boost innovative strategies of companies turned towards a reliable future.

References

1. Kao (1997) The science of creativity. Manag Dev Rev 10(6): 203–204
2. Freeman C, Louçã F (2002) As Time goes By: From the industrial revolutions to the information revolution. Oxford University Press, USA
3. Kapás J (2005) Towards an understanding of the variety of firms. Acta Oecon 55(1):43–61
4. Cavallucci D (2011) A research agenda for computing developments associated with innovation pipelines. Comput Ind 62(4):377–383
5. Binder J (2004) The inventor's key to innovation. Aerosp Am 42(5):24–26
6. Altshuller G (1991) To find an idea: Introduction to the theory of inventive problem solving, 2nd éd. Novosibirsk: Nauka
7. Cascini G, Rissone P, Rotini F, Russo D (2011) Systematic design through the integration of TRIZ and optimization tools. Procedia Eng 9:674–679
8. Johne A (1994) Listening to the voice of the market. Int Mark Rev 11(1):47–59
9. Hertzum M, Pejtersen AM (2000) The information-seeking practices of engineers: searching for documents as well as for people. Inf Process Manage 36(5):761–778
10. Anglei P (1995) Thinking out of the box: a new approach to product development. Bus Horiz 38(3):18–22 (Long Range Plan 28(5):129 Oct 1995)
11. Järrehult BJ (2009) The end of the funnel 9(1):4–16
12. Eastman C, Parker DS, Jeng TS (1997) Managing the integrety of design data generated by multiple applications: the theory and practice of patching. Res Eng Design 9(3):125–145

13. Dodgson M, Hinze S (2000) Indicators used to measure the innovation process: defects and possible remedies. Res Eval 9:101–114
14. Colombo G, Pugliese D (2006) The role of knowledge management in product lifecycle. In: Brissaud D, Tichkiewitch S, Zwolinski P (eds) Innovation in life cycle engineering and sustainable development. Springer, Berlin, pp 397–406
15. Cavallucci D, Rousselot F, Zanni C (2008) Representing and selecting problems through contradictions clouds. In: Cascini G (ed) computer-aided innovation (CAI), vol 277. Springer, Boston, pp 277:43–56
16. Khomenko N. De Guio R (2007) OTSM network of problems for representing and analysing problem situations with computer support In: Leon-Rovira N (ed) Trends in computer aided innovation, vol 250. Springer,Boston, pp 77–88
17. Rousselot F, Zanni-Merk C, Cavallucci D (2012) Towards a formal definition of contradiction in inventive design. Comput Ind 63(3):231–242
18. Mizuyama H, Ishida K (2007) Systematic Decision making process for identifying the contradictions to be tackled by triz to accomplish product innovation. JAMRIS 1(4):21–29, 2007
19. Bultey AA (2007) A substance-field ontology to support the TRIZ thinking approach. Int J Comput Appl Technol 30(1–2):113–124
20. Prickett P, Aparicio I (2012) The development of a modified TRIZ technical system ontology. Comput Ind 63(3):252–264
21. Cavallucci D, Rousselot F, Zanni C (2011) An ontology for TRIZ. Procedia Eng 9:251–260
22. Nonaka I, Takeuchi H (1995) The knowledge-creating company, vol 1. The Free Press, New York, p 995
23. Bondy JA (1976) Graph theory with applications. Elsevier Science Ltd, New York
24. Altshuller G, Shulyak L, Rodman S (1999) The innovation algorithm: TRIZ, systematic innovation and technical creativity. Technical Innovation Center, Inc., USA
25. Zanni-Merk C (2009) An ontological basis for computer aided innovation. Comput Ind 60(8):563–574
26. Vocabulaire européen des philosophies - broché - Collectif - Livre - Fnac.com
27. Cavallucci D, Khomenko N (2007) From TRIZ to OTSM-TRIZ: addressing complexity challenges in inventive design. Int J Prod Dev 4(1–2):4–21
28. Zanni-Merk C, Cavallucci D, Rousselot F (2011) Use of formal ontologies as a foundation for inventive design studies. Comput Ind 62(3):323–336
29. Cavallucci D, Rousselot F (2011) Evolution hypothesis as a means for linking system parameters and laws of engineering system evolution. Procedia Eng 9:484–499, 2011
30. Dubois S, Rasovska I, De Guio R (2008) Comparison of non solvable problem solving principles issued from CSP and TRIZ. In: Cascini G (ed) computer-aided innovation (CAI), vol 277. Springer, Boston, pp 83–94
31. Millon E, Brice JF, Zerardin R, Evrard O, Guillot L, Seurin P, Entringer M. L'émail: un solvant basique et réducteur à l'état fondu. Industrie céramique 824:113–117
32. Pugh S (1990) Total design: integrated methods for successful product engineering. Addison-Wesley Educational Publishers Inc, New York

Chapter 13
The Function-Behaviour-Structure Ontology of Design

John S. Gero and Udo Kannengiesser

13.1 Introduction

Design is one of the profound activities of humans. It is the way humans intentionally change the physical and virtual worlds they inhabit. Society recognises designing as important and privileges defined groups as designers such as engineers and architects, which are long-standing professions, along with relatively new groups such as software designers. It is therefore surprising that formal research into designing commenced relatively recently. Design research has largely adopted the scientific paradigm in which it is assumed that there are regularities that underlie phenomena and it is the role of research to discover and represent those regularities.

The early seminal works in design research in the 1960s and 1970s focused on methods and processes and produced an array of terminologies to describe designing. It was unclear whether the terms used by one group of researchers mapped onto terms used by other researchers or whether they were describing different phenomena. Design appeared to present problems for scientific research in that the results of the acts of designing were always unique and therefore there would be no regularity. This issue has been addressed in two ways. The first way was to look for underlying regularities in designs rather than surface features. The second way was to look for regularities in design processes. The term 'designing' is used to signify the act and the term 'design' is used to signify the result of designing.

This chapter presents the development of an approach to represent such regularities in designs and in designing. It commences with a brief introduction to the historical development of the concepts before expounding the Function-

J. S. Gero (✉)
Krasnow Institute for Advanced Study, George Mason University, University of North Carolina at Charlotte, Charlotte, NC, USA

U. Kannengiesser
Metasonic AG, Pfaffenhofen, Germany

A. Chakrabarti and L. T. M. Blessing (eds.), *An Anthology of Theories and Models of Design*, DOI: 10.1007/978-1-4471-6338-1_13,

Behaviour-Structure (FBS) ontology of design and designing. This is followed by a section describing the situated Function-Behaviour-Structure (sFBS) ontology of design and designing. Empirical studies based on utilizing the FBS ontology are presented and provide experimental evidence in support of the ontology.

In searching for a way to think about designing, an axiom was proposed:

The foundations of designing are independent of the designer, their situation and what is being designed.

This has important consequences as it implies that the differences between design professions and design practices are not foundational to designing notwithstanding the apparent differences. The expectation was that the foundations of designing would not rely on any designing particulars.

Based on this axiom two hypotheses about representing designs and designing were proposed:

1. all designs could be represented in a uniform way, and
2. all designing could be represented in a uniform way.

What was being looked for was a set of irreducible foundational concepts of design and designing. These irreducible foundational concepts should cover the acts of designing and the representation of the design. Further, these irreducible foundational concepts should be distinct and have no overlap. In the 1980s a number of approaches to this were being developed by researchers that were based on the division of the design from the way it worked: Structure (S) for the design and Behaviour (B) for how it worked or performed. Many of these approaches used the term Function (F) to mean the intended behaviour of the design and as a consequence conflated Function and Behaviour and failed the no-overlap requirement.

FBS was developed between 1984 and 1986 and presented as part of a series of lectures on understanding design at Carnegie-Mellon University and at a seminar at Xerox PARC while the senior author was a consultant there in 1987. These presentations honed the understanding of the concepts. The ideas were presented at various conferences and resulted in the paper in a special issue on design in the *AI Magazine* in 1990 as part of a broader set of ideas [8].

Clancey's 1997 book *Situated Cognition* [3] mapped well onto ill-formed concepts about the role the designer's cognitive understanding of the world inside their heads and around them as they designed. This led to the development of a cognitively richer articulation founded on FBS resulting in the situated sFBS framework of design and designing [11–13].

Gruber developed the modern idea of an ontology [16]. The notion of a foundational framework for the field of design mapped well onto the notion of an ontology since they both referred to the meta-level knowledge of a field. Thus, the FBS and sFBS frameworks became ontologies as frameworks for the knowledge in the field of designing.

Up to 1995 the FBS ontology was a conceptual construct that had been used to construct conceptual and computational models. Empirical studies of designing based on verbal protocol analysis [7] had been introduced into design research some years earlier. These early studies and many of those continuing up to this day use project-specific schemes to code the protocol. The effect of this is that the results are incommensurable, i.e. they cannot be compared to each other since the dimensions of what is being measured varies across projects. The FBS ontology offers a project-independent scheme to code the protocols. At the same time the ability of the FBS ontology-based coding scheme to capture the design-related utterances of designers in a protocol provides evidence of its utility if not its validity. This is not to claim that other coding schemes that take a different view of designing are not useful. The section on Empirical Studies demonstrates the wide-ranging applicability and utility the FBS ontology.

13.2 The Function-Behaviour-Structure Framework

The FBS ontology is a design ontology that describes all designed things, or artefacts, irrespective of the specific discipline of designing. Its three fundamental constructs—Function (F), Behaviour (B) and Structure (S)—are defined as follows:

Function is the teleology of the artefact ('what the artefact is for'). It is ascribed to the artefact by establishing a connection between one's goals and the artefact's measurable effects. Table 13.1 shows some examples of function of various artefacts.

Behaviour is defined as the artefact's attributes that can be derived from its structure ('what the artefact does'). Behaviour provides measurable performance criteria for comparing different artefacts. The examples of behaviour in Table 13.1 show that most instances of behaviour relate to notions of quality, time and cost.

Structure is defined as its components and their relationships ('what the artefact consists of'). The various examples of structure in Table 13.1 indicate that this definition can cover any physical, virtual or social artefact.

Humans construct connections between function, behaviour and structure through experience and through the development of causal models based on interactions with the artefact. Specifically, function is ascribed to behaviour by establishing a teleological connection between the human's goals and the observable or measurable performance of the artefact. Behaviour is causally connected to structure, i.e. it can be derived from structure using physical laws or heuristics. There is no direct connection between function and structure [5].

The FBS framework [8] is an extension of the FBS ontology to represent the process of designing as a set of transformations between function, behaviour and structure. The most basic view of designing consists of transformations from function to behaviour, and from behaviour to structure:

(1) $F \rightarrow B$, and
(2) $B \rightarrow S$.

In this view, behaviour is interpreted as the performance expected to achieve desired function. Yet, once a structure is produced, it must be checked whether the

Table 13.1 Examples of function, behaviour and structure of different artefacts

	Dwelling	Editing software	Manufacturing process	Team
Function (F)	Provide safety, provide comfort, provide affordability	Be time efficient, provide affordability	Be safe, be time efficient, provide sustainability, provide affordability	Be time efficient, provide affordability
Behaviour (B)	Strength, weight, heat absorption, cost	Response times, cost	Throughput, accuracy, speed, waste rate, cost	Working speed, success rate, cost
Structure (S)	Geometrically interconnected walls, floors, roof, windows, doors, pipes, electrical systems	Computationally interconnected program components	Logically and physically interconnected operations and flows of material and information	Socially interconnected individuals

artefact's 'actual' performance, based on the structure produced and the operating environment, matches the 'expected' behaviour. Therefore, the FBS framework distinguishes two classes of behaviour: expected behaviour (Be) and behaviour derived from structure (Bs). This extends the set of transformations with which we can describe designing to include:

(1) F $\rightarrow$ Be,
(2) Be $\rightarrow$ S,
(3) S $\rightarrow$ Bs, and
(4) Be $\leftrightarrow$ Bs (comparison of the two types of behaviour).

The observable input and output of any design activity is a set of requirements (R) that come from outside the designer and a description (D) of the artefact, respectively. The FBS framework subsumes R in the notion of function and defines D as the external representation of a design solution:

(5) S $\rightarrow$ D.

Based on the common observation that designing is not only a process of iterative, incremental development but frequently involves focus shifts, lateral thinking and emergent ideas, the FBS framework defines the following additional transformations:

(6) S $\rightarrow$ S',
(7) S $\rightarrow$ Be', and
(8) S $\rightarrow$ F' (via Be).

These three transformations assume an existing structure as the driver for generating changes in structure, behaviour or function.

The eight fundamental transformations or processes in the FBS framework are shown and labelled in Fig. 13.1:

1. Formulation (R $\rightarrow$ F, and F $\rightarrow$ Be)
2. Synthesis (Be $\rightarrow$ S)
3. Analysis (S $\rightarrow$ Bs)
4. Evaluation (Be $\leftrightarrow$ Bs)
5. Documentation (S $\rightarrow$ D)
6. Reformulation type 1 (S $\rightarrow$ S')
7. Reformulation type 2 (S $\rightarrow$ Be')
8. Reformulation type 3 (S $\rightarrow$ F' (via Be)).

The FBS framework represents the beginnings of a theory of designing, through its ability to describe any instance of designing irrespective of the specific domain of design or the specific methods used. Section 13.4 will present how empirical studies provide a validation of the FBS framework in the sense of a theory of designing.

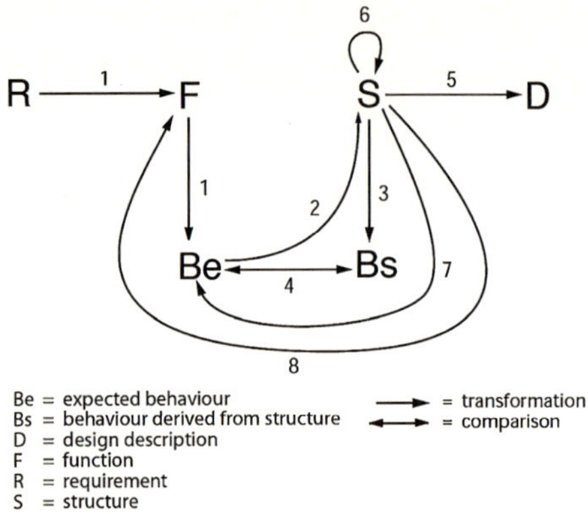

Fig. 13.1 The FBS framework

13.3 The Situated Function-Behaviour-Structure Framework

The situated FBS framework was developed in 2000 as an extension of the FBS framework to include the notion of situatedness [11]. It is founded on the idea that situated designing involves interactions between three worlds: the external world, the interpreted world and the expected world, Fig. 13.2.

The *external world* is the world that is composed of things outside the designer. No matter whether things are 'real' or represented, we refer to all of them as just 'design representations'. This is because their purpose is to support interpretation and communication of designers.

The *interpreted world* is the world that is built up inside the designer in terms of sensory experiences, percepts and concepts. It is the internal representation of that part of the external world that the designer interacts with. The interpreted world provides an environment for analytic activities and discovery during designing.

The *expected world* is the world imagined actions of the designer will produce. It is the environment in which the effects of actions are predicted according to current goals and interpretations of the current state of the world.

These three worlds are related together by three classes of interaction. *Interpretation* transforms variables that are sensed in the external world into sensory experiences, percepts and concepts that compose the interpreted world. *Focussing* takes some aspects of the interpreted world and uses them as goals for the expected world. *Action* is an effect which brings about a change in the external world according to the goals in the expected world.

Figure 13.2b presents a specialised form of this model, with the designer (described by the interpreted and expected world) located within the external world, and with general classes of design representations placed into this nested model. The set of expected design representations (Xe^i) corresponds to the notion of a design state space, i.e. the state space of all possible designs that satisfy the set of requirements. This state space can be modified during the process of designing

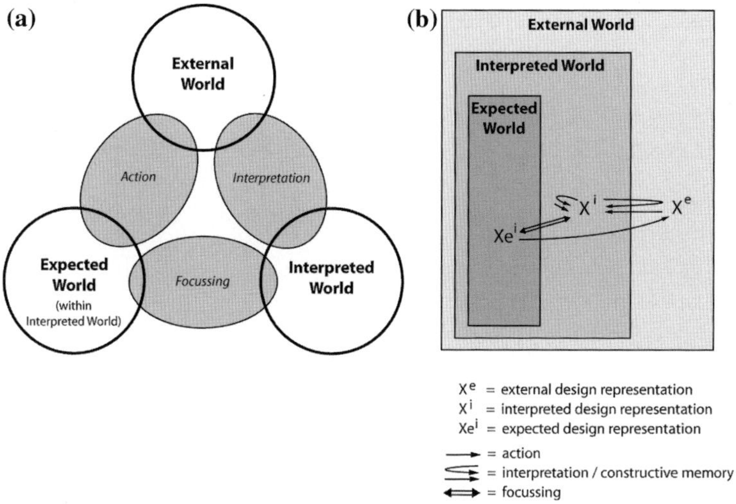

Fig. 13.2 Situatedness as the interaction of three worlds: **a** general model, **b** specialised model for design representations

by transferring new interpreted design representations (X^i) into the expected world and/or transferring some of the expected design representations (Xe^i) out of the expected world. This leads to changes in external design representations (X^e), which may then be used as a basis for re-interpretation changing the interpreted world. Novel interpreted design representations (X^i) may also be the result of memory (here called *constructive memory*), which can be viewed as a process of interaction among design representations within the interpreted world rather than across the interpreted and the external world.

Both interpretation and constructive memory are viewed as 'push–pull' processes, i.e. the results of these processes are driven both by the original experience ('push') and by some of the agent's current interpretations and expectations ("pull") [9]. This notion captures two ideas. First, interpretation and constructive memory have a subjective nature, using first-person knowledge grounded in the designer's interactions with their environment [2, 3, 32, 37]. This is in contrast to static approaches that attempt to encode all relevant design knowledge prior to its use. Anecdotal evidence in support of first-person knowledge is provided by the common observation that different designers perceive the same set of requirements differently (and thus produce different designs). And the same designer is likely to produce different designs at later times for the same requirements. This is a result of the designer acquiring new knowledge while interacting with their environment between the two times.

Second, the interplay between 'push' and 'pull' has the potential to produce emergent effects, leading to novel and often surprising interpretations of the same internal or external representation. This idea extends the notion of biases that simply reproduce the agent's current expectations. Examples have been provided

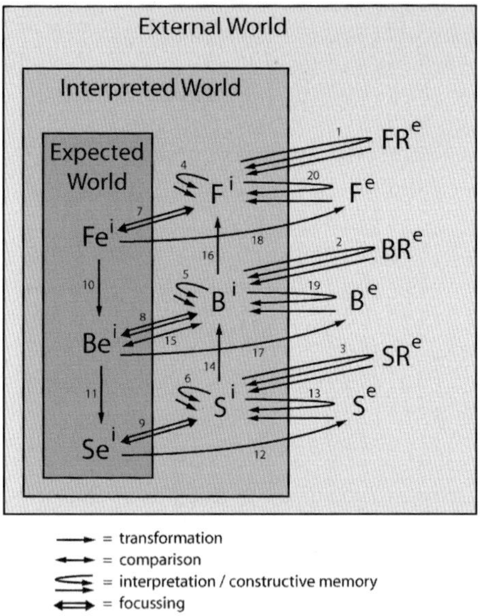

Fig. 13.3 The situated FBS framework [13]

from experimental studies of designers interacting with their sketches of the design object. Schön and Wiggins [30] found that designers use their sketches not only as an external memory, but also as a means to reinterpret what they have drawn, thus leading the design in a surprising, new direction. Suwa et al. [33] noted, in studying designers, a correlation of unexpected discoveries in sketches with the invention of new issues or requirements during the design process. They concluded that 'sketches serve as a physical setting in which design thoughts are constructed on the fly in a situated way'. Guindon's [17] protocol analyses of software engineers, designing control software for a lift, revealed that designing is characterised by frequent discoveries of new requirements interleaved with the development of new partial design solutions. As Guindon puts it, 'designers try to make the most effective use of newly inferred requirements, or the sudden discovery of partial solutions, and modify their goals and plans accordingly'.

Gero and Kannengiesser [11–13] have combined the FBS framework with the model of interacting worlds, by specialising the model of situatedness shown in Fig. 13.2b. In particular, the variable X, which stands for design representations in general, is replaced with the more specific representations F, B and S. This provides the basis of the situated FBS framework, Fig. 13.3. In addition to using external, interpreted and expected F, B and S, this framework uses explicit representations of external requirements, represented as external requirements on function (FR^e), external requirements on behaviour (BR^e), and external requirements on structure (SR^e). The situated FBS framework also introduces the process of comparison between interpreted behaviour (B^i) and expected behaviour (Be^i), and a number of processes that transform interpreted structure (S^i) into interpreted behaviour (B^i), interpreted behaviour (B^i) into interpreted function (F^i), expected

Fig. 13.4 Formulation

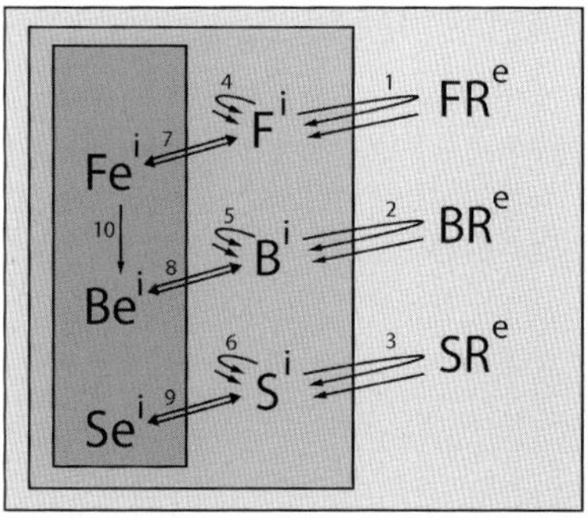

function (Fe^i) into expected behaviour (Be^i), and expected behaviour (Be^i) into expected structure (Se^i). Figure 13.3 uses the numerals 1–20 to label the resultant set of processes; however, they do not represent any order of execution.

The 20 processes in the situated FBS framework map onto the eight fundamental processes in the original FBS framework. The remainder of Sect. 13.3 presents these mappings, and illustrates them using a turbocharger as the artefact.

13.3.1 Formulation

Formulation frames the design task by defining a state space of potential design solutions (structure state space) and a set of criteria for assessing these solutions (behaviour state space). This activity uses a set of goals (function state space) and constraints that are given to the designer by external specification or are constructed based on the designer's own experience. In the situated FBS framework, formulation includes processes 1–10, Fig. 13.4.

Example: A turbocharger designer is provided with a set of requirements by an automobile company that include:

- FR^e: increase the power output of a specific engine of a specific passenger car
- BR^e: air mass flow and efficiency ratio for a range of different engine speeds
- SR^e: maximal spatial dimensions; position of connecting points to other components.

These requirements are interpreted to produce F^i, B^i and S^i (processes 1, 2 and 3) that are complemented with implicit requirements constructed from the designer's memory (processes 4, 5 and 6). These additional requirements include:

- F^i: provide reliability, provide reduced manufacturing cost
- B^i: ranges of values for the pressure ratio of compressor and turbine at the different engine speeds
- S^i: basic components (compressor, turbine, core assembly) and their parameters including geometrical variables and classes of material (e.g., aluminum for

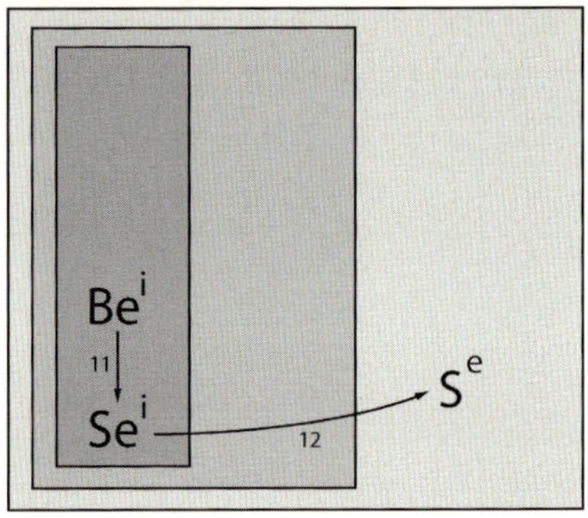

Fig. 13.5 Synthesis

compressor, and cast iron for turbine); ranges of values for inlet and outlet diameters of compressor and turbine.

Processes 7, 8 and 9 represent deciding on a set of turbocharger requirements to form the design state space. Process 10 captures how additional expected behaviour (Be^i) is derived from expected function (Fe^i). For example, expected ranges of thermal strength are derived from the function requirement of reliability.

13.3.2 Synthesis

Synthesis instantiates a design solution in terms of a point in the structure state space. It includes processes 11 and 12, Fig. 13.5.

Example: The designer produces a design by deciding on the values of the formulated structure variables for the turbocharger (process 11). The design is then externalised (process 12) as a drawing on paper, as a computational model using a computer-aided drafting (CAD) tool, or as a physical prototype.

13.3.3 Analysis

Analysis derives the behaviour from the design solution. It includes processes 13 and 14, Fig. 13.6.

Example: The designer uses a range of calculations, simulations and physical prototype tests to analyse the design solution of the turbocharger. This requires interpretation of external structure (process 13), either by the designer or an engineer testing a prototype or visually inspecting iconic or mathematical models, or by an analysis tool that reads CAD files. Behaviour can then be derived (process 14) in one of three ways:

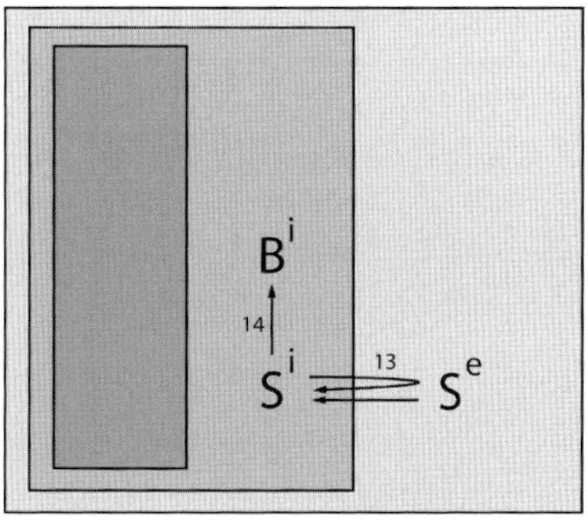

Fig. 13.6 Analysis

- By computation: Specialised tools are used to perform complex calculations and simulations. For example, thermal strength of turbochargers (particularly of their turbine components) is often derived using a finite-element analysis (FEA) tool.
- By physical measurement: Behaviours can be derived from the physical, electrical or chemical effects caused by the interaction of measurement devices and physical prototypes. This is frequently used in turbocharger analysis, to derive pressures and temperatures produced by turbines and compressors under realistic operating conditions.
- By human reasoning: This is done only for very simple derivations of behaviour and usually involves extensive use of external memory aids. Human reasoning is best applied in combination with computation or physical measurement. For example, dividing the compressor's inflow pressure (an exogenous variable) by its outflow pressure (a behaviour measured in a prototype test) is a simple calculation that produces the compressor's pressure ratio (a derivative behaviour).

13.3.4 Evaluation

Evaluation assesses the design solution on the basis of the formulated criteria, i.e. by comparison of the behaviour derived from the design solution and the expected behaviour. Evaluation includes process 15, Fig. 13.7.

Example: The designer compares the air mass flows, pressure ratios, strength, etc., analysed with the ones required (process 15). Based on the outcome of this comparison, the designer decides whether the design of the turbocharger satisfies the requirements. In most cases, changes are needed that lead to further cycles of synthesis, analysis and evaluation. For example, the turbine's pressure ratio may be evaluated as too low to achieve required mass flow rates. The designer may then decide to synthesise a modified structure with larger values for the turbine wheel's geometric variables.

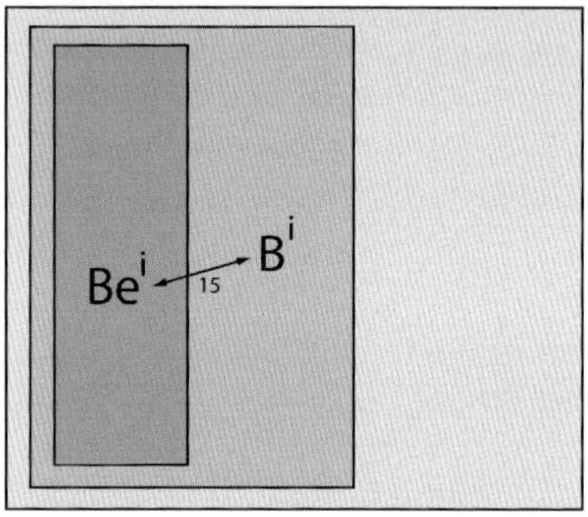

Fig. 13.7 Evaluation

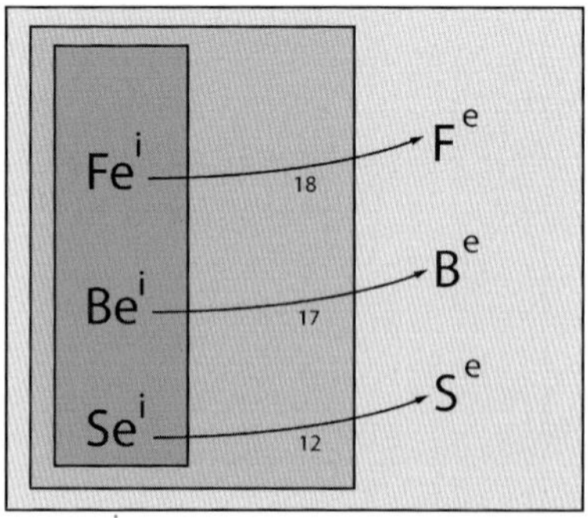

Fig. 13.8 Documentation

13.3.5 Documentation

Documentation produces an external representation of a design solution for purposes of communicating that solution. In most instances of designing ‘physical’ products, this step is required to provide the builder or manufacturer with a ‘blueprint’ for realizing the product. Documentation includes processes 12, 17 and 18 (Fig. 13.8).

Example: After successful evaluation of the turbocharger, a number of drawings and CAD models are produced of the assembly including its individual components (process 12) so that the turbocharger can be manufactured. A number of diagrams documenting some of the behaviour, such as efficiency, air mass flow and pressure ratio, are also generated (process 17) as ‘performance maps’ for the automobile company. Some functions may be documented for purposes of indexing, marketing or explaining design decisions (process 18).

13.3.6 Reformulation Type 1

Reformulation type 1 reframes the structure state space, directly creating a new space of possible designs. This often entails a subsequent modification of the behaviour state space. Reformulation type 1 includes process 9 (Fig. 13.9).

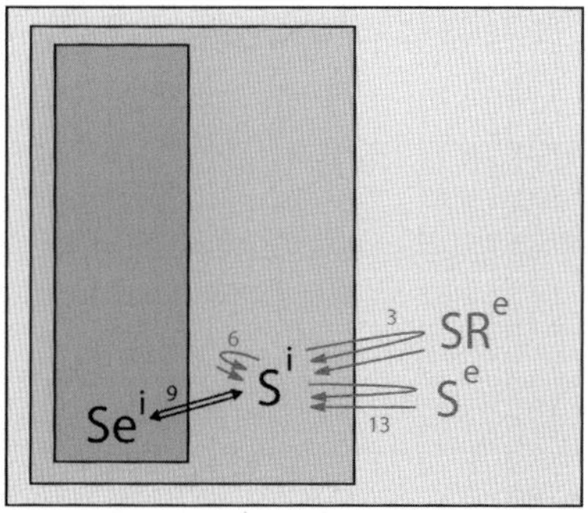

Fig. 13.9 Reformulation type 1

Processes 3, 6 and 13 are the potential drivers of this type of reformulation, as they all have the potential to produce new structure.

Example: The designer may decide to extend the ranges of values of the turbocharger's geometric dimensions (process 9), such that it allows the selection of much smaller values than previously expected. This can be seen as creating a new family of (smaller) turbocharger variants. The decision to reformulate structure may be the result of external drivers, such as new external requirements from the car manufacturer (process 3) or studies of a competitor's product (process 13), or an internal driver, such as reflection on integrating new technologies (e.g., new materials) (process 6).

13.3.7 Reformulation Type 2

Reformulation type 2 reframes the behaviour state space. In most cases, this leads to a modification of the structure state space, and thus to the creation of a new space of possible designs. In some cases, the new behaviour may also drive changes in the set of functions. Reformulation type 2 includes process 8, Fig. 13.10. Processes 2, 5, 14 and 19 are the potential drivers of this type of reformulation, as they all have the potential to produce new behaviour.

Example: The designer may want to introduce a new control behaviour for varying the air mass flow. This leads to the creation of a design state space with new characteristics that become visible through changes in structure. Possible changes of the turbocharger's structure are the addition of variable guide vanes or a variable sliding ring inside the turbine. The reformulation of the turbocharger's control behaviour may be the result of external drivers, such as new external requirements from the car manufacturer (process 2) or the interpretation of ideas articulated in a brainstorming meeting (process 19), or internal drivers, such as reflection on previous experiences regarding variable control (process 5) or analogical derivation of behaviour from structurally related objects (e.g., water turbines with variable inlet nozzle sizes to control water supply) (process 14).

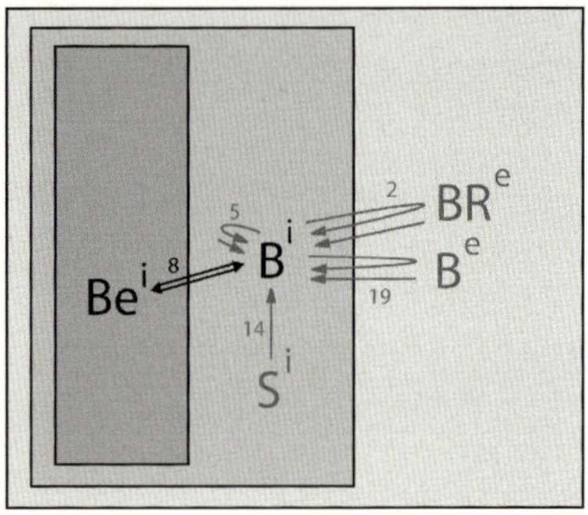

Fig. 13.10 Reformulation type 2

13.3.8 Reformulation Type 3

Reformulation type 3 reframes the function state space. In most cases, this leads to a modification of the behaviour and structure state space, and thus to the creation of a new space of possible designs. Reformulation type 3 includes process 7, Fig. 13.11. Processes 1, 4, 16 and 20 are the potential drivers of this type of reformulation, as they all have the potential to produce new function.

Example: Supporting modified engine characteristics represents new function requirements for the turbocharger. For example, turbocharging an engine with significantly increased exhaust temperature may affect the thermal strength such that a more resistant class of material needs to be chosen for the turbine. The reformulation of the turbocharger's function may be the result of external drivers, such as new external requirements from the car manufacturer (process 1) or the interpretation of alternative functions expressed in a morphological matrix (process 20), or internal drivers, such as reflection on previous experiences with products of high temperature resistance (process 4) or analysis of potential consequences of technological improvements regarding thermal strength (process 16).

The situated FBS framework represents a further development towards a theory of designing, by accounting for the dynamics of the situation within which most instances of designing occur. Section 13.4 will present how empirical studies provide a validation of the situated FBS framework in the sense of a theory of designing.

13.4 Empirical Studies

Verbal protocol analysis is a rigorous methodology for eliciting verbal reports of thought sequences as a valid source of data on thinking. It is a well-developed, validated method for the acquisition of empirical data on thinking (4; 7; 34). The generic process of protocol analysis results in a sequence of codes that represent the cognitive activations during thinking. Using the FBS ontology as the basis of

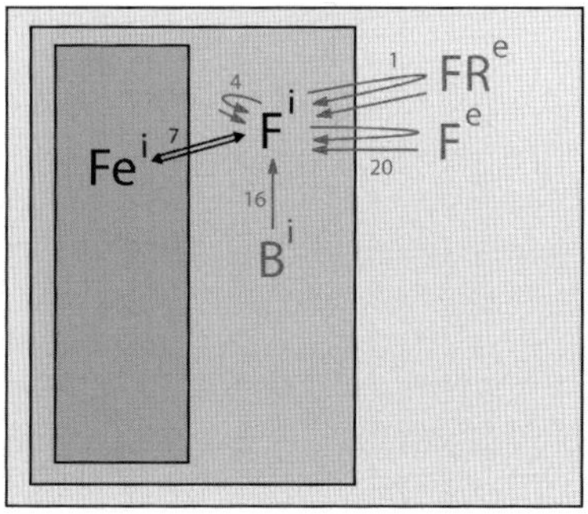

Fig. 13.11 Reformulation type 3

the coding scheme produces results that are commensurable across protocols independent of the designer, the design task and all aspects of the design environment. The FBS codes represent the cognitive activations of the design issues that the designers are thinking about as they are designing. The FBS-based design processes that are a consequence of the transformations of the design issues (Fig. 13.1) are available from the coding of the protocols [20].

Such empirical studies can be used to test the utility of the FBS ontology by measuring the percentage of design-related utterances not covered by the FBS coding as well as being used to characterise the cognition of designing. In a wide range of protocol studies the percentage of design-related utterances not covered in any protocol has been zero or diminishingly small. This does not imply that the FBS ontology-based coding is the only coding scheme that covers empirical data about designing, rather the implication is that the FBS ontology provides a robust foundation for the development of a generic coding scheme. Protocols coded using the FBS coding are commensurable. Results from FBS coded protocols provide insight into designing and confirm the utility of the FBS ontology. A small number of such results is presented below to provide indicate exemplars of what can be found using this approach.

13.4.1 Comparing Different Disciplines Designing

The question of what are the differences between different disciplines as they are designing can be addressed through empirical studies. The results of a set of studies of architects, software designers and mechanical engineers designing in terms of their respective design issue distributions are shown in Fig. 13.12. The use of the FBS coding scheme allows for a direct comparison. These results from these studies indicate that architects spend more of their cognitive effort on the design issue of function than do software designers and mechanical engineers. Mechanical engineers spend more of their cognitive effort on behaviour from structure and less on expected behaviour than do architects and software designers.

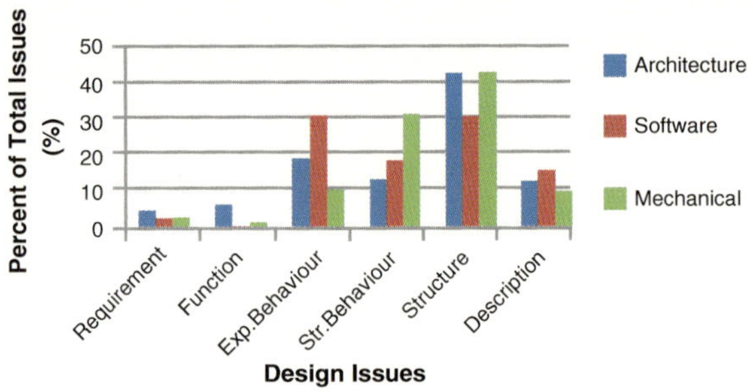

Fig. 13.12 Comparing the design issue distributions of three different design disciplines (after 21])

These results provide evidentiary support for the claim that the FBS ontology can be used independently of design discipline and design task.

13.4.2 Comparing High School and University Students Designing

Do high school and university students design differently? An experiment was conducted where high-school students and sophomore (second year) Mechanical Engineering university students were given the same design task. The results of their design issue distributions are presented in Fig. 13.13.

These results show that university students have a different distribution of their cognitive effort than do high-school students. That difference manifests itself primarily in the differences in both expected behaviour and behaviour from structure.

13.4.3 Comparing Effects of Using Different Design Techniques

Does teaching different concept generation techniques result in different design behaviours? The same cohort of students was taught brainstorming, morphological analysis and TRIZ. After learning each technique the cohort carried out a design task using the technique just learned. The results of measuring the distributions of design issues when utilizing each concept generation technique are presented in Fig. 13.14.

These results indicate that the use of different concept generation techniques produces different cognitive behaviour in the designers. The most significant difference manifests itself in the increased cognitive effort expended on function and expected behaviour when using TRIZ compared to the other two methods.

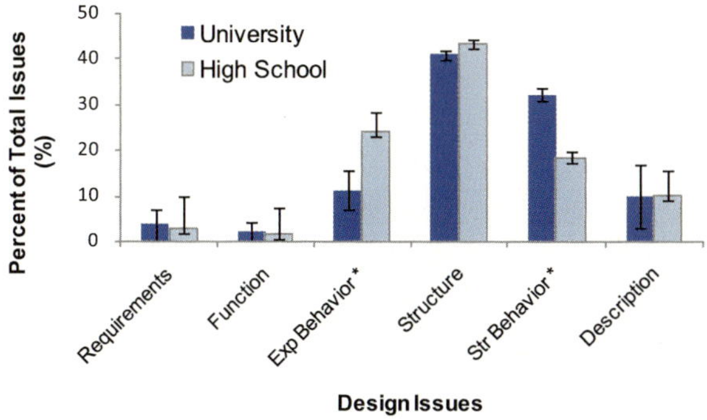

Fig. 13.13 Comparing the design issue distributions of high-school students and university students (after [27])

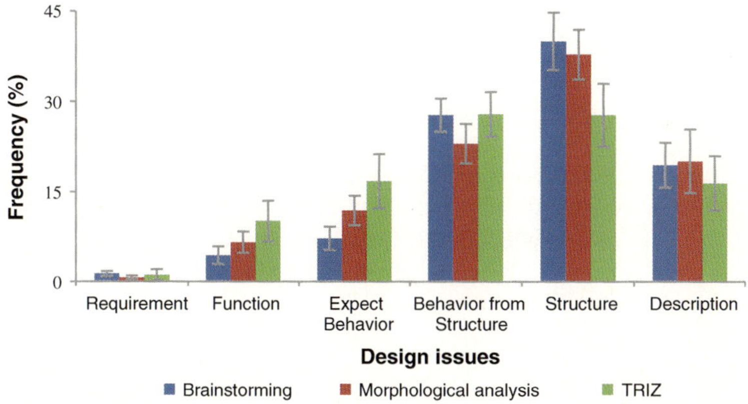

Fig. 13.14 Comparing the design issue distributions when designing with three different concept generation techniques. Data are from undergraduate engineering students (after [14])

13.4.4 Who is Doing What in a Design Team?

As designing is increasingly carried in teams, the behaviour of design teams and the individuals in them become of interest. One characterisation that provides access to the behaviour of teams and individuals in teams is the design process. In the FBS ontology design processes are the transformations from one design to another. This is represented by the semantic linkograph of the protocol [20, 15]. The linkograph of a team of designers provides the basis to extract the design process of individuals and to articulate which members of the team are involved in each process.

The design issue at the end of each link generates the design process and linking the names of the individuals associated with each end of a design process provides

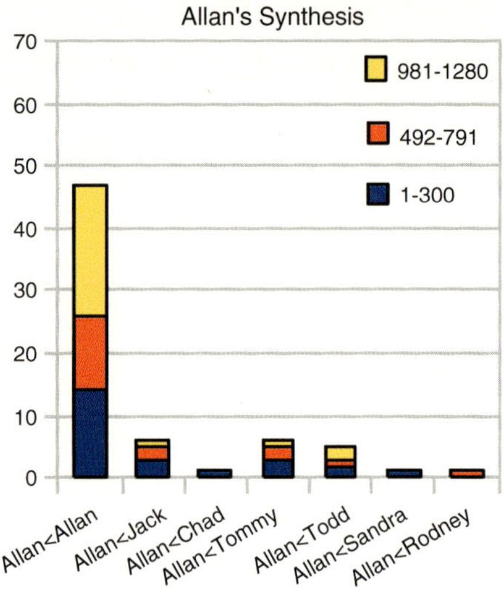

Fig. 13.15 Allan's (the team leader) synthesis process (Be → S) in terms of interactions with other team members, presented as a stacked histogram of the results derived from dividing the design session into three-thirds (defined by their segment number ranges on the right of the graph). This allows for comparing the time behaviour of synthesis interactions across the design session. Data are from a design team in industry (after [22])

a highly detailed description of the design process involvement of each individual during a design session.

In a study of a design team in industry the synthesis process of Allan, the team leader, is extracted from the protocol's semantic linkograph in such a way that his interactions with each member of the 7-person team can be followed (Fig. 13.15).

These results demonstrate the wide-ranging applications of utilizing a coding scheme based on the FBS ontology.

13.5 Discussion

The validation of methods, models and theories in design is a process of building confidence in their usefulness [29]. An increasing number of studies are supporting such confidence for the FBS ontology of design. They provide evidence for its applicability, and for the tools it can offer for understanding designing and designs.

The applicability of the FBS ontology is shown through its large coverage that has been demonstrated conceptually and empirically. Conceptually, the FBS ontology has been used in various design domains including architectural, mechanical, software and business process design [1, 26, 35, 36], Erden et al ([6]) to represent designs and design processes as a basis for methodologies and computational models [24, 28]. Empirically, the FBS ontology has been used for coding hundreds of design protocols representing design processes that varied along multiple dimensions such as the designers' expertise and discipline, the design task, and the size and composition of the design team.

The FBS ontology provides a number of tools for understanding designing and designs. The FBS and sFBS frameworks, in particular their graphical depictions, are tools for understanding designing in terms of its fundamental processes and its situatedness, respectively. They have been used for understanding a process not directly connected to designing: how people construct affordances of a designed object [25]. The FBS-based annotations proposed by Kannengiesser [23] are a tool for understanding a more abstract class of designs, business process designs. Some well-developed analysis tools for design protocols are also based on the FBS ontology [10]; they include the entropy of semantic linkographs for measuring the creation of novel concepts during designing [19] and the problem–solution (P–S) index for measuring the relative cognitive effort spent on either the problem or the solution [18].

A limitation of the high level of generality of the FBS ontology is that some specific aspects of designing are not directly addressed. Articulating the FBS ontology to map onto other framework descriptions of design may demonstrate more detailed areas of coverage that to date are not immediately obvious. For example, subclasses of function, behaviour and structure may be defined to represent different levels in a compositional hierarchy. Transformations between subclasses of the same ontological class but at different hierarchical levels would then represent processes of composition or decomposition, which are commonly described activities in other models of designing [31].

The FBS ontology has been shown to be a robust descriptor of designs and designing. The ontology continues to be widely cited with an average of two to three citations a week for the last decade.

Acknowledgments The research reported here has been supported by multiple grants from the Australian Research Council, the US National Science Foundation (NSF Grant Nos: EEC-0934824, CMMI-0926908, IIS-1002079, CMMI-1161715) and the US Defense Advanced Projects Agency (DARPA). Any opinions, findings, and conclusions or recommendations expressed in this material are those of the authors and do not necessarily reflect the views of the National Science Foundation.

References

1. Bergmann R (2002) Experience management: foundations, development methodology, and internet-based applications. Springer Berlin
2. Bickhard MH, Campbell RL (1996) Topologies of learning. New Ideas Psychol 14(2):111–156
3. Clancey WJ (1997) Situated cognition: on human knowledge and computer representations. Cambridge University Press, Cambridge
4. Crutcher RJ (1994) Telling what we know: The use of verbal report methodologies in psychological research, Psych Sci 5:241–244
5. De Kleer J, Brown JS (1984) A qualitative physics based on confluences. Artif Intell 24:7–83
6. Erden MS et al (2008) A review of function modeling: approaches and applications. AIEDAM 22(2):147–169
7. Ericsson KA, Simon HA (1993) Protocol Analysis; Verbal Reports as Data, MIT Press, Cambridge

theory of design known and applied by Galen; how was it originated? Then we turn to the question of current interest: Does the ancient theory of design have significance still today?

14.2 Origin of Design Theorising

14.2.1 Aristotle as Design Theorist

We contend that the theory of design referred to, applied and further developed by Galen has its origin in Aristotle. In his Nicomachean Ethics, Aristotle (384–322 BC) states that 'the person who deliberates seems to investigate and analyse in the way described as though he were analysing a geometrical construction'. Actually, this short statement, along with the sentences surrounding it (Table 14.1) contains a powerful view on design—however, several layers of interpretation in light of other classical texts are needed for revealing this.

The first question is whether Aristotle means design here. Indeed, he does not use this term for the simple reason that it is of a much more recent origin. Instead, he focuses on deliberation, in the sense of figuring out what to do [4]. In his examples, this deliberation occurs in the framework of production (*poiesis*), which has a wide interpretation: it covers medicine, oratory, shoe making, house building, shipbuilding, agriculture and also artistic activities such as poetry and music.

In the scheme of Aristotle [5], production has two stages: thinking and making:

> Of the productions or processes one part is called thinking and the other making,—that which proceeds from the starting point and the form is thinking, and that which proceeds from the final step of the thinking is making.

He exemplifies this through healing by a doctor; the medical knowledge of Aristotle is obsolete but the concepts and methods are clearly visible [5]:

> The healthy subject is produced as the result of the following train of thought: since this is health, if the subject is to be healthy this must first be present, e.g. a uniform state of body, and if this is to be present, there must be heat; and the physician goes on thinking thus until he reduces the matter to a final something which he himself can produce. Then the process from this point onward, i.e. the process towards health, is called a 'making'.

The similarity of logic (means-ends structure) and subject (healing) in the quoted passage and the focused passage in Nicomachean ethics allows concluding that this thinking equates to deliberation. In the case of producing an object, thinking arguably includes the mental operations required before any making is possible, namely designing (in its colloquial sense, specifying the functional principles, form and material of an object) and planning. Thus, when discussing deliberation, Aristotle covers designing.

Table 14.1 Aristotle's account on deliberation in Nicomachean Ethics [40]

We deliberate not about ends but about means. For a doctor does not deliberate whether he shall heal, nor an orator whether he shall persuade, nor a statesman whether he shall produce law and order, nor does any one else deliberate about his end. They assume the end and consider how and by what means it is to be attained; and if it seems to be produced by several means they consider by which it is most easily and best produced, while if it is achieved by one only they consider how it will be achieved by this and by what means this will be achieved, till they come to the first cause, which in the order of discovery is last. For the person who deliberates seems to investigate and analyse in the way described as though he were analysing a geometrical construction (not all investigation appears to be deliberation—for instance mathematical investigations—but all deliberation is investigation), and what is last in the order of analysis seems to be first in the order of becoming. And if we come on an impossibility, we give up the search, e.g. if we need money and this cannot be got; but if a thing appears possible we try to do it.

14.2.2 Analysing a Geometrical Construction

What, then, does 'analysing a geometrical construction' mean? This refers to the method of analysis in geometry, a sophisticated and well-known procedure already at Aristotle's time, although the only written account of it (Table 14.2), by Pappus (AD c. 290–c. 350), comes from late Antiquity.

It is useful to briefly outline the procedure contained in the method of analysis (Fig. 14.1). In geometry, one typical problem is to construct a given geometrical figure using a ruler and a compass (this is the problematical analysis of Pappus). The starting point of analysis is to assume the sought figure already done, and to consider through which means it can be created, further through which means this can be achieved, until one comes to something well known, such as a theorem generally known to be true (thus, reasoning in analysis consists of inferences backward). This is the end point of analysis, and simultaneously the start point of synthesis. In synthesis, one follows, in a deductive manner, the steps taken in analysis, but in reverse order, and comes finally to the sought figure. Synthesis contains both the construction of the sought figure and its proof. It has to be stressed that the sophistication and richness of the method of analysis is not transmitted in such a brief outline.

14.2.3 Comparison of Deliberation by Aristotle with the Method of Analysis by Pappus

A comparison of Aristotle's account of deliberation and Pappus' description of the method of analysis (Table 14.3) reveals an astonishing similarity, as noted by Hintikka and Remes [6]. Indeed, it can be asked whether Pappus was influenced by Aristotle's account of deliberation, but this is hardly probable as the method of

Table 14.2 Pappus' account on the method of analysis (from Hintikka and Remes [6])

Now analysis is the way from what is sought—as if it were admitted—through its concomitants in order to something admitted in synthesis. For in analysis, we suppose that which is sought to be already done, and we inquire from what it results, and again what is the antecedent of the latter, until we on our backward way light upon something already known and being first in order. And we call such a method analysis, as being a solution backwards. In synthesis, on the other hand, we suppose that which was reached last in analysis to be already done, and arranging in their natural order as consequents the former antecedents and linking them one with another, we in the end arrive at the construction of the thing sought. And this we call synthesis.
Now analysis is of two kinds. One seeks the truth, being called theoretical. The other serves to carry out what was desired to do, and this is called problematical. In the theoretical kind, we suppose the thing sought as being and as being true, and then we pass through its concomitants in order, as though they were true and existent by hypothesis, to something admitted; then, if that which is admitted be true, the thing sought is true, too, and the proof will be the reverse of analysis. But if we come upon something false to admit, the thing sought will be false, too. In the problematical kind, we suppose the desired thing to be known, and then we pass through its concomitants in order, as though they were true, up to something admitted. If the thing admitted is possible or can be done, that is, if it is what the mathematicians call given, the desired thing will also be possible. The proof will again be the reverse of analysis. But if we come upon something impossible to admit, the problem will also be impossible.

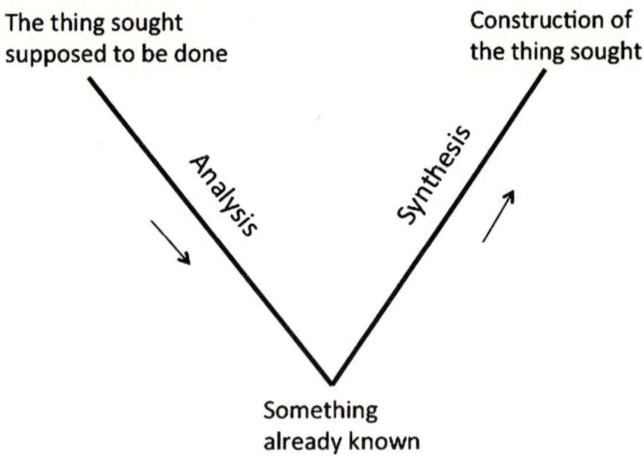

Fig. 14.1 Outline of the method of analysis (problematical analysis)

analysis was the paramount methodological resource for Greek geometers and Pappus, a practitioner of the method of analysis, must have absorbed it from his teachers and prior mathematical treatises. Indeed, Menn [7] argues that the same logical description of analysis that we find in Pappus was already available in Plato's and Aristotle's time.

When describing deliberation, Aristotle refers to all the steps of the method of analysis, as later described by Pappus (and presents one additional step not covered by Pappus). Deliberation starts in the same way as analysis, proceeds through the

Table 14.3 Comparison of deliberation and analysis

Aristotle's description of deliberation	Corresponding parts in Pappus' description of the method of analysis
They assume the end	For in analysis we suppose that which is sought to be already done,
and consider how and by what means it is to be attained;	and we inquire from what it results,
and if it seems to be produced by several means they consider by which it is most easily and best produced,	
while if it is achieved by one only they consider how it will be achieved by this and by what means this will be achieved,	and again what is the antecedent of the latter,
till they come to the first cause, which in the order of discovery is last.	until we on our backward way light upon something already known and being first in order.
For the person who deliberates seems to investigate and analyse in the way described as though he were analysing a geometrical construction (not all investigation appears to be deliberation- for instance mathematical investigations- but all deliberation is investigation),	
and what is last in the order of analysis seems to be first in the order of becoming.	In synthesis, on the other hand, we suppose that which was reached last in analysis to be already done, and [...] we in the end arrive at the construction of the thing sought.
And if we come on an impossibility, we give up the search, e.g. if we need money and this cannot be got; but if a thing appears possible we try to do it.	But if we come upon something false to admit, the thing sought will be false, too.

same steps and even ends similarly. Arguably, Aristotle claims that deliberation and analysis are throughout similar or analogous, from beginning to end. In relation to this claim, several implications and consequential questions arise.

14.2.4 Implications from the Viewpoint of Design

Thus, in proposing that the person who deliberates seems to investigate and analyse as though he were analysing a geometrical construction, Aristotle presents the first theory—proto-theory—of design. It encompasses the claim that design is similar, or analogous, to geometric analysis. This proto-theory was influential still in late Antiquity, as Galen's example shows, but fell then into oblivion.

Did Aristotle mean that just the topics explicitly mentioned by him are similar between design and geometrical analysis, or does the remark imply an overall, deeper structural similarity? The rhetorical figure he used, from beginning to end, would pinpoint to the latter alternative. Likewise, the character of his writings as lecture notes [8] suggests taking the overall similarity as the hypothetical starting point: it falls to us to establish the extent of similarity.

What, according to Aristotle, is the degree of similarity between design and analysis? Not much can be concluded regarding this question. Several stages of deliberation seem to be similar to their counterparts in analysis; however, the wording 'as though he were analysing a geometrical construction' allows a looser analogy, too.

Interpreting the method of analysis as a theory of design leads to the question, whether it is a descriptive or prescriptive theory. Aristotle's wording itself refers to a descriptive account; however, the method of analysis is a prescriptive procedure. Perhaps Aristotle means that expert 'deliberators' naturally drift to similar steps as in analysis.

Why was Aristotle concerned with deliberation as it occurs in production? Science of production (*techne*) was one of the three sciences defined by him, all expected to provide information about causes. Aristotle recognised four different causes: efficient, formal, material and final. Now, this account of deliberation may be interpreted as providing explanation on how the final cause comes to be in production. As final cause had the explanatory priority, this was an important piece in the scientific edifice of Aristotle.

14.2.5 Wider Questions Arising

An account of geometrical analysis and synthesis and of the place given to it by Aristotle in the productive science, as presented above, immediately raises several wider questions. Given the long-standing and wide interest into Aristotle's works, how can the significance of his remark for design theorising have avoided attention up to now? Why has not the understanding of analysis and synthesis as a precisely defined, ancient method been transmitted to the present day?

That this Aristotle's remark is being focused on only now has its explanation both at the 'supply and demand' side. As already mentioned, the interpretation of this remark requires support from other parts of Aristotle's corpus, and its full significance can be grasped only if the method of analysis is understood. This understanding has not been widely diffused through the history. Modern theorising on design started only in the 1960s. Neither now nor 50 years ago has it been usual to look for Aristotle as a source for theorising.

The relative disappearance of the original understanding of geometric analysis is due to several factors, which can be only briefly outlined here. Although the method of analysis had been fundamental for the initial development of geometry, philosophy of science (the scientific method) and productive science, all these disciplines failed to maintain understanding on their roots. In geometry and mathematics, the success of the geometric analysis in stimulating further advances, especially infinitesimal calculus, dwarfed Euclidean geometry and new meanings were given to the term analysis. In natural science, Enlightenment led to an anti-Aristotelian sentiment and the emphasis shifted to empirical studies, from the consideration of classical texts where the geometric roots of the scientific method are explained. In turn, productive science, *techne*, encountered a discontinuation; although productive science was well known in late Antiquity, it was not widely recognised in the Renaissance and fell into oblivion. Altogether, these developments effectively removed understanding on geometric analysis from the public domain.

14.3 The Current Significance of the Proto-Theory of Design

What is the current—rather than historical—significance of this proto-theory of design? This question is examined in four steps. First, we have to settle what we now know about the method of analysis. Second, we compare that to the current theoretical landscape of design. Third, we report a case of utilising understanding on the proto-theory to clarify a current approach to conceptual design. Fourth, based on all this, we evaluate the proto-theory as a theory of design and draw the conclusions regarding its significance.

14.3.1 What is the Method of Analysis, Understood as the Proto-Theory of Design?

The method of analysis was well known and practised in Antiquity, but in modern times, the interest has mostly been towards understanding and reconstructing it. Besides Aristotle's and Pappus' accounts, examples of ancient geometric practice (like those presented by Euclid and Pappus) and the interpretation tradition in the Middle Ages [9, 10] may give insights to this method. Last, current examinations of the method of analysis in mathematics and philosophy of science (for example, [6, 11, 12] provide useful directions.

Drawing from these sources (Table 14.4), although mainly from ancient descriptions, seven features of the method of analysis can be extracted:

1. Two types of analysis: problematical and theoretical.
2. Two stages in analysis: selecting among different means, and completing the analysis regarding the selected means.
3. The qualitative difference between the start point and the end point of analysis.

Table 14.4 Justification for the pinpointed features of the method of analysis

Feature of the method of analysis		Coverage of the features of the method of analysis in different sources				
		Aristotle's account of deliberation	Pappus' account on the method of analysis	Mathematical practice, as presented in the ancient literature	Tradition of interpretation [9]	Modern interpretation
1. Two types of analysis: problematical and theoretical		Problematical analysis implicitly pinpointed	Explicitly described	Commonly found		Described by Polya [12]
2. Two stages in analysis		Explicit description				
3. Start and end points qualitatively different		Explicit description	Explicitly described	Commonly found		
4. Three types of reasoning in two directions	Regression/ deduction	Explicit description		Commonly found	Scholastic tradition	
	Decomposition/ composition	Explicit description	Explicitly described		Scholastic tradition	
	Transformation			Commonly found (auxiliary lines)	Scholastic tradition	Examined by Hintikka and Remes [6]
5. The unity of the two directions of reasoning		Vaguely described	Explicitly described	Commonly found		
6. Two strategies of reasoning		Explicitly described for analysis	Explicit description for analysis and synthesis			Discussed by Polya [12]
7. Two targeted outcomes		Explicit but brief description	Explicit but brief description	*Reductio ad absurdum* as a way of establishing the impossibility of a solution		

4. Three types of reasoning in two directions: in analysis, regressive inferences, decomposition and transformation; in synthesis, deductive inferences, composition and (reverse) transformation.
5. The unity of the two directions of reasoning.
6. Two strategies of reasoning: in analysis heuristic and iterative, in synthesis predetermined.
7. Two targeted outcomes: finding a solution or showing that a solution is impossible.

In the following, each of these is described in more detail. For clarity, the same numbering as above will be used throughout the paper.

1. Two types of analysis
   According to Pappus, there are two types of analysis: theoretical and problematical. Problematical analysis aims at constructing a wished geometrical figure whereas theoretical analysis aims at proving a theorem. These are, in Polya's [12] generalised terms, the problem to find (a certain object, the unknown of the problem) and the problem to prove (an assertion true or false).
2. Two stages in analysis
   Aristotle states: 'if it seems to be produced by several means they consider by which it is most easily and best produced, while if it is achieved by one only they consider how it will be achieved by this and by what means this will be achieved'. This suggests that in the common case of several means, there are two stages: first selecting the best means among different alternatives and then completing the analysis regarding the selected means, through a chain of inferences. This feature is related to the general tendency towards economising. That it is not mentioned by Pappus may be explained by the fact that there is no specific mathematical method in play, rather the question is about a judgment; furthermore, in mathematical problem solving, the need for economising is not a central issue. Nevertheless, Aristotle seems to have thought that this step of deliberation is also part of the method of analysis.
3. The qualitative difference between the start and end points of analysis
   Pappus' description implies that the start and end points of analysis are qualitatively different. Regarding the start point in theoretical analysis, that is the 'thing sought', we do not know whether it exists and is true, but assume that. Instead, the end point consists of something admitted, that is, already known. In geometry, axioms and theorems already proven provide a body of admitted things. In turn, synthesis provides the proof that the 'thing sought' is existent and true. Correspondingly, in problematical analysis we do not know the 'desired thing', but assume it to be known.
4. Three types of reasoning in two directions
   Pappus makes it clear that reasoning in analysis involves inferences backwards and also Aristotle refers to this kind of reasoning in the passage from Nicomachean Ethics. Such inferences backwards are called regressive analysis.
   Two other types of reasoning are evident in the interpretation tradition [9] and they can be also deduced from ancient practice. Thus, analysis also comprises

transformational aspects, where the original problem is transformed into another form for facilitating its solution [9]. In geometric analysis, the use of auxiliary lines is the main form of this type of procedure. Moreover, a decompositional (or configurational) analysis is usually also involved [6, 13]. In geometry, the question is about investigating from which parts (lines, angles, points, etc.) a figure is made up, and which relations exist between those parts. Regarding synthesis (called proof by him), Pappus says that it is the reverse of analysis. Thus, the three types of reasoning of synthesis are carried out in reverse order compared to analysis.

5. Unity of the two directions of reasoning
   According to Pappus, both directions of reasoning are needed: in analysis, backwards for the solution, and in synthesis, forwards for the proof or the construction of the desired figure.
6. Two strategies of reasoning
   In Pappus' description, the method in itself does not provide detailed guidance on which particular moves one should carry out in analysis, except regarding the targeted end point; something admitted which is true or possible or can be done. Thus, analysis is heuristic rather than algorithmic and obviously often leads to an iterative approach of trial and error. In contrast, the synthesis stage is predetermined in the sense that it mirrors the (successful steps of) analysis, even if in reverse order.
7. Two targeted outcomes
   Obviously, the main target of an analysis is to find a solution. However, Pappus also claims that the analysis stage can end up showing that a solution to the problem at hand is impossible. Although his wording is laconic and vague, it is reasonable to assume that he refers to *reductio ad absurdum,* a well-known technique in ancient geometry. It creates a proof of a thesis by argumentation that derives a contradiction from its negation [14].

In summary, it can be stated that these seven features of the method of analysis provide guidance and a flexible methodical arsenal for the geometer on how to approach the task, how to structure it, where to start and where to stop, which are the possible reasoning strategies and moves as well as what to target. It is in this sense that the method of analysis is suggested to provide a proto-theory of design; the thesis is that this guidance and methodical arsenal would apply also to design.

14.3.2 How Does the Proto-Theory Compare to the Current Theoretical Landscape on Design?

In view of the arguments just made, the crucial question addressed is: Do the features of the proto-theory have similar or analogous counterparts in the current methodical and theoretical landscape of design? Namely, it can be assumed that if the features of the method of analysis are relevant to design, those features would have surfaced in the recent methodical and theoretical design literature.

1. Two types of analysis
   Briefly stated, the two types of analysis include finding (a solution) and proving (an assertion, say, on the validity of a proposed solution). The main difference of these is that in the former, one endeavours to create a chain of inferences from the problem towards a solution, whereas in the latter a solution is first guessed and then analysed for its validity. In the design literature, analogous approaches have been called problem-oriented and solution-oriented strategies, and it is recognised that completing a design requires the application of both [15]. The many stage models of design, some positing that analysis precedes synthesis [16], some that synthesis precedes analysis [17], have tried to accommodate and clarify both types, with varying success (it is important to note that in these models the meanings of analysis and synthesis have drastically drifted from the sense these terms are used in geometry).
2. Two stages in analysis
   The two stages in analysis, of selecting a means among different alternatives and completing the analysis regarding the selected means, of course correspond to the dichotomy between conceptual design and embodiment/detail design (for example, [18]. In the former, one tries to find the best solution in principle; in the latter, one endeavours to translate it into a practical solution. Morphological analysis [19] and parameter analysis [20] provide examples of approaches that have endeavoured to develop methods for conceptual design.
3. The difference between the start and end points of analysis
   The philosopher Schütz [21] proposed the concept of future perfect in relation to the theory of action: 'So we have to place ourselves mentally in a future state of affairs which we consider already as realized…' This proposal, which has been used in the design domain, is similar to Pappus' general description of the start point of analysis: 'For in analysis we suppose that which is sought to be already done, and we inquire from what it results…'
   However, there is a newer proposal in the design field that comes near this feature. Hatchuel and Weil (2002) take it as their 'fundamental proposition' that design reasoning must always make a distinction between two related spaces: the space of concepts and the space of knowledge. A concept (C) is defined as a proposition that has no logical status, i.e. we cannot know whether it is true or false. In turn, propositions in the knowledge space (K) have a logical status, and clearly the most interesting knowledge is what is known to be true. Design is defined as a process by which a concept generates other concepts or is transformed into knowledge. Thus, in the C–K theory, design is conceptualised by its start (C) and end points (K), which have similar characteristics as the start and end points in analysis. The C–K theory expands the knowledge about what occurs between these points.

4. Three types of reasoning in two directions
Regressive and deductive inferences equal, respectively, to backward and forward reasoning, as widely identified in the design domain. As an example, Quality Function Deployment embodies the chain of regressive inferences from the requirements to the product design. Decompositional and compositional inferences refer to breaking down and putting together. Such types of reasoning are often argued to exist in design (e.g. [22]). Indeed, Product Breakdown Structure is a pure application of decomposition. In transformational inferences, the problem is transformed into another problem for facilitating its solution. This idea is used in TRIZ [23], where a particular problem to be solved is abstracted to a more general level, at which the knowledge about inventive opportunities lies.
5. The unity of the two directions
The Vee model, developed in the framework of systems engineering [24] and recently diffused in software engineering and project management [25], similarly implies two directions of reasoning.
6. Two strategies of reasoning
The view that the design process is heuristic and iterative, as in analysis, is now widespread (e.g. [22, 26, 27]). However, this was a new idea in the 1980s, as evident from the observations of many who reported that, in practice, the designers unpredictably move between goals and means, instead of a linear, one-way process (e.g. [26, 28]). The axiomatic design approach as presented by Suh [29] represents an attempt to facilitate this heuristic and iterative search through rules. The predetermined deductive process of synthesis, in turn, is present in the right wing of the Vee model.
7. Two targeted outcomes
As all other parts and aspects of the method of analysis are geared towards finding a solution, the interest here in the impossibility of it. In engineering design, it has been found that requirements set based on customer wishes may be unrealistic [30]; engineering models are proposed as a means for pinpointing the impossibility of a solution. More generally, a feasibility analysis stage has been suggested [16] for dealing with this issue.

Several interesting observations and insights stand out. First, for all the features explicitly, or implicitly, contained in the method of analysis, we can pinpoint modern, corresponding ideas, concepts and methods, many very recently rediscovered (Table 14.5). This adds to the validity of the method of analysis as a theory of design. Second, without exception, the modern concepts and practices have been forwarded by their originators without any reference to the ancient counterparts—clearly, due to ignorance of them. Further, insights into the breadth and depth of the proto-theory as well as into its use for analysing the evolution of design methodology can be made—these are discussed below.

Table 14.5 Proto-theory versus current theories and methods

Features of geometric analysis		Embodied in current methods	Expanded in current methods
1. Two types of analysis		Problem-oriented and solution-oriented approaches	
2. Two stages in analysis		Distinction between conceptual design and embodiment/detail design	Morphological analysis; parameter analysis (expansion of conceptual design)
3. Difference between start and end point of analysis		Future perfect approach	C–K theory
4. Three types of reasoning in two directions	Regression	Quality function deployment	
	Decomposition	Product breakdown structure	
	Transformation	TRIZ	
5. Unity of the two directions		Vee model	
6. Two strategies of reasoning			Axiomatic design
7. Two targeted outcomes			Engineering models (expansion of 'proof of impossibility of a solution')

14.3.3 Use of the Proto-Theory for Clarification of a Current Approach to Design

In [31], research endeavouring to interpret the parameter analysis (PA) methodology of conceptual design [20] through the reconstructed proto-theory of design is reported.

From the viewpoint of parameter analysis, the notions of the proto-theory are found to create added clarity when applied to this contemporary design approach. Especially, they allow interpreting each of the parameter analysis steps separately in terms of the types of reasoning involved. Also, it is clarified that reasoning backwards towards a solution and reasoning forwards towards the proof are integrated into one process.

In turn, from the viewpoint of the proto-theory, it is of interest that most features of the proto-theory can be connected to steps or aspects in PA. This, for its part, empirically adds to the validity of the proto-theory. Second, the proto-theory is helpful in pinpointing aspects or parts of a suggested design process that remain implicit or not fully elaborated. Arguably, this is related to the prevailing relative lack of precise notions to describe design reasoning in detail. Third, the examination of PA provides evidence on the role of the proto-theory as a useful reference, for example, a novel strategy of reasoning in PA (focus on those parts of the problem where uncertainty can be most steeply reduced) could readily be identified when it was compared to the corresponding feature of the proto-theory.

In addition, this research highlighted certain differences of design reasoning in comparison to geometric reasoning. For example, in design, reasoning is more often based on informal logic than in geometry. Furthermore, there seem to be steps in PA that do not nicely fall into the proto-theory. Comparison of alternatives belongs to such steps. This may indicate that for some aspects and stages of design, notions and explanations that go beyond the proto-theory are needed. This is discussed in the next section.

14.3.4 Evaluation of the Proto-Theory as a Theory of Design

In [32], it has been argued that general functions of a theory comprise explanation, prediction, direction (for further progress) and testing (for validity). Furthermore, especially in a managerial science, a theory should also provide the functions of providing tools for decision and control, communication, learning and transfer (to other settings). Due to the novelty of the topic, only a part of these functions can be used as a basis of evaluation: explanation, direction, testing, tools for decision and control and communication.

14.3.4.1 Explanation

Does the proto-theory provide an explanation on how design can effectively be carried out? If yes, is that explanation better than prior explanations?

The close correspondence of the features of the proto-theory to topics in the current theoretical landscape of design arguably indicates that the proto-theory provides an explanation on design. Some further remarks on the quality of that explanation can be provided.

First, regarding the conceptualization of design, the proto-theory seems to provide a broader explanation than recent design theory proposals, even those with practical success, such as axiomatic design [29] or the C–K theory [33]. The former deals with the strategy of regressive and decompositional reasoning, for which heuristic (non-proven) rules have been developed. The latter is oriented around the start (C) and end (K) points of analysis [34]. In it, the dynamic and interactive nature of the start and end points, concepts and knowledge, is accentuated. By formally representing the start and end points, the space of concepts (C) and the space of knowledge (K), it has been possible to model the revision of object identities in C and the expansion of knowledge in K through four operators, three of which arguably are new. However, despite these advances, both of these approaches orderly cover only one or at most two of the several features of the method of analysis.

Second, this proto-theory can be claimed to be point wise deeper than the present body of knowledge on design. It shows the intellectual origin of such practically used and popular methods as Vee model and Product Breakdown

Structure, and gives them an initial explanation by way of a geometric analogue. The theoretical basis for these methods has hitherto been totally missing.

Third, at the outset, the explanation provided by the proto-theory is constrained by any intrinsic differences between the two areas: geometry and design. However, this does not seem to be a serious limitation; in such cases, it is possible to proceed through analogical reasoning towards finding the nearest design counterpart for a geometrical feature as well as to clarify the difference between the two.

14.3.4.2 Direction

Does the proto-theory give direction for further progress? When comparing current design theories and methods to the proto-theory, an interesting pattern of evolution is initially revealed (Table 14.5). Many design theories and methods seem to be based on descriptive but somewhat shallow knowledge on some aspect of the design process, equalling one feature of the proto-theory. However, only in such cases where a single feature of the proto-theory has been expanded and operationalised, manifest advances in design theory and/or methodology seem to have been made. Axiomatic design and the C–K theory, as discussed above, provide examples of this.

This leads to the hypothesis that the development of design theory and methodology should concentrate on expanding each feature of the proto-theory, focusing first on those where the theoretical and methodical advances have been meagre, such as 'unity of two directions'. Thus, indeed direction for further progress seems to flow from the proto-theory.

14.3.4.3 Testing

Does the proto-theory allow for testing its validity? As the method of analysis is relatively precisely defined, this should be generally possible. In principle, it would be feasible to empirically test the proto-theory both as a descriptive account (do designers use the seven features in their design activities; to which extent there are activities or aspects that are beyond the seven features?) and as a prescriptive guide (are the design outcomes or the design process improved through the implementation of the proto-theory?). Such studies have not yet been carried out.

However, based on analysis of Aristotle's seminal remark, the mentioned application of the proto-theory to clarify a current approach, and generally discussed features of design, something can be said regarding the question: Are there limitations related to the method of analysis as a proto-theory of design? At least four important gaps can be discerned.

First, the examples used by Aristotle are on design by one individual, analysis by one mind. However, design is very rarely an activity that is embodied within one individual. Rather, outcomes of design have to be presented to the client, to the producer and to other designers. Each designer needs to persuade others that his output is the best possible in the situation. This interaction, communication and persuasion are not covered by the method of analysis.

Second, the need for plausible (rather than logical) reasoning in design becomes evident, for example, in the comparison of alternative concepts and solutions. The method of analysis has no means to cover this type of reasoning.

Third, the starting point of analysis is either the task of proving a mathematical theorem or the task of drawing a certain geometrical figure. In both cases, the question is about a self-contained starting point, presented through unambiguous mathematical concepts. Both tasks are universal in the sense that their results are applicable and true everywhere and always. Instead, design is about a particular need of particular user(s). Thus exploration of that particular case is required at the outset of design. Also, it is implied that our understanding of a particular case is never complete. The stage of making sense of the particular case in question is missing from geometric analysis.

Fourth, geometric analysis is about the existence and production of a solution or about the proof. Design, when it comes to aesthetical aspirations, is about influencing the audience. This is also missing from geometric analysis.

Interestingly, these gaps (except perhaps the third one) exist also in the theoretical landscape of (engineering) design, if not absolutely so, at least as weaknesses. However, they all point to the need of embracing another ancient discipline into design theorising, namely rhetoric [35–37].

14.3.4.4 Decision and Control/Communication

Does the proto-theory provide practical tools for design? Does it help in communication? As evidenced in the clarification of the parameter analysis method through the proto-theory, the proto-theory seems to provide a more precise terminology in the field of design, than is currently available; this supports communication. But the interpretation of the parameter analysis through the concepts of the proto-theory as such shows that the latter have value in contributing to decision and control in design.

14.3.4.5 Concluding Remark

The conclusion is that the proto-theory fulfils several of the functions of a theory in a superior and fertile way. All in all, it can be contended that this proto-theory is not only of historical interest, but also provides a contribution to the theoretical and methodical knowledge on design.

14.4 Conclusion: The Lessons for Design Theorising

Perhaps the most important conclusion for design theorising is about the significance of history; there has been a fertile legacy for understanding design but it has not been embraced by the movement towards design theorising that started in the 1960s. It is tempting to draw an analogy to the general history of sciences and philosophy; the forgotten and lost legacy from the Antiquity was reintroduced in Europe during Renaissance and this crucially triggered development towards Enlightenment and the modern period. The intriguing question arises whether we will witness a late Renaissance in the discipline of design. Indeed, if we accept that the proto-theory of design has been rediscovered, we are compelled to see the evolution of design science under a new light. The core theory of design has been missing, and although scholars of design have endeavoured to discover it, the progress has been painfully slow and results fragmented. This missing of the core theory has arguably contributed to the maintenance of disciplinary fragmentation around design. This situation invites sounding whether the proto-theory of design, for its part, could still be used for advancement and unification of design science.

The argument for unified design science is not new. At the outset, there was a unified approach to design and making: the Aristotelian science of production, *techne*. No valid rationale is visible for the current fragmented disciplinary situation in this field, with engineering design, industrial design, systems engineering, new product development and project management having their own communities and knowledge bases. It should be possible to unify the many design disciplines around their common theoretical basis, or at least pinpoint their connections. Indeed, the task ahead is to compile a common conceptual and theoretical core for the various design and production sciences, and to develop associated ways of contextualising it to specific situations. In this regard, it is interesting to note that the view on the ubiquitous nature of design, as recently highlighted by the initiatives to establish management as a design science [38], is fully compatible with the original wide scope of *techne*.

Another challenge posed by the proto-theory to current theorising on design is about the scope of the phenomenon of design. Most current design theories seem narrow in comparison to the proto-theory. On the other hand, the proto-theory cannot be claimed to cover the whole area of design; from Antiquity onwards, it has been contended that there are aspects and stages in design that are best approached through rhetoric. The task of agreeing on the boundaries of the phenomenon of design seems still to be in front of us.

Lastly, the proto-theory renders the terminological problems plaguing the discipline of design visible. The terms analysis and synthesis have maintained a long-standing prestige, and as the understanding of their original meaning has been corrupted, new meanings have been given to them in different knowledge domains. This has led to a fundamental confusion of the role and meaning of analysis and synthesis in design. The current popular understanding in the design literature of analysis as a rational stage and synthesis as a creative stage is in direct contradiction with the ancient understanding. In his account on synthesis, Roozenburg [39] recognises those ancient meanings for analysis and synthesis as they have been transmitted in the philosophy of science, but comments that this use of words is rather confusing. Unfortunately, it is rather the design field that is using words in a way that are detached from their historical roots. It is opportune to clarify this confusion for the sake of the advancement in the field.

Acknowledgments Discussions with Dr. Ehud Kroll have been useful for clarifying the design interpretation of the features of the method of analysis. The helpful comments by two anonymous reviewers are also gratefully acknowledged.

References

1. Cross N (1993) Science and design methodology: a review. Res Eng Design 5(2):63–69
2. NSF Report on Engineering Systems Design Workshop (2010) Report on the NSF engineering systems design workshop held on 22–24 Feb 2010
3. Galen (1997) The affections and errors of the soul. In: Selected works. Singer PN (transl) Oxford University Press, Oxford, pp 120–149
4. Cooper JM (1975) Reason and human good in Aristotle. Harvard University Press, Cambridge
5. Aristotle (1924) Aristotle's metaphysics, a revised text with introduction and commentary by WD Ross. Clarendon Press, Oxford
6. Hintikka J, Remes U (1974) The method of analysis: its geometrical origin and its general significance. Boston Studies in the Philosophy of Science, Dordrecht
7. Menn S (2002) Plato and the method of analysis. Phronesis XLVII(3):193–223
8. Barnes J (2000) Aristotle: a very short introduction. Oxford University Press, Oxford
9. Beaney M (2009) Analysis, The Stanford Encyclopedia of Philosophy, Summer 2009 edn, Zalta EN (ed). http://plato.stanford.edu/archives/sum2009/entries/analysis/
10. Raftopoulos A (2003) Cartesian analysis and synthesis. Stud Hist Philos Sci 34:265–308
11. Hintikka J (2011) Method of analysis: a paradigm of mathematical reasoning? Hist Philos Logic 33(1):49–67
12. Polya G (2004) How to solve it: a new aspect of mathematical method. Princeton University Press, Princeton
13. Byrne PH (1997) Analysis and science in Aristotle. State University of New York Press, Albany
14. Rescher N (2005) Reductio ad absurdum. In: Internet Encyclopedia of Philosophy
15. Wynn D, Clarkson J (2005) Models of designing. In: Clarkson J, Eckert C (eds) Design process improvement: a review of current practice. Springer, London, pp 34–59
16. Asimow M (1962) Introduction to design: fundamentals of engineering design, Prentice-Hall, Englewood Cliffs
17. Hall AD (1962) A methodology for systems engineering. Princeton, New Jersey
18. Roozenburg NFM, Eekels J (1995) Product design: fundamentals and methods. Wiley, Chichester
19. Ritchey T (2006) Problem structuring using computer-aided morphological analysis. J Oper Res Soc 57(7):792–801
20. Kroll E, Condoor SS, Jansson, DG (2001) Innovative conceptual design: theory and application of parameter analysis. Cambridge University Press, Cambridge
21. Schuetz A (1943) The problem of rationality in the social world. Economica (new series) 10(38):130–149

22. Pahl G, Beitz W (1996) Engineering design: a systematic approach. Springer, Berlin
23. Cavallucci D (2002) TRIZ, the Altshullerian approach to solving innovative problems. In: Chakrabarti A (ed) Engineering design synthesis: understanding, approaches, and tools. Springer, London
24. Stevens R, Brook P, Jackson K, Arnold S (1998) Systems engineering: coping with complexity. Prentice-Hall, London
25. Forsberg K, Mooz H, Cotterman H (2005) Visualizing project management, 3rd edn. Wiley, Hoboken
26. Cross N (2000) Engineering design methods: strategies for product design. Wiley, Chichester
27. Hubka V, Eder WE (1987) Scientific approach to engineering design. Des Stud 8(3):123–137
28. Rasmussen J, Pejtersen AM, Goodstein LP (1994) Cognitive system engineering. Wiley, New York
29. Suh N (2001) Axiomatic design. Oxford University Press, Oxford
30. Ramaswamy R, Ulrich K (1993) Augmenting the house of quality with engineering models. Res Eng Design 5(2):70–79
31. Kroll E, Koskela, L (2012) Interpreting parameter analysis through the proto-theory of design. In: Proceedings of the 12th international design conference DESIGN 2012, Cavtat, Croatia, 21–24 May 2012
32. Koskela L (2000) An exploration towards a production theory and its application to construction. VTT Technical Research Centre of Finland
33. Hatchuel A, Weil B (2003) A new approach of innovative design: an introduction to C–K theory. In: Proceedings of the international conference on engineering design (ICED-03), Stockholm
34. Hatchuel A, Weil B (2009) C–K design theory: an advanced formulation. Res in Eng Des 19(4):181–192
35. Ballard G, Koskela L (2013) Rhetoric and design. In: International conference on engineering design (ICED13), Seoul, 19–22 Aug 2013
36. Buchanan R (1985) Declaration by design: rhetoric, argument, and demonstration in design practice. Des Issues 2(1):4–22
37. Koskela L, Ballard G (2013) The two pillars of design theory: method of analysis and rhetoric. In: International conference on engineering design (ICED13), Seoul, 19–22 Aug 2013
38. Boland RJ, Collopy F (2004) Managing as designing. Stanford University Press, Stanford
39. Roozenburg NFM (2002) Defining synthesis: on the senses and the logic of design synthesis. In: Chakrabarti A (ed.) Engineering design synthesis: understanding, approaches, and tools. Springer, London
40. Aristotle (s.a.) Nicomachean ethics [electronic resource]. http://classics.mit.edu/Aristotle/nicomachaen.html

Chapter 15
Perspectives on Design Models and Theories and Development of an Extended-Integrated Model of Designing

B. S. C. Ranjan, V. Srinivasan and A. Chakrabarti

15.1 Introduction

In the first part of this chapter (Sect. 15.2), we describe our perspective on the following questions:

- What are design models and theories?
- What are their characteristics?
- What are their purposes?
- What criteria should they satisfy to qualify as design models or theories?

In the second part of this chapter, we explain two models of designing, where the latter is an extension of the former:

- IMoD, developed earlier to explain task clarification and conceptual design (Sect. 15.3), and
- its further development into E-IMoD, developed to extend the scope of the model beyond task clarification and conceptual design (Sect. 15.4).

15.2 Theory and Model of Designing

In this section, we first report findings from literature, and based on these findings, present our perspective on the characteristics of design models and theories, their purposes, and the criteria they should satisfy to qualify as design models and theories.

In Merriam-Webster dictionary [19, p. 1223], the following alternative meanings are provided for the word 'theory': (a) analysis of a set of facts in their relation to one

B. S. C. Ranjan · V. Srinivasan · A. Chakrabarti (✉)
IDeaS Lab, Centre for Product Design and Manufacturing, Indian Institute of Science, Bangalore, India

A. Chakrabarti and L. T. M. Blessing (eds.), *An Anthology of Theories and Models of Design*, DOI: 10.1007/978-1-4471-6338-1_15,

another, (b) abstract thought: speculation, (c) the general or abstract principles of a body of facts, a science or an art, (d) a belief, policy, or procedure proposed or followed as the basis of action, (e) an ideal or hypothetical set of facts, principles or circumstances often used in the phrase 'in theory', (f) a plausible or scientifically accepted general principle or body of principles offered to explain phenomena, (g) a hypothesis assumed for the sake of argument or investigation, (h) an unproved assumption: conjecture, (i) a body of theorems presenting a concise, systematic view of a subject. Mautner [18, p. 426] defines a theory as a set of propositions which provide principles of analysis or explanation of a subject matter. In Dictionary of Ideas [9, p. 507], a theory is defined as a set of ideas, concepts, principles or methods used to explain a wide set of observed facts. Sutherland [28, p. 9] describes a theory as an ordered set of assertions about a generic behaviour or structure assumed to hold throughout a significantly broad range of specific instances.

Anderson [1] uses the term 'model' to refer to any way of visualising or conceiving of a structure or a mechanism that can account for observable phenomena. According to Friedman [12], a theory in its most basic form is a model. A theory illustrates how something works by showing its elements in relationship to one another. Some models show the elements in a dynamic relationship by describing process or action. Others, such as taxonomy, describe relationships without describing process or action. The dynamic demonstration of working elements in action as part of a structure or the demonstration of relationship is what distinguishes a model from a simple catalogue.

ASME [2] defines the field of design theory and methodology as '... an engineering discipline concerned with process, understanding and organised procedures for creating, restructuring and optimising artefacts and systems'. Rabins et al. [23] use design theory to mean systematic statements of principles and experientially verified relationships that explain the design process and provide the fundamental understanding necessary to create a useful methodology for design. The act of designing is certainly a natural phenomenon; so it seems reasonable that developing theories about the design process is a worthy goal [29]. Evbuomwan et al. [10] state that 'design models are the representations of philosophies or strategies proposed to show how design is and may be done. Often, models are sketched as flow diagrams, showing the iterative nature of designing through feedback links'. According to Ullman [29], design theories have one of the following foci: product, process and interactions between people and tools.

From the above literature, theory or model, in general, can be defined as a set of theorems, abstract principles, beliefs or propositions, to analyse or explain a wide range of observed facts or their relationships. We do not distinguish between a model and a theory of design. Note that the term 'model of design', as used in this chapter, means models explaining designing, and not models used for supporting designing. Designing involves multiple facets—product, process, people, tools, environment, micro- and macro-economy [3]. Designing happens in several stages, starting from an often abstract stage with less information and culminating in a detailed stage with more information. Therefore, a model or a theory of designing should be able to explain characteristics of one or more facets of designing, including relationships between the facets, at one or more stages of designing, including the transitions from one stage to another, of a design process. A model or theory of designing consists of:

- constructs and their definitions; and
- a set of propositions, expressed as descriptive relationships among the constructs, as statements about designing;

These propositions are meant to be used to describe or explain:

- various characteristics of the facets of designs and designing,
- relationships among the facets, or
- relationships among these and various characteristics of design success.

Observed facts, in the case of designing, can take a variety of forms, e.g. audio and video recordings of designing sessions, contents of diaries of designers, information about outcomes of designing (e.g. product and their descriptions), etc.

What are the purposes of a model or theory of designing? Design research aims to improve the chances of producing successful products by making designing more effective and efficient. To realise this aim, a two-stepped approach is proposed as follows by Blessing and Chakrabarti [3]: (a) formulate and validate an understanding of current designing and (b) develop and validate a support, founded on the understanding, in order to improve the current designing.

In designing, the characteristics of the facets can have positive or negative influence on the outcomes of designing. Apart from helping to describe phenomena associated with designing, a model or theory of designing should be used as a basis to identify the positive and negative characteristics influencing design.[1] Further, design models or theories can be used as a basis to improve the design process by enhancing its positive traits and suppressing its negative traits. A prescriptive support for designing, in the form of guidelines, methods, tools, etc., based on a model or a theory of designing, should help describe or explain how influences on a given design or designing should be changed in order to improve its strengths and reduce its weaknesses.

What are the criteria to qualify as model or theory of designing? According to Popper [22], the criterion for the scientific status of a theory is its falsifiability, refutability or testability. A design theory or model should satisfy the following criteria:

(a) A theory or model of design should contain a set of propositions to describe or explain some characteristics of (one or more facets of) designing (and design success);
(b) The propositions should be testable using empirical data, or using logical consistency with other theories or models, that are already validated.

[1] In our research, we use the term 'design' to mean both design (representing the product facet) and designing (representing the process facet).

15.3 Developing and Validating IMoD

The model described in this chapter has been developed in two stages. This section explains the first stage, i.e. development of IMoD. IMoD is developed with the intent of describing task clarification and conceptual design, and, for explaining how various characteristics of these stages relate to one another, e.g. how levels of abstraction at which design outcomes are developed relate to novelty of outcomes [26, 27]. The goal is to use this understanding as a basis for developing support to enhance design outcomes.

Activities are defined here as deeds of problem-finding and problem-solving. Outcomes are defined here as properties of designs at various levels of abstraction. Requirements are defined here as expressions of what designs should have at various levels of abstraction. Solutions are defined here as means to satisfy requirements. IMoD combines the views of activity, outcome, requirement and solution.

15.3.1 Research Methodology

The following research methodology has been used in the development of IMoD:

1. Survey literature on existing design models, approaches and theories, to determine to what extent the views of activity, outcome, and requirement-solution are considered, separately or together, in the literature.
2. Identify the constructs of the three views, based on the literature survey, and develop models of the views from the identified constructs. Develop an integrated model of designing by combining the individual models of the views.
3. Test the models of the individual views and the integration, to check the following: (a) whether or not all the constructs of the models can be used to describe the instances in designing and (b) whether or not all the instances in the designing can be described using the constructs of the models.

The following methodology has been used for validating IMoD:

1. Using existing protocol studies: Protocol studies of six design sessions from earlier research [5] are used to evaluate the proposed model. The research is undertaken before the proposed model is developed. Tables 15.1 and 15.2 show data about these protocol studies.
2. *Coding of the transcribed data*: The transcriptions of the verbal utterances in these designing sessions are coded to check the presence of activities (generate, evaluate, modify, select), outcomes (action, state-change, input, phenomenon, effect, organs, parts), requirements and solutions.

Table 15.1 Designers and teams [26, 27]

Teams	Designers	Education		Nature
		Bachelors	Masters	
T1	D11	Mechanical	Product design and engineering	Novice
	D12	Mechanical	Product design and engineering	Novice
	D13	Mechanical	Product design and engineering	Experienced
T2	D21	Mechanical	Product design and engineering	Novice
	D22	Mechanical	Product design and engineering	Novice
	D23	Architecture	Product design and engineering	Novice

Table 15.2 Design sessions [26, 27]

Problem	Task	Team	Methods
P1	To develop conceptual solutions for an efficient means of keeping the university campus free from dry leaves	T1	M1 [FA]
		T2	M3 [IDA]
		T2	M2 [ISQ]
P2	To develop conceptual solutions for a locking system that does not require any physical key or numbers to remember	T2	M1 [FA]
		T1	M2 [IDA]
		T2	M3 [ISQ]

FA Functional analysis, *ISQ* Innovation situation questionnaire, *IDA* Ideal design approach

3. *Analysis of the coded transcriptions*: The coded transcriptions are analysed in the following way:
   a. It is checked whether or not all the constructs can describe the instances in designing.
   b. It is checked whether or not all the instances in designing can be described using the constructs.

15.3.2 Activity Model: GEMS Model

From the literature on design models, theories and approaches, four activities—Generate, Evaluate, Modify and Select (GEMS)—are identified. The activity model—GEMS model (see Fig. 15.1)—is developed from the four constructs. An episode is defined as an event in designing that involves an exploration of an outcome. Generate is the activity that brings an outcome into an episode. Evaluate is the activity that judges the quality, importance, amount or value of an outcome in an episode. Modify is the activity that changes an outcome in an episode. Select is the activity that decides an outcome as acceptable or unacceptable in an episode.

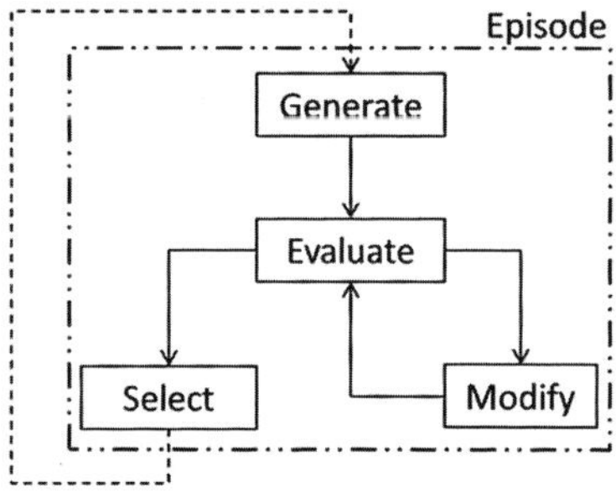

Fig. 15.1 GEMS activity model [26]

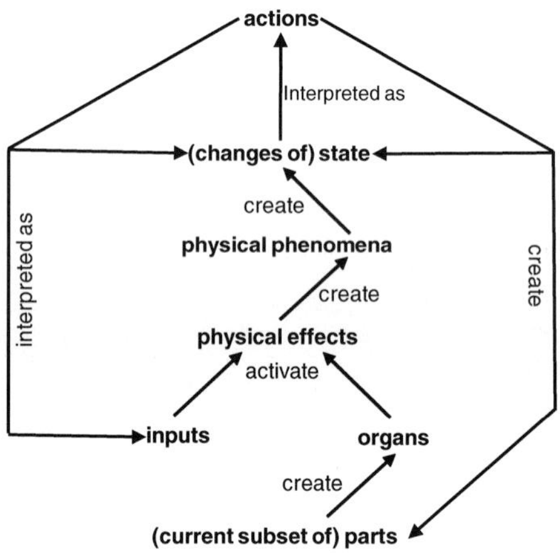

Fig. 15.2 SAPPhIRE model of causality [6]

15.3.3 Outcome Model: SAPPhIRE Model of Causality

The SAPPhIRE model of causality (see Fig. 15.2) [6] has been used and empirically validated as a model of outcomes [26]. The constructs are defined in [25] and are as follows: *Phenomenon* is an interaction between an entity and its surroundings. *State change* is a change in a property of an entity and its surroundings involved in an interaction. *Effect* is a principle of nature that governs an interaction. *Action* is an abstract description or high-level interpretation of an interaction. *Input* is a physical quantity, taking the form of material, energy, or information that comes from outside an entity's boundary and is essential for an interaction. *Organ* is a set of properties and conditions of an entity and its surroundings that is also required for an interaction. *Part* is a set of physical components and interfaces that constitutes an entity and its surroundings. *Entity* is defined as a subset of the universe under consideration, and is characterised by its boundary; *surroundings* are defined as all the other subsets of the universe; *interaction* is a communication between an entity and its surroundings, to reach equilibrium. The model of outcomes is based on the model of causality, and helps describe designing as starting at a high level of abstraction (e.g. actions) and culminating in a low level of abstraction (e.g. parts).

15.3.4 Requirement-Solution Model

A design process is initiated with the recognition of a need, leading to the establishment of requirements for the intended product [21]. Therefore, capturing requirements is essential and a central issue in design research [4]. Cooper [7] and Nidamarthi et al. [20] measure the success of a product in terms of how well it satisfied its requirements. The co-evolving model of requirement solution has been developed and empirically validated [26]. Requirement is defined as an expression of what a design should have at a level of abstraction. Solution is defined as a means to satisfy requirements. Associated information, which is added to the model during Stage 2 of the development but mentioned here for completeness of the view as it stands in its recent most form, is defined as information that is related to requirement or solution, but is neither a requirement nor a solution.

15.3.5 Integrated Model of Designing

IMoD is developed by combining the individual models of activity, outcome and requirement-solution [26]. The main proposition, according to this model, is the following: In task clarification and conceptual design, *GEMS activities are performed on SAPPhIRE outcomes, which co-evolve as requirements and solutions.*

15.3.6 Evaluation of IMoD

For the sake of brevity, the results of the evaluation of IMoD are not shown here, but the results are explained in detail in [26].

15.4 Extended-Integrated Model of Designing

In our attempts to use IMoD to describe the stages of designing beyond conceptual design, we felt that the model needed to be extended. Literature gave strong indications that system-environment view is essential for (a) explaining the entire design process, (b) explaining various characteristics of design such as requirement-solution co-evolution and (c) for obtaining greater clarity on how constructs and propositions developed in earlier research can be mapped against one another.

Therefore, IMoD is extended by combining the system-environment view with the existing views of activity, outcome and requirement-solution, with the intents as explained above. The following methodology is used for developing E-IMoD:

1. Survey literature of existing design models, approaches and theories, to determine to what extent the views of activity, outcome, requirement-solution and system-environment are considered, separately or together, in the literature.
2. Identify the constructs of the four views, based on the literature survey, and develop models of the views from the identified constructs, and integrate these views into E-IMoD.
3. Evaluate the models of the individual views and E-IMoD, to check the following: (a) if all the constructs of the models can be used to describe the instances in designing and (b) if all the instances in the designing can be described by the constructs of the models.

15.4.1 System-Environment view

Hall [13], in his systems engineering process, uses a system-environment view consisting of system, environment, sub-systems and objects. He considers initial and final environment in the systems engineering process, but environment is not considered in the processes in between. Deng et al. [8] consider environment as an explicit element in their Function-Behaviour-Working Environment-Structure model of designing. INCOSE [15] for the systems engineering process uses a system view with the following constructs: system, element or segment, subsystem, assembly, subassembly, components and parts. Hubka and Eder [14] use a system-environment view that consists of: system, sub-system, elements, components, environment and active environment.

The constructs of the system-environment view proposed by Ranjan et al. [25] are relationships, elements, sub-system, system and environment. In this view, both system and environment are potentially *evolvable* constructs in the process of designing. The constructs are defined as follows. A system is the overall product (the artefact to be transacted) being designed, at any level of abstraction. A subsystem (SS) is a subset of a system that can be further divided. An element (El) is a subset of a system or a sub-system, which cannot be further divided. An environment refers to all the subsets of the universe other than the system. Relationships are means of interlinking these constructs.

The constructs can be illustrated as follows. At the level of abstraction of parts, a ballpoint pen is a system made up of a refill (SS), body or cover (SS) and a cap (El). The refill is a sub-system consisting of elements such as nib, ink and ink reservoir. The environment for the ballpoint pen includes but is not limited to papers on which the user has to write and an agent that uses it to write. This example can also be extended to the action level, as follows. The action at the system level is 'to write on paper'; the actions at the sub-system level are 'to store and supply ink to write' and that of the sub-system body or cover along with the element cap is 'to minimise smudging due to ink leakage on agent or paper'. This system-environment view is adopted for extending IMoD, as it allows environment to be an explicit and evolvable construct.

15.4.2 Development of Extended-Integrated Model of Designing

From an extensive literature survey, it is found that the existing design models, approaches and theories have been based, implicitly or explicitly, on one or more of the above views: system-environment, activity, outcome and requirement-solution. However, these are not integrated into a single design model, theory or approach [25, 26]. Therefore, a model of designing that integrates all these views is not only novel, but also has the potential to serve as a platform with substantially more explanatory power that can be used as a basis for understanding and supporting various complex characteristics requiring consideration of multiple views. Examples of such complex characteristics include comparing and benchmarking design models, developing a detailed understanding of the different stages of designing, various forms of co-evolution that characterise designing, etc.

The views of system-environment, activity, outcome and requirement-solution, identified in the earlier sections, are based on their unique traits. The constructs in system-environment, requirement-solution and outcome views are all evolvable, i.e. outcomes in any of these views evolve during designing. The GEMS activity view consists of a generic set of activities as constructs. In [26], the compatibility among the three views: activity, outcome and requirement-solution, was empirically established. In [25], the major hypothesis is that there is compatibility between these three views and the system-environment view. The model is represented in a 4D space with the four views being the four co-ordinate dimensions of any event in designing (see Fig. 15.3). The hypothesis amounts to the following: any event of engineering designing can be represented using a combination of constructs from the four views, and each such event should require at least one construct from each of the four views for it to be represented.

This model is referred to as the E-IMoD, which combines the (constructs of the) views of activity, outcome, requirement-solution-information and system-environment. E-IMoD is intended to describe engineering design including but not limited to task clarification and conceptual design. The main proposition according to this model is the following: *GEMS activities are performed on SAPPhIRE or other outcomes, which evolve as requirements or solutions or associated information of relationships, element, sub-system, system or environment.*

15.4.3 Evaluation of E-IMoD

The following approach is used to empirically validate E-IMoD:

1. Using existing protocol studies: Protocols from six design sessions from [16] are used to evaluate the proposed model. The sessions were conducted before the proposed model is developed. Each designing session recorded in these

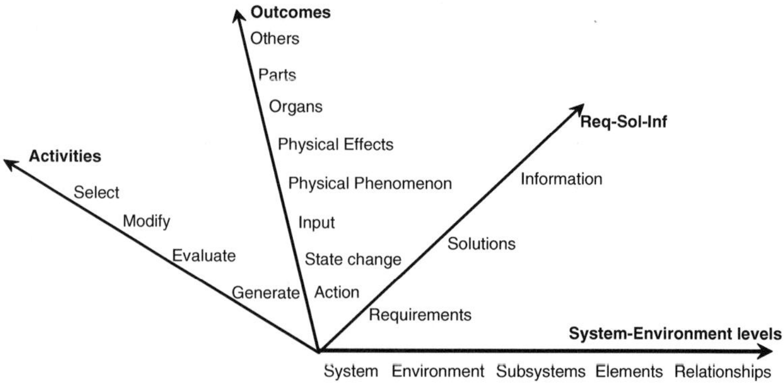

Fig. 15.3 E-IMoD

protocols involves an individual designer solving a design problem in laboratory conditions. The designers are trained in think-aloud protocol [11] and instructed to use this during designing. The problem briefs, details of the designers involved, videos and audio recordings, associated sketches and transcriptions (i.e. written descriptions of utterances, associated transactions and gestures) are available from earlier research. Further details of the protocol studies are provided in Tables 15.3 and 15.4.

2. *Coding of the transcribed data:* Transcribed data is coded for the constructs of activities (generate, evaluate, modify, select), outcomes (action, state-change, input, phenomenon, effect, organ, part), requirement-solution, and system-environment view (system, sub-system, element, relationship and environment). In the early stage of coding, it is found that the outcome view needed an extra construct—others—as current constructs do not address factors like cost, ergonomics, or safety, all of which became important as designing progressed beyond conceptual design. Also, a new construct 'associated information' is identified and incorporated into the requirement-solution view, since descriptions are identified that are found to be related to requirements or solutions, but as such are neither requirements nor solutions (Table 15.5).

The following instructions are given to designers:

"At the end of designing you are expected to provide drawings and a bill of materials and any other detail necessary for the production of the product. You may consider all the life cycle phases of the device/product for your design.

Life Cycle Stages

Every product/device passes through several stages in its life from birth to death. These are called product life cycle stages. The following are the main life cycle stages of a product: raw material, production, distribution, usage and after-use."

3. *Analysis of coded transcriptions:* The transcriptions are analysed in the following way:

Table 15.3 Problem brief given to designers

	Problem-briefs
P1	India has a large number of people with transferable jobs. They need to shift frequently from one place to other (every 1–2 years). And often face problems transferring present types of furniture, which are bulky and heavy. It is not economical for them to buy furniture and sell it before shifting to other place. This furniture occupies lot of space and this is an additional problem since they live in small houses. It takes more time to pack the furniture and it damages during transport if it is not packed properly
	Your task is to design a portfolio of furniture which will help in sleeping and storing things while taking the above problems mentioned. Setup time and effort on the part of users should be minimal
P2	Many modern executives find it difficult to spend spare time from their busy schedule to go to the gym for workout. On the other hand, they are often reluctant to spend money on expensive gymnasium for personal use. There is some equipment available for personal use, but it is expensive. There is no privacy in gymnasium. Current equipments occupy a lot of space and are usually not portable
	Your task is to design a product that will help in solving these problems. Users should be able to use it without any difficulty in setting up the equipment. It should be portable and should help in complete workout of the body
P3	India has a large number of people with transferable jobs. They need to shift frequently from one place to other (every 1–2 years). And often face problems transferring present types of furniture, which are bulky and heavy. It is not economical for them to buy furniture and sell it before shifting to other place. This furniture occupies lot of space and this is an additional problem since they live in small houses. It takes more time to pack the furniture and it damages during transport if it is not packed properly
	Your task is to design portfolio of furniture which will help to sit, write and dine while taking the above problems mentioned. Setup time and effort on the part of user should be minimal
P4	There are many problems associated with the increase in temperature during summer. A huge number of people die because of heat waves. A number of products such as umbrella, hat, etc. are available to help alleviate part of these problems; however, these are useful only for blocking the direct sunlight. These are not able to protect a person from high temperature and heat waves. Air conditioners are available and solve this problem; however they are expensive and work only in a fixed setting. No mobile and portable equipment is available that can be carried around while in transit. There is a need for a product that will help in maintaining body temperature within a comfortable range
	Your task is to design a product that will help in solving these problems. It should help the user in avoiding direct sunlight and maintaining body temperature. The user should be able to use it without any difficulty in setup and it should be portable

Table 15.4 Designing Sessions

Problems	Designers	Education background		Designer background		Field
		Bachelors	Masters	Experience	In years	
P1	D1	Mechanical	MDes	Industry designer	5	Transport
P2	D2	Mechanical	MDes	Industry designer	2	Furniture
P3	D3	Mechanical	MDes	Industry designer	2	Automobile
P1	D4	B.Arch	MDes	Student designer	Novice	Novice
P2	D5	Mechanical	MDes	Student designer	Novice	Novice
P4	D6	Mechanical	MDes	Student designer	Novice	Novice

Table 15.5 Codes for different constructs

Views	Constructs	Codes
Activity	Generate	G
	Evaluate	E
	Modify	M
	Select	S
Outcome	Action	a
	State-change	sc
	Input	i
	Physical phenomenon	ph
	Physical effect	e
	oRgan	r
	Part	p
	Others	o
Req-sol	Requirement	Req
	Solution	Sol
	Associated information	Info
System-environment	System	Sy
	Subsystem	SS
	Elements	El
	Relationship	Rel
	Environment	En

i. It is checked if all the proposed constructs can describe the instances in designing in the transcriptions.
ii. It is checked if all the instances of designing in transcriptions can be described using the constructs.

4. *Comparison of results with earlier research*: The designing sessions used in the research on IMoD [26] were focused on task clarification and conceptual design only. These results are compared with the new, E-IMoD results, which are intended to describe/explain designing beyond the conceptual stage.

15.4.4 Evaluation of E-IMoD

The following results are obtained in the analysis of the coded transcriptions in Ranjan [24]:

1. the individual view of activity,
2. the individual view of outcome,
3. the individual view of requirement-solution,
4. the combined views of activity and outcome,
5. the combined views of activity and requirement-solution,
6. the combined views of outcome and requirement-solution,

7. the individual view of system-environment,
8. the combined views of system-environment and activity,
9. the combined views of system-environment and outcome, and
10. the combined views of system-environment and requirement-solution.

While the first six results re-evaluate IMoD [GEMS of SAPPhIRE as req-sol] using the present set of protocol studies; the other four results are related to system-environment view, and evaluates the integration of system-environment view to IMoD. Thus, together the above results evaluate E-IMoD. Due to lack of space, only the last three results are outlined in this chapter; see Srinivasan and Chakrabarti [26] for the first six results, and Ranjan [24] for the remaining results (all of which include the system-environment view).

15.4.5 Evaluation of Combined System-Environment and Activity view

Here, we discuss the percentage of the constructs of the system-environment view over the activity view, for all the six designing sessions.

The following observations are made on the percentage distribution of 'Generate' activity in the system-environment view for all the six designing sessions:

1. A high percentage of descriptions at the system and element levels is observed in all the designing sessions.
2. A low percentage of descriptions at the environment, sub-system and relationships level is observed in all the designing sessions.

The following observations are made on the percentage distribution of 'evaluate' activity in the system-environment view for all the six designing sessions:

1. A high percentage of descriptions at the system level is observed in all the designing sessions.
2. A low percentage of descriptions at the environment, sub-system, element and relationships level is observed in all the designing sessions.

The following observations are made on the percentage distribution of 'modification' activity in the system-environment view for all the six designing sessions:

1. A high percentage of element-level descriptions is observed in all designing sessions except D1-P1.
2. A high percentage of descriptions at the sub-system level is observed in D2-P2, D3-P3 and D5-P2, while such descriptions are absent in the other sessions.
3. A high percentage of system-level descriptions is observed in three of the six designing sessions [D1-P1, D4-P1, D6-P4].
4. No modifications to descriptions at the environment-level are observed in any of the designing sessions.

The following observations are made on the percentage distribution of 'selection' activity in the system-environment view, for all the six designing sessions:

1. A high percentage of system-level descriptions is observed in all the designing sessions.
2. A significant percentage of element-level descriptions is observed in all the designing sessions.
3. A very low frequency of environment and sub-system-level descriptions is observed in all the designing sessions except D6-P4. Relationship-level descriptions are observed only in three [D1-P1, D2-P2, D4-P1] of the designing sessions.

The overall observation from these findings is that, all activities (except modification on environment) are performed at all the levels of the system-environment view. The process is dominated by GEMS of system-level descriptions, followed by GEMS in the element level. This may be because of the relatively low complexity of the systems designed that required relatively few sub-systems to be designed. No explicit example of modification of environment is found in the current protocols. However, the designer in D1-P1 generates an environment level solution by proposing a slot in walls. It is not inconceivable that for implementing this, the designer may have to evolve the characteristics of the slot, thereby introducing modifications to the environment. The overall inference is, therefore, *(almost) all the activities are performed at every level of the system environment view.*

15.4.6 Evaluation of Combined System-Environment and Requirement-Solution views

This section focuses on the distribution of the number of descriptions for constructs of the system-environment view over requirements, solutions and associated information. The distribution of the number of descriptions for constructs of the system-environment view over requirements shows the following:

1. A high percentage of system-level descriptions is noticed in all the designing sessions.
2. A significant percentage of environment level is noticed in the designing sessions D1-P1, D2-P2, D4-P1 and D5-P2.
3. A low percentage of sub-system, elements and relationships is observed in all the designing sessions except D6-P4.

The distribution of the number of descriptions for constructs of the system-environment view over solutions shows the following:

1. A high percentage of system-level and element-level descriptions is noticed in all the designing sessions.

2. A significant percentage of sub-system and relationships-level descriptions is noticed in all the designing sessions.
3. A very low percentage of environment-level descriptions is observed in all the designing sessions except D6-P4.

For these examples, the activity is dominated by system-level requirements and solutions. While a significant percentage of requirements are produced (which could be due to the nature of the problem brief), these are used more as constraints on the system, with relatively little effect on changing the environment. The overall inference from the above section about the validation of the model is that, *requirements and solutions exist at all system-environment levels.*

15.4.7 Evaluation of Combined System-Environment and Outcome views

This section discusses the distribution of the number of descriptions for constructs of the system-environment view over SAPPhIRE outcomes. The following observations are made:

1. A very high frequency of part-level descriptions is noticed under all the system, environment-levels.
2. A very high frequency of organ-level description is noticed at all the system-environment levels.
3. A very high frequency of action-level descriptions is noticed under all the system, environment, sub-system and element-levels.
4. A low frequency of state change, input, effect and phenomenon-level descriptions is noticed at all the system levels.
5. A high frequency of other-level descriptions is noticed at all system-environment levels.

For these examples, the outcome levels of action, organ and part dominated proceedings at all levels of abstraction. The overall inference relevant for validation of the model is that, *all constructs exist at all outcome levels, for all system-environment levels.*

What is the implication of these findings on supporting improvement in design? As found in earlier work by Srinivasan and Chakrabarti [26], the exploration shows a "bathtub" profile in the plot along the x-axis of the number of outcomes at various SAPPhIRE levels; exploration at the initial action level and the last organ and part levels are high, while exploration is low in the middle. This is counter-intuitive, as abstract outcomes are expected to allow germination of a larger number of less abstract outcomes; this indicates scope for better support for exploring designs at the levels where the frequency of exploration is found wanting.

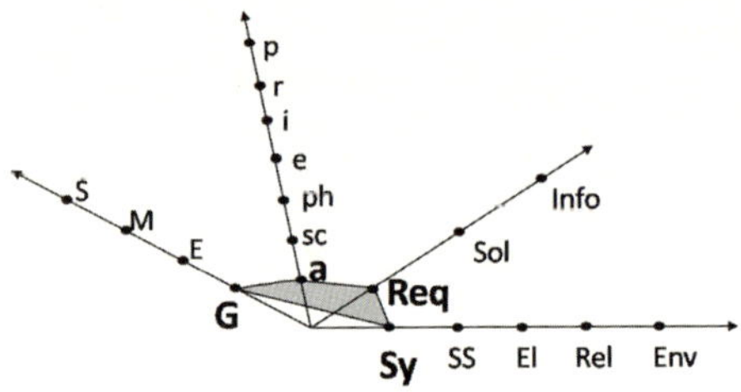

Fig. 15.4 *Generate action* level *requirement* at *system* level

15.4.8 Illustrative Examples

In this section, examples are used to illustrate how the proposed model E-IMoD can be used to describe designing. Where used, the figures following the description of each event (utterances in double quotes) provide a representation of the event, using the views of E-IMoD.

Example 1: This example is taken from designing session D5-P2. "Development of personal use gymnasium equipment" is the core of the problem brief for P2 (see Table 15.3 for details) given to designer D5.

1. Based on this problem brief, D5 *generates* an *action* level *requirement* at *system* level as follows: "body building for which basic exercises are weight-lifting,…" (Fig. 15.4).
2. For this requirement, D5 *generates* a *phenomenon* (Fig. 15.5a) and *part* (Fig. 15.5b) level *solution* at *system* level: 'we can think of magnetic springs or magnetic material which can create some kind of force or which act as some kind of spring'. (Fig. 15.5).
3. In the next step, D5 *generates effect* level *solution* at *element* level (Fig. 15.6a) and *input* level *solution* at *system* level (Fig. 15.6b): 'If we incorporate that kind of thing (in magnetic springs) there we have to control the complete current which is passing through that depends on what force you should apply and what are various spring properties or weight properties' (Fig. 15.6). Here one can notice that D5 also plans to *generate* the *organ* level *solution/s* at e*lement* level: 'various spring properties or weight properties' (see Fig. 15.7).
4. D5 *evaluates* the *solution* against another at *system* level (Fig. 15.8a) (for the purpose of weightlifting): 'we can directly control with some electrical or by changing these some electrical basic properties of that given instrument. It is very easy when compared to mechanical springs. So we have to work on these two first of all; but still we are not aware of that kind of material. Whether that kind of material is existing or not right now. It is not possible to go for study and so we fix upon mechanical springs'. As can be seen from the last line above, he *selects* 'mechanical springs' of the *part*-level *solutions* at *system* level (Fig. 15.8b).

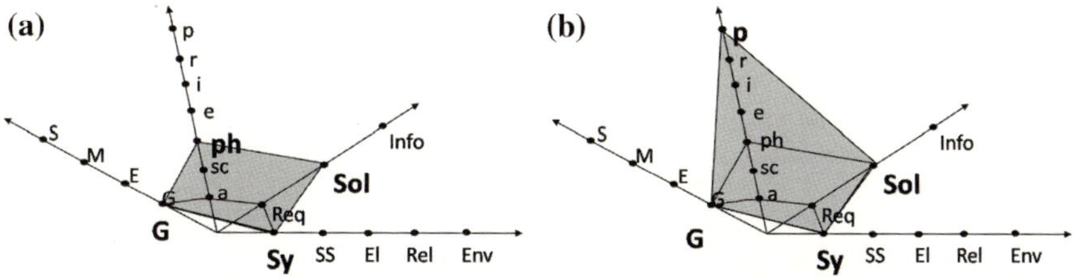

Fig. 15.5 **a** *Generate phenomenon* and **b** *part* level *solution* at *system* level

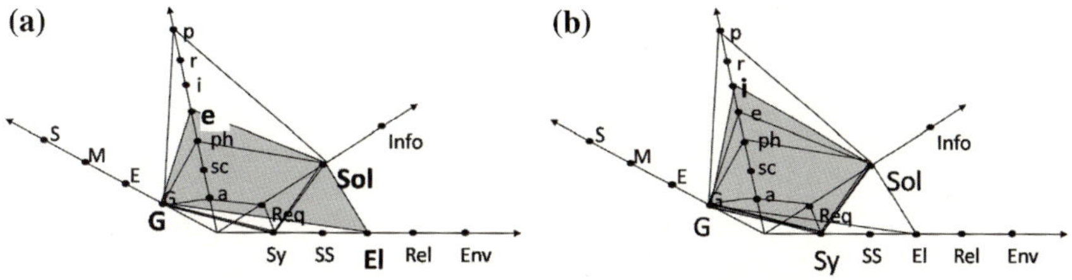

Fig. 15.6 **a** *Generates effect* level *solution* at *element* level and **b** *input* level *solution* at *system* level

Fig. 15.7 *Generate organ* level *solution* at *element* level

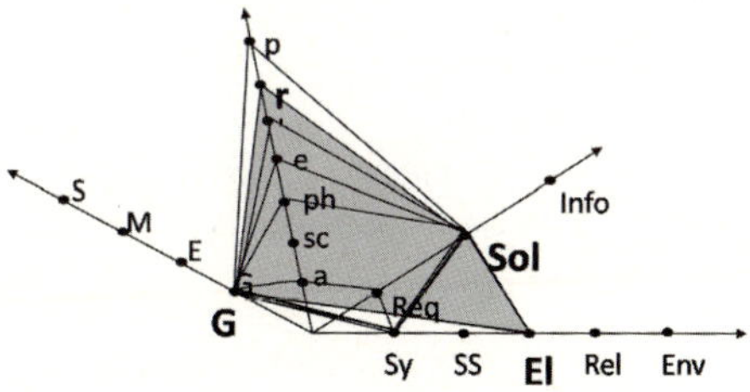

Example 2: This example is from designing session D1-P1.

1. For the problem brief for P1, the designer at an early stage *generates* an *action*-level solution: '…we have seen somewhere the furniture is already fixed; you don't need to bring any furniture from other place; you hire that house all the furniture is fixed; now you keep all your bedding material storing material; you don't need to buy a new furniture or something like that; already it is there and you can use that for your daily purposes'.
2. The researcher clarifies that it is not a popular idea in India. This clarification is received by D1 as follows: 'culture is not there, for the existing culture; Cultural Part, Culture and Use—In India, already fixed furniture is not popular, that

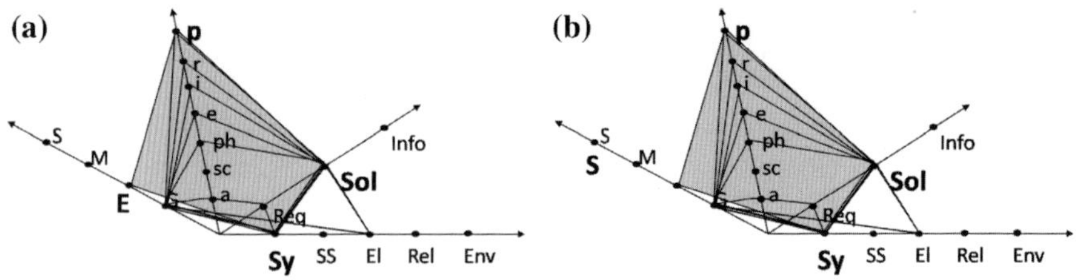

Fig. 15.8 **a** *Evaluate* and **b** *select part*-level *solution* at *system* level

means, people will like to carry their furniture'. This is classified as *other* level outcome which is *information* at *environment* level.

Example 3: This example is from designing session D6-P4.

1. In D6-P4 session, after developing some concepts designer D6 *selects* a concept that he named 'Capsule' and starts detailing: 'concept is done we will detail it so. Detail design: so now we are detailing (and starts drawing and dimensioning and naming of parts)'.
2. During this episode, D6 generates a *part* level *solutions* at *sub-system* and *element* levels in the sequence as follows:

    a. '…this is the area where we have motor, fan, plus water tank';
    b. '…compartment made of fabric netlon fabric';
    c. '…base is sturdy…';
    d. drawing–adding to sketch—'…wheels which will be castor wheels of 15 cm diameter'. D6 than gives overall dimensions: 'This is of 2 ft (adding in the detail design sketch to the width); and this is 7 ft (adding height in the detail design sketch)'.

3. At this stage D6 generates another *part*-level *solution*: 'telescopic pillars made of Stainless Steel'. D6 explains that the telescopic pillars are to change the height of the 'capsule' from 'fully open position' and "fully closed position (for which D6 draws a new sketch)". This is a *state change* level *solution* at *system* level; D6 *generates* this *state change* level *solution* to satisfy the compactness *requirement* in the problem-brief of P4.
4. While *generating* the sketch for the 'fully closed position of capsule', D6 makes a series of modifications as shown in Fig. 15.9a–d. D6 utters the following words while drawing the sketch in Fig. 15.9a: 'we have a pulley connected'. After this *modifies* the location of the 'pulley' as in Fig. 15.9b and the location of the 'castor wheel' as in Fig. 15.9c. Designer D6 then reveals the reasons for the addition of pulley: 'ease of use' and 'portability'. We can also observe that D6 defines the *relationship* between 'pulley' and the 'capsule', in no detail other than that the two are connected together.

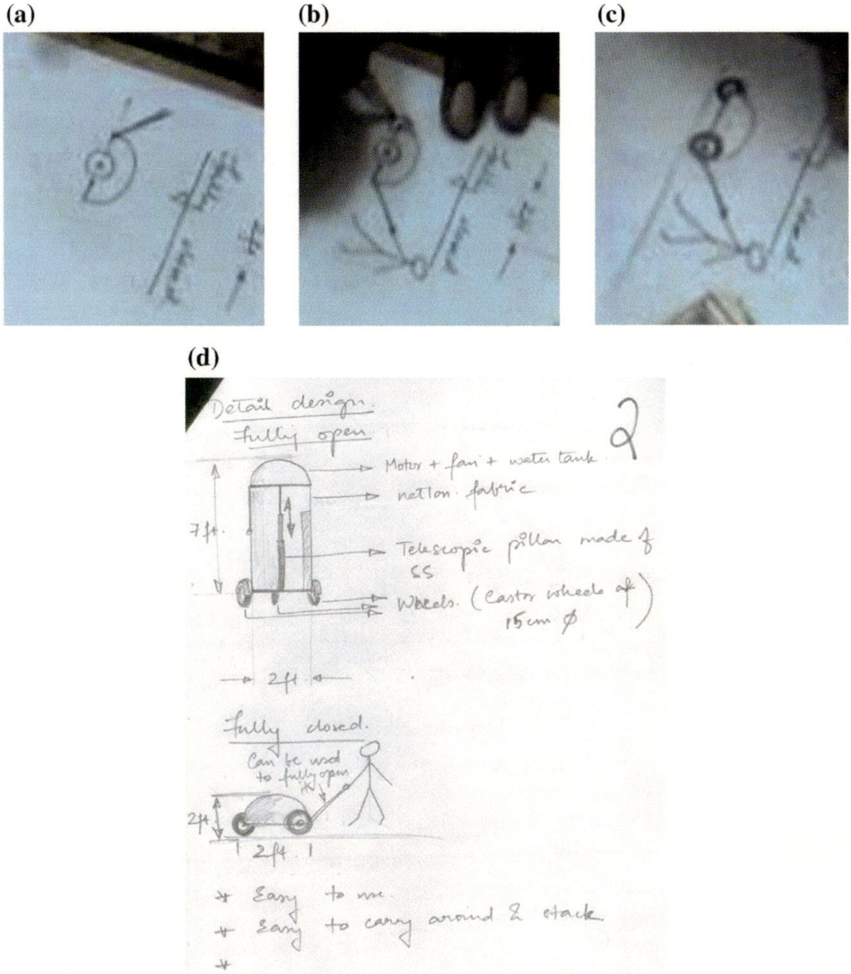

Figs. 15.9 a–d

The activity modification is usually observed during the sketching or drawing, and because of this, the reasons for the modifications made are usually not captured unless the designer makes them verbally explicit.

15.4.9 Summary

All the constructs of the proposed model of designing are found to be present in the designing sessions (see [24, 26] for more details). Most of the instances of the designing sessions can be coded using the constructs of the proposed model.

However, in the outcomes view, 11 % of the instances could not be coded using the constructs of SAPPhIRE; these were related to issues of manufacturing, cost and so on, and were coded as 'others'.

The findings are summarised as follows [24]:

1. Activities are performed on the outcomes.
2. Outcomes evolve as requirements, solutions and associated information.
3. All the activities are performed on both requirements and solutions.
4. Almost all the activities are performed at all the levels of system-environment.
5. Requirements and solutions exist at all the levels of system-environment.
6. All the outcomes are present at all the levels of system-environment.

15.5 Conclusions

In this research, E-IMoD is proposed. The model combines four views:

- an outcome view, which can explain the working of a given system with a common terminology developed; the outcome view consists of state change, action, part, phenomenon, input, organ, effect and others,
- a requirement-solution view consisting of requirements, solution and associated-information,
- an activity view consisting of: generate, evaluate, modify and select, and
- a system-environment view consisting of system, sub-system, elements, relationships, and environment.

Using protocol data from design sessions, the model is evaluated as follows:

- Each construct of the above views is empirically validated in terms of its ability to describe designing. All design events were possible to be explained using the above constructs.
- The co-existence of the above views in a design event is empirically validated by showing that:
  - every activity is performed on every design outcome,
  - every activity is performed on requirement, solution and associated-information,
  - every activity is performed on every construct of system-environment view,
  - every outcome exists at requirement-solution-associated information view,
  - every outcome exists at every construct of system-environment view and
  - requirement, solution and associated information exist at every level of the system-environment view.

From the empirical observations of six designing sessions in [26] and six design protocols in [24] (which are also presented in this chapter), the proposition that *GEMS activities are performed on SAPPhIRE outcomes, which evolve as requirements or solutions* is satisfactorily corroborated using another set of empirical studies. Also, within the constraints of the six design protocols in [24] and presented in this chapter according to E-IMoD, *GEMS activities are performed on SAPPhIRE or other outcomes, which evolve as requirements, solutions or associated information of element, relationship, sub-system, system or environment* is validated.

The above statements provide a concise description of the main propositions in the models of designing proposed in this work. Popper [22] uses testability as the criterion for assessing whether or not a theory is scientific. All the statements made in the proposed model of designing are verifiable or testable, as demonstrated in the chapter using an empirical approach to validation. It is found through this empirical evaluation that the model is capable of explaining almost all the events in the designing sessions. Thus, the proposed model consists of a set of views and the well-defined constructs, capable of explaining designing reasonably well. However, the number of design sessions is limited, and more studies need to be conducted to have a greater degree of confidence in the model. Further, the construct 'other' provides an estimate of what cannot be explained by the remaining constructs in the outcome view, and opens up possibility for further improvement of the model by providing more distinct constructs to account for what is now categorised under this construct.

In a Lakatosian sense, the development of IMoD and E-IMoD provides an interesting illustration of how scientific theories follow the general pattern of adjusting and extending a theory, when entities are encountered that either contradict, or cannot be explained by the theory [17].

References

1. Anderson (1964) The discovery of the electron: the development of the atomic concept of electricity. Commission on College Physics, Van Nostrand, Princeton, NewJersey
2. ASME (1986) Goals and priorities for research in engineering design. ASME, New York
3. Blessing L, Chakrabarti A (2009) DRM a design research methodology. Springer, London
4. Chakrabarti A (1994) Requirements identification: a central issue in design research. In: Proceedings of the international east–west conference on IT in design, Russia, pp 108–117
5. Chakrabarti A (2003) Towards a measure for assessing creative influences of a creativity technique. In: Proceedings of the international conference on engineering design ICED03, Stockholm, Sweden
6. Chakrabarti A, Sarkar P, Leelavathamma B, Nataraju BS (2005) A functional representation for aiding biomimetic and artificial inspiration of new ideas. AI EDAM 19(2):113–132
7. Cooper R (1991) New products: what distinguishes the winners? Res Technol Manag 33(6):27–31
8. Deng YM, Tor SB, Britton GA (2000) Abstracting and exploring functional design information for conceptual mechanical product design. In: Engineering with computers, vol 16. Springer, London, p 36–52
9. Dictionary of Ideas (1994) Brockhampton Press, London
10. Evbuomwan NFO, Sivaloganathan S, Jebb A (1996) A survey of design philosophies, models, methods and systems. Proc Inst Mech Eng Part B: J Eng Manuf 210(4):301–320

11. Eckersley M (1988) The form of design processes: a protocol analysis study. Des Stud 9(2):88–94
12. Friedman K (2003) Theory construction in design research: criteria: approaches, and methods. Des Stud 24:507–522
13. Hall AD (1962) A methodology for systems engineering, 6th edn. Van Nostrand, Princeton
14. Hubka V, Eder WE (1988) Theory of technical systems: a total concept theory for engineering design. Berlin and New York, Springer-Verlag
15. INCOSE (2004) Systems engineering handbook—a "What To" guide for all systems engineering practitioners. INCOSE-TP-2003-016-02 Version 2a. Released by Technical Board, International Council on Systems Engineering
16. Kota S, Chakrabarti A (2009) Understanding the needs of designers for developing environmentally friendly products. In: Chakrabarti A (ed) Research into design—supporting multiple facets of product development, p 128-135
17. Lakatos I (1976) Proofs and refutations. Cambridge University Press, Cambridge. ISBN 0-521-29038-4 & 978-0-521-29038-8
18. Mautner T (1996) A dictionary of philosophy. Blackwell, Oxford
19. Merriam-Webster's Collegiate Dictionary (1990) 9th edn. Merriam-Webster Inc, Springfield, MA
20. Nidamarthi S, Chakrabarti A, Bligh T (1997) The significance of co-evolving requirements and solutions in the design process In: Proceedings of the international confer-ence in engineering design (ICED97), Tampere, pp 227–230
21. Pahl G, Beitz W (1996) Engineering design: a systematic approach. Springer, London
22. Popper KR (1968) Conjectures and refutations. Harper & Row, New York
23. Rabins M, Ardayfio D, Fenves S, Seireg A, Nadler G, Richardson H, Clark H (1986) Design theory and methodology—a new discipline. Mech Eng 108(8):23–27
24. Ranjan BSC (2012) An extended integrated model of designing. M.Sc. Engineering thesis, Indian Institute of Science, Bangalore
25. Ranjan BSC, Srinivasan V, Chakrabarti A (2012) An extended, integrated model of designing. In: Horváth I, Albers A, Behrendt M, Rusák Z (eds) Proceedings of TMCE, Karlsruhe, 7–11 May 2012
26. Srinivasan V, Chakrabarti A (2010) An integrated model of designing. In: Goel AK, de Silva Garza AG (eds) Special issue on knowledge based design. ASME J Inf Sci Eng (JCISE). doi: 10.1115/1.3467011
27. Srinivasan V, Chakrabarti A (2010) Investigating novelty–outcome relationships in engineering design. In: Maher ML, Bonnardel N, Kim YS Special issue on creativity: simulation, stimulation, and studies. AI EDAM 24(2):161–178
28. Sutherland JW (1975) Systems: analysis, administration and architecture. Van Nostrand, New York
29. Ullman DG (1991) The status of design theory research in the United States. Des Stud 12(4):204-208. ISSN 0142-694X, doi: 10.1016/0142-694X(91)90032-R
30. Zavbi R, Duhovnik J (2001) Analysis of conceptual design chains for the unknown input/known output pattern. In: Proceedings of the international conference on engineering design ICED01, UK

Chapter 16
Modelling Products and Product Development Based on Characteristics and Properties

Christian Weber

16.1 Introduction

In the last decades, considerable effort has been invested into theories and models of technical products and product development/design processes (Design Theory and Methodology, DTM). Developing/designing products is a very complex and diversified activity that is still not fully understood.

DTM became an independent topic of research after World War II. As shown by Heymann [31], the overall task was (and is) to explore how much of designing is art (i.e. based on intuition which supposedly could not be taught and trained) and how much can be systemised, maybe even automated, based on scientific findings and concepts which make the activity teachable and trainable.

Several approaches to DTM have been published and discussed, some of them considered incompatible with one another. In the last 10–20 years, a wealth of new ideas have come up, sparked off by limitations of the existing approaches, by more complex (and increasingly heterogeneous) products, by new tools (e.g. computer support in development processes) and new procedures (e.g. globalised work distribution).

Therefore, the question of what is an appropriate model or theory of designs and designing has to be discussed and answered from time to time again. Based on an earlier contribution [73] the answer could be:

- A model or theory of designs and designing, like any scientific statement or theory, must explain and predict observations in its field. Since Popper [49] published his famous 'Logik der Forschung' (re-written English version Popper [50]) we want a model or theory falsifiable rather than verifiable.

C. Weber (✉)
Institute of Design and Precision Engineering, Engineering Design Group, Technische Universität Ilmenau, PO Box 10 05 65 98694 Ilmenau, Germany
e-mail: christian.weber@tu-ilmenau.de
URL: http://www.tu-ilmenau.de/konstruktionstechnik/

A. Chakrabarti and L. T. M. Blessing (eds.), *An Anthology of Theories and Models of Design*, DOI: 10.1007/978-1-4471-6338-1_16,

- A model or theory of designs and designing should have two sides: collecting and systematising knowledge about 'what is' (descriptive part) as well as collecting and systematising knowledge about actions and skills that can change the present state into another, previously not existing state (prescriptive).

It should be noted that this approach is not in contradiction with current concepts of 'science'—even if we can observe a strong bias towards the 'descriptive' or 'purely analytical' part in other disciplines. It should also be noted that the requirement of 'falsifiability' is often problematic in the prescriptive part of a model or theory.

- Any science deals with certain objects in its field. In the case of designs and designing, there are two different 'objects' to be considered: the designs (as artefacts) and the designing (as a rationally captured process to create artefacts).
- A model or theory of designs and designing is 'situated' in the sense that external influences (knowledge in other fields, society, markets, new technologies, time, ...) have to be considered as they evolve, resulting in modified or new models and theories.
- There may be different 'stakeholders' who pose requirements (maybe: 'demands') on models and theories of designs and designing. In Weber and Birkhofer [73] four groups of 'stakeholders' were identified and their requirements/demands investigated: scientists, designers in practice, students (including PhD students) and tool/software developers.
- We may state that an 'appropriate model or theory of designs and designing', beyond the usual criteria of a descriptive science (e.g. truth, completeness, level of detail), has to meet criteria like 'usefulness' (for different stakeholders!) and 'timeliness'.
- The author's conviction is: Developing/designing products is such a complex process that not one model alone can explain every aspect; several models may exist in parallel. However, an integrating framework would be beneficial.

This article presents an approach that has been developed by the author and his team during the last ca. 12 years. It comprises two parts:

- "Characteristics-Properties Modelling" (CPM) as the *product* modelling side.
- "Property-Driven Development/Design" (PDD) explaining the *process* of product development/design.

CPM/PDD does not claim to be a new or alternative design methodology; its concern is to provide a framework that can integrate many existing approaches and to deliver some explanations of phenomena in product development/design that have been insufficiently understood so far.

16.2 Goals of the CPM/PDD Approach

Within the space of this article, it is impossible to present all existing approaches in DTM. In accordance with the editors of this book, a brief overview is given at the end of this article as a separate appendix.

As already stated before, the overall goal of the CPM/PDD is to provide a framework that can integrate many of the existing approaches and to deliver some additional explanations of phenomena in product development/design. In more detail, sub-goals are:

- CPM/PDD shall build upon existing findings and knowledge in DTM. Many existing models, methods and tools shall be integrated. This includes seemingly incompatible views such as the concepts of the 'European/German-speaking schools' (e.g. represented by Pahl and Beitz [47] or VDI 2221 [63]) and Suh's Axiomatic Design Theory [57, 58].
- CPM/PDD shall bring DTM closer to the way practitioners think and proceed in product development/design. This has two aspects:
  - Many of the 'traditional' approaches of DTM concentrate on original design (new product development). With view to engineering practice, development/design processes based on existing solutions (variant and adaptive design) must be addressed and explained with equal intensity.
  - If possible, differences between development/design processes in different branches of industry and/or different companies shall be explained.
- CPM/PDD shall redefine the role of computer methods, tools and architectures in product development/design, based on a more solid scientific foundation. Hints for the further development of software for product modelling and process support should be derived.

16.3 Modelling Products and Product Development (Core of CPM/PDD)

16.3.1 Fundamentals

The CPM/PDD approach is based on the distinction between the characteristics (in German: *Merkmale*) and properties (*Eigenschaften*) of a product:

- ***Characteristics*** (***C_i***) are made up of the parts structure, shape, dimensions, materials and surfaces of a product (*Struktur und Gestalt*, *Beschaffenheit*). They can be directly influenced or determined by the development engineer/designer.
- ***Properties*** (***P_j***) describe the product's behaviour (*Verhalten*), e.g. function, weight, safety and reliability, aesthetic properties, but also things like manufacturability, assemblability, testability, environmental friendliness and cost. They can *not* be directly influenced by the developer/designer.

The characteristics are very similar to Hubka and Eder's [35] 'internal properties' and what Suh [57, 58] calls 'design parameters'. The properties, as introduced here, are related to the 'external properties', as defined by Hubka and Eder [35], and to the 'functional requirements', according to Suh [57, 58]. Recently, Birkhofer and Wäldele [8] proposed alternative terms that focus more strongly on what can and cannot be directly influenced: What is 'characteristics' here is called 'independent properties'; 'properties', as used here, becomes 'dependent properties'. It may also be noted that the property concept of Hubka and Eder [35] was considerably extended in Eder and Hosnedl [16, 17].

The concept of properties, as used here, is also related to the 'affordance' approach [40, 41].

For reasons not to be discussed here, the author prefers to use Andreasen's [4] nomenclature 'characteristics/properties'.

Characteristics and properties are two different concepts for describing products and their behaviour, respectively. Similar concepts have been used in DTM for a long time. The only new aspect of CPM/PDD is that this duality is put into the *centre* of modelling products and product development/design processes.

To handle characteristics and properties—literally thousands of them in complex products—and to keep track of them in the development process they have to be structured. Figure 16.1 shows the basic concept, as discussed in CPM/PDD:

- On the left, a proposition for the (hierarchical) structuring of characteristics is given, following the parts' structure of a product. It complies with standard practice, and links considerations to the data structures of CAx-systems.
- On the right, a first proposition for the structuring properties is presented, based on life-cycle criteria, and reflecting frequently discussed issues in product development/design.

While in Fig. 16.1 the characteristics are structured quite deeply, on the properties side only top-level headings are displayed. Of course, also the properties have to be decomposed further in order to be usable in product development/design: Which (measurable) parameters define function, strength, safety, etc.? However, the author is convinced that any further structuring of properties as well as an assessment of their importance are always specific to individual industries (product classes), often even specific to individual companies and, in addition, also time-dependent. As discussed in Weber [71], listing the relevant properties and decomposing them into (measurable) sub-properties is already the first step of deriving an application-specific development/design methodology.

This statement stands quite opposed to most approaches to DTM. Two consequences, one negative and one positive, arise:

- Negative: There is no hope of being able to find one generic procedural model (methodology) for the development/design of any product.
- Positive: Here is the explanation of what makes the differences between development/design processes in different branches of industry and companies—in each case another set of properties is relevant and controlling the process.

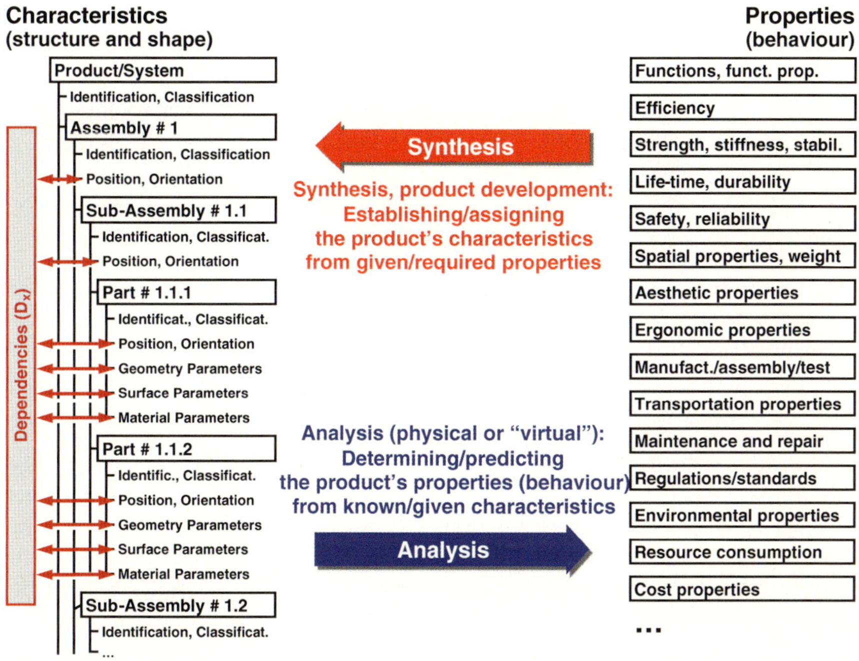

Fig. 16.1 Characteristics and properties, and their two main relationships

Another difference between the characteristics side and the properties side according to the model in Fig. 16.1 may be noted:

- For characteristics, there is usually no problem to check and maintain 'completeness'. We are used to hierarchical structures, and we can decompose the parts structure down to the last detail (e.g. faces, edges, vertices of the CAD-model).
- For properties, there is no way to achieve completeness. In this respect we can only go for the criterion 'relevance'. The reason is that every product always has more properties (behavioural parameters) than the ones considered in the development/design process. These considerations form an interesting link to the discussions about affordance-based design [40, 41].

On the characteristics (left) side of Fig. 16.1, an additional block is drawn that represents dependencies (D_x) between characteristics. Development engineers and designers are familiar with these types of dependencies, e.g. geometric or spatial dependencies, as well as those concerning fits, surface and material pairings, even conditions of existence. It should be noted that the existence of these dependencies is a great advantage: Each one of them reduces the number of characteristics (and, thus, the number of design degrees of freedom) that the product developer/designer has to take care of by one. Without the (implicit or explicit) use of these dependencies finalising a complex design would probably be impossible, especially in the detailing stage of the process where literally thousands of characteristics (or 'design parameters') have to be assigned.

Today, geometric and spatial dependencies can be captured and administered by parametric CAD- or PDM-systems. One hint for the further development of CAx-systems is that *all* relevant types of dependencies should be covered.

Finally, Fig. 16.1 shows the two main relationships between characteristics and properties:

- *Analysis:* Based on known/given characteristics (structural parameters, design parameters) of a product, its properties are determined (and therefore, its behaviour), or—if the product does not yet exist—predicted.
- *Synthesis:* Based on given, i.e. required, properties, the product's characteristics are established and appropriate values assigned.

Synthesis is the main task of product development/design. The requirements list is, in principle, a list of required properties; the task of the development engineer/designer is to find appropriate solutions, i.e. an appropriate set of characteristics that meet the requirements to the customer's or user's satisfaction. In many practical cases the requirements already contain certain characteristics—but this means that some partial solutions (or solution elements, solution patterns, see Sect. 16.3.2.3) are set from the beginning.

As will be discussed further down, during product development/design also many analysis operations are required. They serve the purpose of checking whether the as-is properties of the solution actually meet the required properties.

16.3.2 Modelling Products (CPM)

In the CPM/PDD approach, analysis and synthesis as the two main relationships between characteristics and properties, are now modelled in more detail. In these considerations the following symbols are used:

$\mathbf{C_i}$	Characteristics (*Merkmale*)
$\mathbf{P_j}$	Properties (*Eigenschaften*)
$\mathbf{PR_j}$	Required properties
$\mathbf{D_x}$	Dependencies (constraints) between characteristics
$\mathbf{R_j}$	Relations between characteristics and properties; for analysis operations
$\mathbf{R_j^{-1}}$	Relations betw. properties and characteristics; for synthesis operations ('inverse rel.')
$\mathbf{EC_j}$	External conditions

As a simplification, from now on both characteristics and properties are displayed as simple lists (no hierarchical or other structure). These lists of

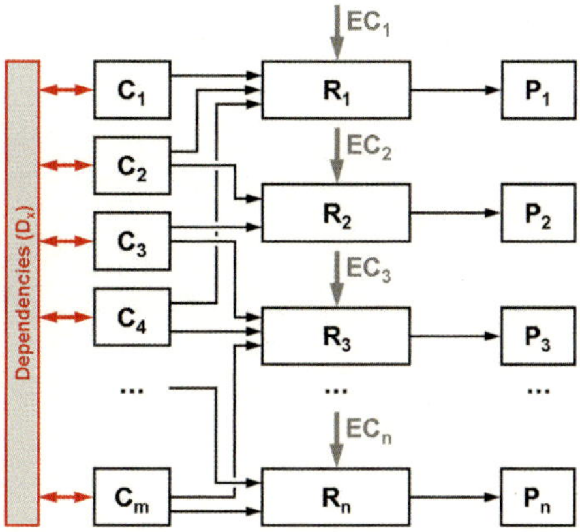

Fig. 16.2 Basic model of analysis

characteristics and properties, respectively, could also be noted as vectors $\vec{\mathbf{C}}$ and $\vec{\mathbf{P}}$—similar to the approach proposed by Suh [57, 58]. In this way, the whole approach can be quite easily formalised—not explained in detail in this article, see Weber [69] for more information.

16.3.2.1 Analysis Model

Figure 16.2 shows the basic model of analysis. It is a network-like model in which the characteristics ($\mathbf{C}_i$) determine the properties ($\mathbf{P}_j$). The only assumption in the model is that the properties can be analysed independently from each other (which does *not* mean that they *have* to be independent!) which complies with the usual practice of analysis.

The core content of Fig. 16.2 is that for a product with given characteristics (analysis!) they determine all relevant properties; however, for each individual property a different combination of characteristics is constitutive.

Once the product exists (i.e. when the product's characteristics $\mathbf{C}_i$ are physically realised, 'materialised') and operates, the analysis of its properties/behaviour ($\mathbf{P}_j$) can be performed by testing and measuring. In this case the product itself is the representation of the relations ($\mathbf{R}_j$).

During product development, however, when there is not yet a finished product, its properties can only be analysed by means of appropriate methods and tools which are based on—physical or non-physical—models. The relation-boxes ($\mathbf{R}_j$) in Fig. 16.2 stand for these methods and tools; their purpose is to tell about the influences of relevant characteristics ($\mathbf{C}_i$) on the respective properties ($\mathbf{P}_j$), thus *predicting* the properties given at that moment.

Models, methods and tools to realise the relation-boxes ($\mathbf{R}_j$) shown in Fig. 16.2 can be—here sorted from informal ('soft') to strongly formalised ('hard'):

- Guesswork, estimation
- Experience
- Interrogation of experts or potential customers (e.g. for hardly measurable properties)
- Physical tests/experiments (using physical models/mock-ups/prototypes of components or the whole product)
- Tables, diagrams (= formalised experience and experimental knowledge)
- Conventional (which usually means: simplified) calculations
- Computer-based methods and tools (which can be based on many different concepts: physics-based models turned into mathematical models and numerically solved—the most common case–, but also rule-based strategies, 'fuzzy logics', semantic or neural networks, case-based reasoning, ...).

As will be shown later (Sect. 16.3.3), along the whole product development process in order to analyse the same property different methods/tools are needed—depending on the development stage.

The basic model of analysis according to Fig. 16.2 displays one more element: The determination/prediction of every product property via an appropriate model, method or tool must be performed with respect to certain external conditions ($\mathbf{EC_j}$). They define the framework in which the statement about the respective property is valid. Examples are: analysing the load capacity or the life-time of a design solution with respect to the external load conditions (and their distribution over time); statements on the manufacturability are always linked to the manufacturing system as an external condition; even assessing the aesthetic properties of a design may be dependent on the assumed cultural background of the customers.

As was shown in Weber [70], these external conditions are particularly important for 'Design for X' (DfX) and correspondent DfX-strategies.

16.3.2.2 Synthesis Model

Formally, the synthesis model emerges from the analysis model by inversion: Based on given properties—i.e. required properties ($\mathbf{PR_j}$)—the characteristics of the solution are to be determined and values assigned. Figure 16.3 shows the model of synthesis which basically is a re-drawn analysis model (see Fig. 16.2) with all arrows reversed.

Now, appropriate synthesis methods and tools are required, in Fig. 16.3 symbolised by the 'inverted relations' ($\mathbf{R_j^{-1}}$). Again sorted from 'soft' to 'hard', these can be:

- Human genius.[1]
- Association (transfer of patterns seen in another context—can be technical or biological patterns).

[1] According to findings of psychologists involved in empirical design studies: probably the same as (quick) association.

Fig. 16.3 Basic model of synthesis

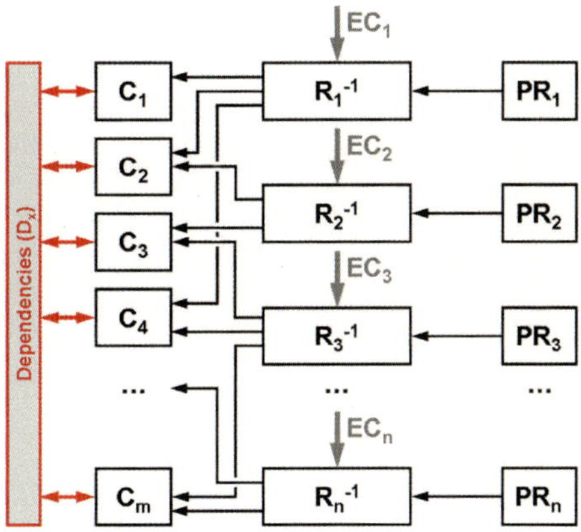

- Experience (= association based on many patterns seen in the past).
- Use of catalogues, standard solutions (e.g. machine elements).
- Set of rules, methodical/systematic approaches (often combining several of the above).
- Inverted calculations (usually only possible in very simple or—for the sake of becoming invertible—simplified cases).
- Computer-based methods and tools (which can again be based on many different concepts—for synthesis even more so than for analysis).

Already, the very simple synthesis model shown in Fig. 16.3 displays the nature of conflicts: Different required properties influence the same characteristic(s) and demand changes into different directions (C_2 … C_m in Fig. 16.3). The classical example is maximising the stiffness—which requires the cross-section to be increased—against minimising the weight—which requires the cross-section to be reduced.

16.3.2.3 Solution Elements, Solution Patterns

The definition and utilisation of solution elements, solution patterns, etc. are extremely important in practical product development/design: Solution elements/patterns enable the reuse of knowledge and are the base of product modularisation. Seen from the perspective of the CPM/PDD approach:

A solution pattern is an aggregation of characteristics ($\mathbf{C_i}$) and properties ($\mathbf{P_j}$) with known relations ($\mathbf{R_j}$, $\mathbf{R_j^{-1}}$) between the two, Fig. 16.4.

The use of solution elements/patterns does not only enhance reuse, standardisation, modularisation, etc., but is also attractive for another reason: If characteristics ($\mathbf{C_i}$), properties ($\mathbf{P_j}$) and relations between them ($\mathbf{R_j}$) are all known, then this 'knowledge' can be used in *both* directions: They can help determining the properties based on given characteristics (analysis), but they can also be used to find appropriate characteristics after searching for required properties (synthesis).

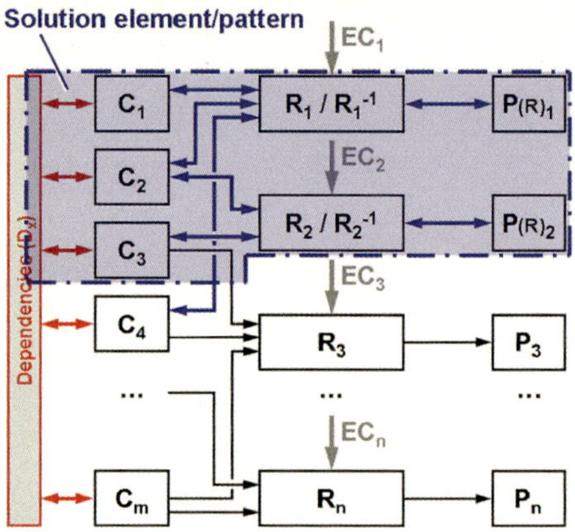

Fig. 16.4 Schematic representation of a solution element/pattern

Solution elements/patterns can be:

- Physical objects: Typical examples are machine elements, where we usually find the link between characteristics and properties in the form of catalogues, tables, diagrams, calculation algorithms.
- 'Virtual' objects: There is a lot of terms for digitised solution elements/patterns, e.g. variant programmes, pre-defined features and feature libraries, templates and—as a recent extension—'Knowledge-Based Engineering' (KBE).

It should be noted that the term "solution element/pattern', as introduced here, is *not* confined to the functional view (as is the case in traditional DTM approaches). Instead, there are solutions elements/patterns with respect to all properties, e.g. manufacturing or assembly elements/patterns, safety elements/patterns, even aesthetic elements/patterns.

The author is more and more convinced that a large proportion of product development/design processes in practice take place by piling known solution elements/patterns on top of each other.

Example: For years, compact-class motor cars bind together solution elements/patterns such as four-cylinder internal combustion engine (Otto or Diesel/turbocharged); multi-gear transmission; front engine, transversally mounted; front-wheel drive; design concept 'two-box' design, fast-back, three or five door; body construction sheet steel, spot-welded, some components bolted, … Changing one single pattern (e.g. replace the IC engine by electrical drive, change steel body to aluminium or fibre-reinforced plastic, …) will cause considerable notion. The company leaving the normal patterns will gain profile in innovation, but only when successful.

16.3.3 Modelling Processes (PDD)

16.3.3.1 Synthesis-Analysis-Evaluation Cycles

Based on the considerations on the new approach to modelling products, now the consequences for product development/design processes are presented.

Product development/design is explained as a cyclic process which, in principle ('strategically'), follows the synthesis model (see Fig. 16.3), but is complemented by analysis operations (see Fig. 16.2); synthesis and analysis are linked by a two-step evaluation procedure that controls the process. During the process—in every synthesis step—ever more characteristics of the product are established and their values assigned; by means of the analysis steps ever more and ever more precise information of the product's properties/behaviour is generated. Figures 16.5 and 16.6 give a schematic overview over two subsequent synthesis-analysis-evaluation cycles.

A1 The product development process starts with a list of requirements. This list is in PDD represented by the required properties ($\mathbf{PR_j}$, *Soll*-properties). Step 1 is a synthesis step and starts from some of the properties and establishes the first major characteristics ($\mathbf{C_{iA}}$). This is often done by adopting partial solutions from previous designs (= solution elements/patterns, see Fig. 16.4).

A2 In step A2 the current properties ($\mathbf{P_{jA}}$, *Ist*-properties) of the present solution state are analysed, based on the characteristics currently established. In this analysis step not only those properties that went into the first synthesis step are considered, but also all of the relevant properties (if possible—in very early stages it may be difficult to reason on some of the more complex properties).

A3 In this step, the analysis results of the previous step are used to determine the deviations of the individual *Ist*-properties against the required (*Soll*-) properties; the result of the comparison ($\mathbf{\Delta P_{jA}}$) represents the shortcomings of the design in its current stage.

A4 The development engineer/designer now has to run an overall evaluation: Extract the main problems and decide how to proceed, i.e. pick out the property or properties to attack next and select appropriate methods and tools for the subsequent synthesis-analysis-evaluation cycle.

It should be noted that this overall evaluation (A4) is the actual driver of the whole product development/design process: If the overall view on the current deviations delivers entirely satisfactory results, the process may end (see Sect. 16.3.3.2). If the results are not satisfactory, another cycle has to be started—with the evaluation results pointing to what to do next.

The first cycle (A, Fig. 16.5) is followed by further cycles (B, C, …) which are all similar. Figure 16.6 demonstrates this for parts of the next cycle (B).

B1 The last step of the previous cycle (A4) determines which property/properties to start from. Appropriate synthesis methods ($\mathbf{R_B^{-1}}$) are required that help modifying the characteristics in a sensible way ($\mathbf{C_{iA}} \rightarrow \mathbf{C_{iB}}$).

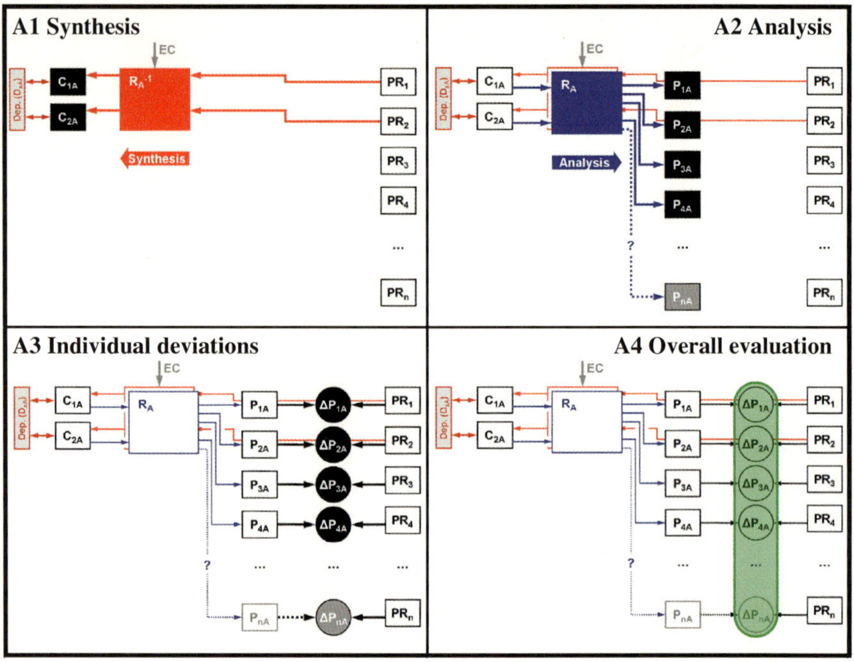

Fig. 16.5 PDD-scheme of steps in the first cycle ('cycle A') of product development

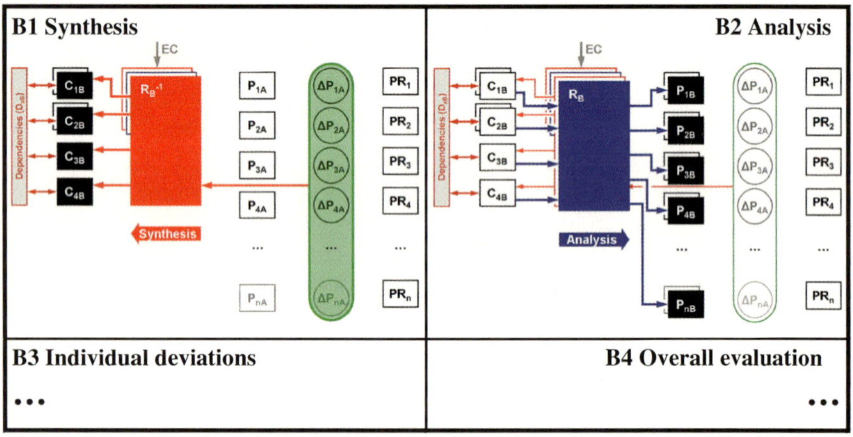

Fig. 16.6 PDD-scheme of steps in the subsequent/second cycle ('cycle B') of product development

'Modifying the characteristics' can either mean changing the values of characteristics that are already established or it means establishing additional characteristics; the last case, in turn, is either adding details to existing components or adding new components.

B2 The next step is an analysis step again. In cycle B a new (changed/extended) set of characteristics is given. Because of this the properties have to be newly determined/analysed ($\mathbf{P_{jA}} \rightarrow \mathbf{P_{jB}}$). Appropriate analysis methods and tools are required again ($\mathbf{R_B}$); in many cases these will be different from the ones used in (the) previous cycle(s) because on the characteristics side more detailed information is available which, in turn, may allow for the use of more detailed, and therefore more exact analysis methods/tools.

It should be noted that in this—actually: in every—step always the same properties are analysed—there is nothing more to analyse than what is given as required properties. However, because of the characteristics becoming ever more detailed from one cycle to the next, therefore, enabling the use of ever more detailed/exact analysis methods and tools, the analysis results will change from one cycle to the next—usually to the better.

Among other things, this means that along the whole product development/design process for each property different analysis methods and tools are needed: In early stages, analysis methods must be able to deliver statements about properties based on only a few given characteristics—with compromises regarding the accuracy of the results. Only in late stages—where the characteristics side is already quite detailed—does it make sense to utilise exact (and often complicated) analysis methods/tools.

Not shown explicitly in Fig. 16.6:

B3 Like in cycle A, the next step is confronting the—revised—*Ist*-properties ($\mathbf{P_{jB}}$) with the requirements or *Soll*-properties ($\mathbf{PR_j}$) in order to determine the individual the deviations between the two ($\mathbf{\Delta P_{jB}}$). While the requirements usually remain the same, compared to the previous cycle there is a new (hopefully improved) situation with regard to the *Ist*-properties; therefore, revised (hopefully reduced) values for the deviations are to be expected.

B4 Like cycle A, cycle B (and all subsequent cycles) ends with an overall evaluation of the—newly determined—individual deviations ($\mathbf{\Delta P_{jB}}$) between *Ist*-properties and *Soll*-properties This is again base of conclusions regarding the next cycle (actual driver of the product development/design process).

In a strongly abstracted representation, the product development process can be seen as a control circuit, as described in Weber [70].

The concept of the product development/design process as presented here can be mathematically formalised—similar to, but extending the statements of Suh's Axiomatic Design Theory [57, 58], see Weber [69] for more details.

It should be noted that the PDD concept implies a view on 'early' and 'late' stages of product development/design that is quite different from traditional DTM:

- The difference between early and late stages of product development/design is ***not*** defined by content (like in the 'European/German-speaking school' of DTM: functional and principle considerations = early stages, layout and detailing = late stages).

- Instead, only the number of characteristics already established and assigned defines what is an early and what is a late stage (with continuous transition).

In every synthesis step, solution elements/patterns (see Sect. 16.3.2.3) may be used in order to bring not yet satisfactory properties closer to the requirements. In an extreme case, the complete design of a predecessor is taken over as the start of the new development/design process. The process then would 'jump' immediately to a late stage, the product developer/designer would 'only' have to look after those properties that are not yet satisfactory against the new requirements.

That said, the CPM/PDD approach provides a relatively simple explanation of original, adaptive and variant design: These types of development/design tasks differ in the extent of a priori given solution elements/patterns. However, the basic PDD concept of product development/design remains the same: the process is entirely controlled by not (yet) satisfactorily realised properties. Also the scheme of synthesis-analysis-evaluation cycles remains unchanged. Therefore, PDD does not need to define different procedures for different types of development/design.

16.3.3.2 Termination of the Product Development/Design Process

A successful product development/design process terminates if and when

1. all characteristics needed for manufacturing and assembly of the product are established and assigned ($\mathbf{C_i}$),
2. all (relevant) properties can be determined/predicted ($\mathbf{P_j}$),
3. with sufficient certainty and accuracy, and
4. all determined/predicted properties are close enough to the required properties ($\mathbf{\Delta P_j \rightarrow 0}$).

It should be noted that most of the traditional approaches to DTM are confined to conditions 1 and 4. Conditions 2 and 3 are closely linked to the use of appropriate, i.e. 'sufficiently certain and accurate' analysis methods and tools—currently a big topic in industry.

16.3.3.3 Integrating Other Models and Approaches

An important goal of the CPM/PDD approach is to integrate existing models, methods and approaches (see Sect. 16.2). This shall be outlined by a couple of selected examples in this section.

Figure 16.7 shows graphically how the well-known design methodology approaches of the 'European/German-speaking school' (e.g. according to VDI-guideline 2221 [63]) and DfX-methods integrate themselves into the PDD process model:

- The traditional approaches concentrate on *function-related properties*; their main task is to provide support for the development of so-called principle

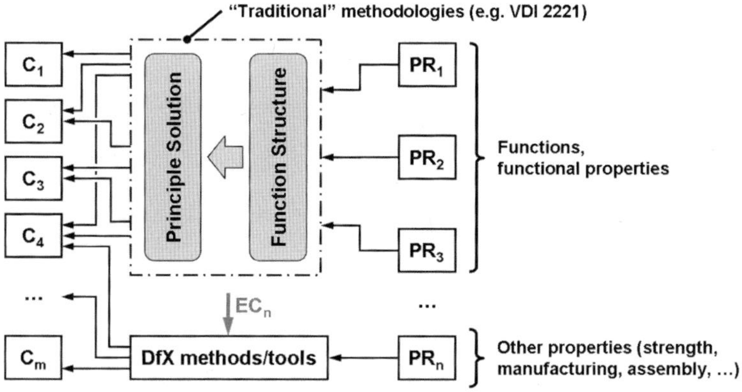

Fig. 16.7 Integration of traditional design methodologies and DfX-methods/-tools (schematic)

solutions. These are usually documented in form of a sketch—containing some, but not many characteristics of the solution. Terms like 'function structure' and 'principle solution' stand for particular intermediate models—that are purely mental constructs serving as 'stepping stones' to bridge the gap between function-related requirements and the solution.

- Other required properties (strength, manufacturability, assemblability, safety, ergonomic or aesthetic properties, …) require different methods and tools, usually summarised under DfX-methods and -tools. Design ***for*** X (DfX), but also Design ***of*** X (i.e. the simultaneous development of product and technology), seen from the perspective of CPM/PDD, have been discussed in Weber [70].

As was shown in Weber [69], the CPM/PDD product and process model includes the Design Equation introduced by Suh [57, 58] as a special case if some pre-requisites are fulfilled or assumed:

- Linear or linearised relations $\mathbf{R_j}$ and $\mathbf{R_j^{-1}}$.
- If the Design Equation is to be inverted (synthesis!), the number of characteristics (in Suh's terminology: design parameters) must be equal to the number of relevant properties (functional requirements).

The PDD process concept is similar to the Function-Behaviour-Structure (FBS) model of designing (Gero [22], Gero & Kannengiesser [23])—if Gero's term 'function'[2] is translated into '*Soll*-properties', 'behaviour' into '*Ist*-properties' and 'structure' into "characteristics'.

A very similar process concept was also presented in Roozenburg [52].

[2] The term 'function' is always difficult for linguistic reasons: While the 'European/German-speaking school' sees the term restricted to the transfer of input to output values, in the Anglo-American (and Australian) literature 'function' can be every sort of requirements.

Finally, the integration of computer-based methods and tools into the product development/design process has always been an important topic in (if not driver of) the development of the CPM/PDD approach. This topic was in the focus of the very first article on CPM/PDD [76]; since then, several subsequent publications have presented additional conclusions for the classification, further development and application of computer-based methods and tools [60, 72, 74, 77].

16.4 Critical Résumé

16.4.1 What Was Achieved?

Modelling products and development processes based on product characteristics and properties could—at least in the author's view—provide new explanations of phenomena that have previously been insufficiently understood. They have been addressed in more detail in the previous sections, to summarise:

- Differences between product development/design processes in different branches of industry and different companies were explained by different profiles, different structures and different weighting of relevant properties.
- The product development/design process as a whole was explained by repeated synthesis-analysis-evaluation cycles.
- The process is controlled ('driven') by product properties—more exact: by the current deviations between *Ist*-properties and *Soll*-properties (requirements).
- Linked to this concept: evaluation as important tool to control the process.
- For the termination of the product development/design process extended conditions could be formulated.
- Based on this: New ideas to define and measure 'product maturity'—not explained in this article, see Weber [70].
- A new and extended explanation of the term 'solution element/pattern' was given, their role in the process was explained.
- Following from that, a new view on different types of development/design tasks (original/adaptive/variant design) was presented—the transition between the types only depending on the extent of solution elements/patterns (re-) used.
- It was shown that the CPM/PDD approach is able to integrate other models, methods, tools, methodologies.
- Several findings and proposals have been presented with regard to computer-based methods and tools in product development/design in other publications. Please refer to [60, 69, 72, 74, 76, 77] for more details.

Further topics that the author and his group investigated from a CPM/PDD point of view shall only be listed in brief:

- Modelling and development of Product-Service Systems [75].
- Controlling the product development/design process not by (technical) properties, but by customer value and cost [12].
- In cooperation with scientists from manufacturing systems planning, transfer of the CPM/PDD approach into this area [13].
- A new approach to engineering change management, based on matrix representations derived from CPM/PDD [37].
- Compilation and structuring of digital catalogues based on the CPM/PDD concept of solution elements/patterns [68].
- Investigation of the term 'design knowledge' from the perspective of the CPM/PDD approach [11].
- Selection of solution principles and solution elements—here in the field of actuator design—based on characteristics and properties [19].

All of these investigations also had the purpose to confront the 'own' CPM/PDD approach with a multitude of questions in order to fathom its limits or even to find—according to Popper [49, 50]—at least one falsification. So far, no real falsification could be found; however, not in all cases spectacular new findings could be achieved either (see the Sect. 16.4.2).

The CPM/PDD approach has found some reception by other scientists and is referenced quite often. It was intensively looked at in a major interdisciplinary research cluster (engineering design, information management, computer science), financed by the Bavarian state, on 'process and workflow support of planning and control of product development' (FORFLOW), see Meerkamm et al. [45].

16.4.2 What Remains Open?

In terms of industrial applications, so far only minor parts of the CPM/PDD concept could be implemented and tested—often in projects re-engineering the company's software architecture. The one big transfer to industry—e.g. along the lines described in Weber [71]—has not taken place. Probably, CPM/PDD shares the fate of most DTM approaches: Effort and risk of a full implementation are too high; furthermore, benefits of methodical/procedural changes are extremely difficult to evaluate.

Originally it was hoped that CPM/PDD, in comparison with traditional DTM approaches, would be able to broaden the scope. However, until today (also) CPM/PDD remains in the area of mechanical engineering—electrical/electronic and software components, therefore mechatronic systems are not well covered.

The author suspects that a transfer to mechatronic systems would require a modified concept of the characteristics: In particular, the hierarchical structure of the characteristics (see Fig. 16.1), which was, in fact, taken over from (mechanical) CAD- and PDM-systems, may have to be re-designed because it cannot capture logical/functional links between the elements. They are, however, indispensable to model the structures of electrical/electronic and software components.

An approach to come to better solutions could be linking CPM/PDD with the Contact and Channel Model (C&CM), see [2, 43]; options are being be discussed with these colleagues.

A similar conclusion has to be drawn for the area of Product-Service Systems (PSS): Despite several attempts, CPM/PDD did not (yet) bring convincing new concepts—again due to the fact that there are problems in capturing and structuring the characteristics of PSS.

On the side of the product properties there are some open questions, too. The biggest problem is the decomposition of properties: How to assign measurable parameters (sub-properties) to properties? So far, both in science and in industry there are only empirical procedures, often specific for a particular branch of industry or even a particular company.

A not identical, however related question is: How to split product properties into properties of the product's components? An answer to this question—apart from its scientific value—could even be interesting for co-operative development/design processes where it is essential to deduce component requirements (required properties) from system requirements.

Finally, there is no decisive concept of how to link several hierarchically structured objects by means of their characteristics-properties networks. It would be most interesting to have such a concept because development processes of several successive systems, even domains (e.g. development of material—component—technical product—technical system—socio-technical system) could be structured accordingly.

16.4.3 Closing Remark

Probably, it is more luck than merit if a scientist is able to contribute a little bit to the theoretical foundations of his/her discipline and finds some positive reception in the 'scientific community'. In this, the most satisfying experience is having intensive discussions with colleagues: In particular, the author owes thank for interesting, partly controversial but always constructive discussions about the CPM/PDD approach as well as DTM in general to A. Albers (Karlsruhe), M.M. Andreasen (Copenhagen), H. Birkhofer (Darmstadt), L. Blessing (Luxemburg), W.E. Eder (Ontario), U. Lindemann (Munich), S. Matthiesen (Karlsruhe), C. McMahon (Bristol), H. Meerkamm (Erlangen), K. Paetzold (Munich), S. Vajna (Magdeburg), S. Wartzack (Erlangen) and K. Zeman (Linz) and, of course, many research assistants as well as participants of the annual Summer School on Engineering Design Research (where the author is allowed to present and discuss his ideas).

It is the big merit of the International Workshop on Models and Theories of Design to provide space and time for further discussions of this type—what a scientist should actually do in the first place.

A.1 16.5 Appendix: Existing Approaches to Design Theory and Methodology—A Brief Overview

In this appendix, a brief overview over Design Theory and Methodology (DTM) findings and approaches is presented. Although this overview has a slight bias towards those approaches that influenced and/or are related to the work presented here, it might be of interest for others. Therefore, the editors of this book agreed to put it into an appendix.

Quite dominant in DTM are the early procedural models; almost all of them stem from mechanical engineering and were developed in Europe, mainly but not entirely in German-speaking countries. Groundbreaking work has to be attributed to authors such as P.J. Wallace, Hansen, Kesselring, Eder & Gosling, Rodenacker, French, Hubka, Koller, Pahl & Beitz, Roth, Ehrlenspiel, Roozenburg & Eekels [15, 18, 21, 27–29, 32, 33, 36, 38, 47, 51[3], 53, 54, 67]. Hubka's work was later extended into [16, 17, 34, 35]. The book of Pahl and Beitz was very competently and quite early translated into the English language [48], therefore is well-known in the DTM community world-wide.

Although the initial DTM concepts displayed some differences in background and focus, the researchers succeeded quite early in formulating a common view under the auspices of the VDI (Verein Deutscher Ingenieure, German Association of Engineers). The outcome was VDI-guideline 2222 [65], later developed into the VDI-guideline 2221 as a basic framework for design processes [63]. VDI 2221 was also translated into English [64] and is—alongside the book of Pahl and Beitz—an important reference for many international research activities to this day.

The core of all these approaches is a generic phase model of product development processes which starts with the task clarification, goes through functional and principle considerations and ends with layout and detail design. The aim of the approaches is to provide support for product developers, in particular for the systematic development of solution alternatives, their evaluation (and subsequent decision making) and the systematic detailing of the chosen solutions. In most cases, the books and guidelines focus on original design (or 'new product development')—despite the fact that this type of development task is not very common in practice where so-called variational and adaptive tasks prevail.

All approaches provide models of the development/design ***process***; models of the ***product*** are not explicitly covered, with the notable exception of Hubka [32], later integrated into Hubka and Eder [34, 35] and nowadays Eder and Hosnedl [16, 17]. Methods and tools to model products and support processes using computers—today of great importance in practice—are not discussed; this is, of course, natural in the older publications, but even more recent articles on DTM do not go very deep into the issue.

---

[3] Most of the books referenced here have had several editions. In these cases two dates are given: Year of the first/last edition.

Hubka and Eder [35], Blessing [9], Wallace and Blessing [66] and Heymann [31] present very profound insights into these European/German-speaking approaches to DTM and their impacts.

Although the approaches mentioned above basically share similar phase models there are several different 'schools' of DTM; they differ in focus, but also in terminology. Therefore, the unification or consolidation still is a frequently discussed topic in DTM (see e.g. [7, 24]).

Based on findings of Hubka [32] and Andreasen [4] introduced, among other things, the concept of distinguishing between structural characteristics, which define or specify the constituents of a technical system, and behavioural properties.

While the models and procedures of developing products/designing according to the European/German-speaking 'schools' proved very successful in teaching, in the 1990s a—still ongoing—discussion started why there is less acceptance and application in engineering practice (see e.g. [6, 14, 20]).

A completely different approach to DTM was presented by Suh [57, 58]. His Axiomatic Design Theory is intensively discussed in the academic as well as the practical world of product development/design, especially in the USA. The core of Axiomatic Design is to see designing as a transformation (or 'mapping') of information from the 'functional space' (represented by a list/vector of functional requirements, **FRs**) into the 'physical space' (represented by a list/vector of design parameters, **DPs**).

Suh also proposed a very elegant mathematical formalisation of his approach—not covered here. Finally, Suh formulates certain axioms (plus corollaries and theorems deduced from them) that define an optimal design solution; the most commonly known axiom is the independence axiom which requires that in an optimal case each FR shall only be dependent of one DP (leading to a 'decoupled design solution').

The 'European School' of product development/design and Suh's Axiomatic Design approach usually are considered as rather incompatible.

Already originating in 1990, John Gero proposed the Function-Behaviour-Structure (FBS) model of designing [22, 23]. In this model, product development/design is a transformation of requirements (function F) into the description of the solution (structure S). Between the two there is the category of 'behaviour B' which can be split up into expected behaviour (Be) and structure behaviour (Bs)—i.e. (actual) behaviour as derived from actual structure. A fifth relevant element is the documentation of the result (D). Between these five categories (F, S, Be, Bs, D), there are 8 basic relations of which the activity of developing products/designing is constituted. As the FBS model was developed from considerations towards Artificial Intelligence (and first published in an AI journal) it was only later recognised in the DTM community.

Sándor Vajna, later together with Tibor Bercsey coined the term 'Autogenetic Design Theory' (first publication Vajna and Wegner [61], advanced and more detailed version in Vajna et al. [59]). This approach describes product development in analogy to evolution processes in nature, in particular as a continuous co-development of objects, techniques and technologies. Developing products is considered as a constant optimisation process, starting with one or more base solutions (the original population) that are defined by chromosomes (i.e. product characteristics); the process then consists of varying the chromosomes (mutation), thus producing new solutions that have to be evaluated against the requirements (selection).

In the 1980s and 1990s—basically decades ***after*** the first approaches to and concepts of product development/design had been published—so-called empirical design studies emerged as a new branch of design research. In laboratory experiments (sometimes using experienced practitioners, more often using students as test persons), the actual process of designing was analysed and studied; experiments in real industrial settings would be even more desirable, but often face many practical and methodical problems. Studies of this type originate in the UK (University of Cambridge), in Germany (TU Darmstadt, TU München) and the Netherlands (TU Delft); they are often performed in co-operation with psychologists and social scientists who can contribute vast methodical know-how in conducting experiments involving humans and in interpreting the results. The outcomes of these studies have always been interesting and challenging, sometimes surprising. One of the earliest contributions came from Hales [25]; the relative large number of publications in this field since that time cannot be cited here in detail.

Before empirical design research, research in DTM had the habit of inventing new procedures, methods, tools and methodologies, but very rarely measuring or proving their necessities and impacts. Blessing and Chakrabarti [10] saw this deficit and presented their Design Research Methodology (DRM). DRM sees the traditional prescriptive activities (i.e. developing procedures, methods, tools, methodologies) framed by two descriptive, often empirical studies: One up front to find out what is needed, one at the end to check whether the measures taken show any improvement. By using the DRM framework, design research is given more rigour, its results become refutable in a scientific sense.

In industrial practice, during the 1980s companies—large and small—started to equip themselves with computer tools to support design, simulation and product data management (CAx-systems). This process is still ongoing, the number of tools increasing, the tools getting more and more complex, gaining considerable influence on product development processes. Already in the late 1990s, Spur and Krause [56] coined the term of the (completely) virtual product and (completely) virtual product development.

However, apart from very few exceptions (e.g. [44, 46, 55]) research in DTM on one hand and in computer support of product development/design processes on the other hand have been very weakly interlinked. DTM tended (and tends) to concentrate on the 'early phases' of product development (e.g. functional and principle reasoning), while CAx-systems are particularly successful in the 'late phases' (embodiment design, numerical simulation, optimisation). The separation of DTM and CAx development is negative both ways: DTM has largely bowed out of discussing computer methods and tools, therefore has lost competences in this field. At the same time computer methods and tools are being developed, introduced and applied without comprehensive methodical background—not always to the benefit of product developers/designers in practice.

Complementing trends in industrial practice, DTM—originally coming from mechanical engineering—widened its focus: The products considered became more 'mechatronic', existing expertise and experiences were transferred to the development of multi-domain products and systems (see e.g. [62]). Next, the development of combinations of material products and services was (and still is) considered (so-called Product-Service Systems, PSS).

The issue of Design for X (DfX) has been broadened considerably in the last couple of decades. In addition to 'traditional' topics like Design for Manufacturing and Assembly (DfM, DfA, sometimes combined to DfMA) new aspects were introduced, e.g. Design for Quality (DfQ) and Design for Environment (DfE). However, also here we find rather weak links between general DTM research and the development of DfX guidelines.

Since 2000, the area of DTM has gained new impetus, maybe due to a new generation of scientists having taken over. A remarkable number of new approaches have been introduced and are being discussed in the community.

In 2000, the author and H. Werner presented the approach of modelling products and development processes based on product characteristics and properties for the first time [76] which was only later called CPM/PDD (Characteristics-Properties Modelling, Property-Driven Development). The focus of this first article was looking at support tools for product development/design (CAx-systems) from a new perspective—still an important, but not anymore the major topic of CPM/PDD.

Andreasen and his group developed earlier views [4] into the so-called Domain Theory [5, 26]. 'Domains' were defined as a set of dedicated views onto a product (in particular: the domains/views of activities, organs and parts) that are used as the skeleton of a procedure for product development/design.

In 2001 Maier and Fadel formulated the concept of affordance-based design [40], more extensively explained and put into context in Maier and Fadel [41]: 'Briefly stated, an affordance is what one system (say, an artefact) provides to another system (say, a user). The concept of affordance is relational because of the complementarity entailed between two interacting systems.' Thus, the user becomes integral part of considerations in a product development/design process. Thus, product and user is seen in context. At the same time, the concept of affordances opens extended views on requirements and properties of products and systems. Finally, the relation concept is extended in order to map affordances against product/system components, using a matrix approach derived from Design Structure Matrix practices [42].

Albers und Matthiesen introduced the Contact and Channel Model (C&CM), see [2, 43]. C&CM takes up earlier work (dating back to Hubka [32]) on working surface pairs as carriers of functions, but concentrates on extensions in two dimensions: First, based on working surface pairs ('contacts') and the structures to connect them ('channels') new design methods are presented. Second, the new approach is not confined to mechanical contacts and channels (like in the past), but broadens the view to fluidic, electrical, even information flows.

In addition, [1, 3] described the so-called SPALTEN methodology as a comprehensive approach to handle problems of different boundary conditions and levels of complexity. From that, Albers recently developed the approach of 'advanced systems engineering'.

Hatchuel and Weil [30] introduced their Concept-Knowledge (C-K) Theory; it explains product development as mutual interplay between extending the 'concept space' (i.e. simplified: generating solutions) and the 'knowledge space' (generating knowledge about the concepts' behaviour via analysis).

Lindemann [39] presented the Munich Procedural Model (MPM) for product development processes. Among other new ideas, it propagates procedural flexibility and the use of a multitude of methods (including computer-supported methods and tools). Another new focus is on the management of product development processes.

References

1. Albers A, Burkardt N, Meboldt M, Saak M (2005) SPALTEN problem solving methodology in the product development. In: Samuel A, Lewis W (eds) Proceedings of ICED 05, full paper on CD-ROM. The Design Society, Melbourne
2. Albers A, Matthiesen S, Ohmer M (2003) An innovative new basic model in design methodology for analysis and synthesis of technical systems. In: Folkeson A, Gra-lén K, Norell M, Sellgren U (eds) Proceedings of ICED 03. The Design Society and the Royal Institute of Technology, Glasgow
3. Albers A, Meboldt M (2007) SPALTEN Matrix—product development process on the basis of systems engineering and systematic problem solving. In: Krause F-L (ed) Proceedings of the 17th CIRP design conference, The future of product development. Springer, Berlin, pp 43–52
4. Andreasen MM (1980) Synthesis methods as system basis—a contribution to a design theory (in Danish). Dissertation, Lund University
5. Andreasen MM (2011) 45 years with design methodology. J Eng Des 22(1):1–40
6. Birkhofer H (2004) There is nothing as practical as a good theory—an attempt to deal with the gap between design research and design practice. In: Marjanovic D (ed) Proceedings of DESIGN 2004, Faculty of mechanical engineering and naval architecture. University of Zagreb, pp 7–14
7. Birkhofer H (2006) The consolidation of design science—a critical review of the status and some proposals to improve it. In: Hosnedl S, Vanek V (eds) Proceedings of the AEDS workshop 2006 (Applied engineering design science). University of West Bohemia, Pilsen, pp 13–22
8. Birkhofer H, Wäldele M (2009) The concept of product properties and its value for research and practice in design. In: Proceedings of ICED 09, vol 2. The Design Society, Glasgow, pp 227–238
9. Blessing LTM (1995) Comparison of design models proposed in prescriptive literature. In: Perrin V, Vinck D (eds) Proceedings of COST A3/COST, "The role of design in the shaping of technology", Social sciences series, vol 5. Lyon, pp 187–212
10. Blessing LTM, Chakrabarti A (2009) DRM, a design research methodology. Springer, London

11. Conrad J, Köhler C, Wanke S, Weber C (2008) What is design knowledge from the viewpoint of CPM/PDD? In: Proceedings of DESIGN 2008, vol 2, Faculty of mechanical engineering and naval architecture. University of Zagreb, pp 745–752
12. Deubel T, Steinbach M, Weber C (2005) Requirement- and cost-driven product development process. In: Proceedings of ICED 05, (executive summary), paper no. 357.46 (full paper, on CD-ROM). The Design Society, Melbourne, pp 166–167
13. Deubel T, Zenner C, Bley H, Weber C (2006) Adaptation of a new design method for the requirement-driven planning of manufacturing systems. In: Proceedigs of the 16th CIRP international design seminar, paper no. 10061. Kananaskis, pp 529–534
14. Eder WE (1998) Design modeling—a design science approach (and why does industry not use it? J Eng Des 9(4):355–371
15. Eder WE, Gosling W (1965) Mechanical systems design. Pergamon Press, Oxford
16. Eder WE, Hosnedl S (2008) Design engineering. CRC Press, Boca Raton
17. Eder WE, Hosnedl S (2010) Introduction to design engineering—systematic creativity and management. CRC Press, Boca Raton
18. Ehrlenspiel K (1995/2009) Integrierte produktentwicklung—methoden für die prozessorganisation, Produkterstellung und Konstruktion. Hanser, München. 1st edn 1995, 4th edn 2009
19. Erbe T, Weber C, Paetzold K (2011) Actuation principle selection—an example for trade-off assessment by CPM-approach. In: Proceedings of ICED 11, vol 4. The Design Society, Glasgow, pp 222–229
20. Franke H-J (1999) Ungelöste probleme der konstruktionsmethodik. In: Franke H-J, Krusche T, Mette M (eds) Konstruktionsmethodik – Quo vadis? Symposium on the occasion of the 80th birthday of Prof. Dr.-Ing. Karlheinz Roth. Institut für Konstruktionslehre, TU Braunschweig, Bericht Nr. 56, pp 13–30
21. French MJ (1971/1998) Engineering design—the conceptual stage, 1st edn. Heinemann, London (Conceptual Design for Engineers), 3rd edn. Springer, London, 1998
22. Gero JS (1990) Design prototypes—a knowledge representation schema for design. AI Mag 11(4):26–36
23. Gero JS, Kannengiesser U (2004) The situated function-behaviour-structure framework. Des Stud 25 (2004):373–391
24. Grabowski H, Rude S, Grein G (eds) (1998) Universal design theory. In: Proceedings of the workshop "Universal design theory". Shaker, Aachen
25. Hales C (1987) Analysis of the engineering design process in an industrial context. Gants Hill Publications, Hampshire
26. Hansen CT, Andreasen MM (2002) Two approaches to synthesis based on the domain theory. In: Chakrabarti A (ed) Engineering design synthesis—understanding, approaches and tools. Springer, London, pp 93–108
27. Hansen F (1955) Konstruktionssystematik—eine Arbeitsweise für fortschrittliche Konstrukteure. VEB Verlag Technik, Berlin. 2nd edn also 1955. Available online www.tu-ilmenau.de/konstruktionstechnik/
28. Hansen F (1966) Konstruktionssystematik. VEB-Verlag Technik, Berlin. 3rd edn 1968. Available online www.tu-ilmenau.de/konstruktionstechnik/
29. Hansen F (1974) Konstruktionswissenschaft – Grundlagen und Methoden. Hanser, München-Wien. 2nd edn 1976. Available online www.tu-ilmenau.de/konstruktionstechnik/
30. Hatchuel A, Weil B (2003) A new approach of innovative design—an introduction to C-K theory. In: Folkeson A, Gralén K, Norell M, Sellgren U (eds) Proceedings of ICED 03. The Design Society and the Royal Institute of Technology, Glasgow/Stock-holm
31. Heymann M (2005) Kunst und Wissenschaft in der Technik des 20. Jahrhunderts—zur Geschichte der Konstruktionswissenschaft. Chronos, Zurich
32. Hubka V (1973) Theorie der Maschinensysteme. Springer, Berlin
33. Hubka V (1976) Theorie der Konstruktionsprozesse. Springer, Berlin
34. Hubka V, Eder WE (1992) Einführung in die Konstruktionswissenschaft. Springer, Berlin
35. Hubka V, Eder WE (1996) Design science. Springer, London

36. Kesselring F (1959) Technische Kompositionslehre. Springer, Berlin
37. Köhler C, Conrad J, Wanke S, Weber C (2008) A matrix representation of the CPM/PDD approach as a means for change impact analysis. In: Proceedings of DESIGN 2008, vol 1. Faculty of mechanical engineering and naval architecture, University of Zagreb, pp 167–174
38. Koller R (1976/1998) Konstruktionsmethode für den Maschinen-, Geräte- und Apparatebau. Springer, Berlin. 1st edn 1976, 4th edn 1998
39. Lindemann U (2005/2009) Methodische Entwicklung technischer Produkte – Methoden flexibel und situationsgerecht anwenden. Springer, Berlin. 1st edn, 2005, 3rd edn 2009
40. Maier JRA, Fadel GM (2001) Affordance: the fundamental concept in engineering design. In: Proceedings of ASME design theory and methodology conference, Pittsburgh, PA. Paper no. DETC2001/DTM-21700
41. Maier JRA, Fadel GM (2009) Affordance based design: a relational theory for design. Res Eng Des 20(2009):13–27
42. Maier JRA, Mocko G, Fadel GM (2009) Hierarchical affordance modeling. In: Norell M, Grimheden M, Leifer L, Skogstad P, Lindemann U (eds) Proceedings of ICED 09, vol 5. The Design Society, Glasgow, pp 397–404
43. Matthiesen S (2002) Ein Beitrag zur Basisdefinition des Elementmodells „Wirkflächen-paare & Leitstützstrukturen"zum Zusammenhang von Funktion und Gestalt technischer Systeme. Dissertation Universität Karlsruhe, IPEK Research Report vol 6, Karlsruhe
44. Meerkamm H (1993) Design system mfk – an Important Step towards an engineering workbench. In: Proceedings of the institution of mechanical engineers (IMechE), vol 207. pp 105–116 (Journal of Engineering Manufacture)
45. Meerkamm H, Henrich A, Jablonski S, Krcmar H, Lindemann U, Rieg F (eds) (2009) Flexible Prozessunterstützung in der Produktentwicklung – Prozesse, Daten, Navigation. Abschlussbericht des Bayerischen Forschungsverbundes FORFLOW 2006-2009. Shaker, Aachen
46. Mortensen NH (2000) Design modelling in a designer's workbench. Dissertation, Technical University of Denmark
47. Pahl G, Beitz W (1977/2007) Konstruktionslehre. Springer, Berlin (1st edn 1977, 7th edn (by Pahl, G.; Beitz, W.; Feldhusen, J.; Grote, K.-H.) 2007)
48. Pahl G, Beitz W (1983/2007) Engineering design. Springer, London (1st edn 1983, 3rd edn 2007; all english editions translated and edited by Wallace K, Blessing LTM)
49. Popper KR (1935) Logik der Forschung. Julius Springer, Wien
50. Popper KR (1959) Logic of scientific discovery. Hutchinson and Co, London
51. Rodenacker WG (1970/1991) Methodisches Konstruieren. Springer, Berlin. 1st edn 1970, 4th edn 1991
52. Roozenburg NFM (2002) Defining synthesis. In: Chakrabarti A (ed) Engineering design synthesis—understanding, approaches and tools. Springer, London, pp 3–18
53. Roozenburg NFM, Eekels J (1995) Product design—fundamentals and methods. Wiley, Chichester
54. Roth K (1982/1994) Konstruieren mit Konstruktionskatalogen. Springer, Berlin. 1st edn 1982, 2nd edn (in two volumes) 1994
55. Schulte M, Stark R, Weber C (1993) Feature structure for the representation of design objects as combinations of technical functions and geometry in CAD. In: Sodhi RS (ed) Proceedings of CARs & FOF'93, published in series "Manufacturing research and technology", vol 22. Advances in manufacturing systems—design, modeling and analysis, 1994. Elsevier, Amsterdam, pp 335–340
56. Spur G, Krause F-L (1998) Das virtuelle Produkt. Hanser, München
57. Suh NP (1990) The principles of design. Oxford University Press, New York
58. Suh NP (2001) Axiomatic design. Oxford University Press
59. Vajna S, Clement S, Jordan A, Bercsey T (2005) The autogenetic design theory—an evolutionary view of the design process. J Eng Des 16(4):423–440
60. Vajna S, Weber C, Zeman K, Bley H (2009) CAx für Ingenieure – eine praxisbezogene Einführung, 2nd edn. Springer, Berlin

Historically, design methodology has been developed in response to intuitive approaches with the rationale that intuition often involves arbitrary choices of problem decomposition and solution generation [16]. On the other hand, several empirical studies over the last 50 years have shown that designers do use other "strategies" than structured design methods to arrive at new ideas, solutions, products, etc. (e.g., Gunther and Ehrlenspiel [12]). However, design methodology seems to ignore intuition as an elementary part of the thinking and acting of the designer.

Intuition plays an important role for planning and decision making in other fields as well. For instance, the mechanisms of intuition are being studied in organizational management, philosophy, and psychology. However, so far, the widespread interest has not led to a more comprehensive empirical investigation on this phenomenon in design research (Kathri and Ng [18]). Even the basic question, "does a designer relying on his/her intuition exhibit good or bad design behavior" is not being asked in a rigorous manner, and, thus, not being answered in the scientific literature.

In addition to the lack of answers provided by design research, there seems to be an even more conflicting situation in practice: On the one hand, management of product development organizations pursue process standardization certifications such as ISO 9000/9001, and, on the other hand, self-reports of designers imply that they do not follow those processes by frequently relying on their intuition. Which approach is the desirable one? Who would feel safe on an airplane where key design decisions have been made by intuition and "gut feeling" instead of explicit and structured reasoning?

In the following sections, we first integrate results from the literature to shed some light on previous research on intuition and define the process of intuition. We then illustrate the use of design intuition in practice by analyzing a set of interviews with professional designers of varying degrees of experience in different fields.

Our goal is to understand the role intuitive processes play in the thinking and acting of designers. Although we are particularly interested in the connection between intuition and decision making, we recognize that design intuition encompasses more, and will consider its multiple facets.

Before we begin our exploration, we should stress that we do not see any evidence for the somewhat common assumption that intuitive processes follow completely different rules compared to the rational decision making process [19] (see Sect. 17.2 for a detailed discussion).

Our exploration will enhance our understanding of how intuition is used in different stages of the design process, the characteristics of the individual designer, and the contextual factors relating to the team and the organization. The driving research questions are:

- What is the nature of design intuition?
- How does intuition influence design processes and performance?
- How do designers conceptualize that influence?
- Should intuitive processes in design thinking be supported? If so, how?

17.2 Intuition and Rational Decision Making: Complementary or Conflicting Behavior?

As the data we will present during our empirical consideration also indicate, when a practicing designer is asked how he/she arrives at a decision while dealing with a complex problem, he/she is likely to reveal that his/her way of working is not necessarily "rational," and that he/she does not always use explicit criteria to compare the utility of generated alternatives when arriving at a decision.

Many decades of research on human decision making model the human being as a "rational decision maker." This model explains the behavior as a probabilistic approach: The decision maker chooses the alternative that delivers the highest utility, which is estimated based on the available information. Thus, the chosen option is optimal in a probabilistic sense. However, the view of human decision making as a purely rational process has been promptly questioned. The most well-known "critique" of the rational problem solver was brought up by Simon [24], who introduced the notion of "bounded rationality," and within that concept, the claim that humans make satisficing rather than optimizing decisions due to limitations associated with unreliable information about alternatives and their consequences, human memory, and resources (also claimed by Kahneman [17]).

Moreover, many studies yielded results about human fallacies in a variety of tasks requiring judgment and decisions—the main reason being that humans base judgments on beliefs and intuition rather than a logical reasoning process [8, 28]. On the other hand, it has been argued that intuition can be a successful element in human's acting and decision making. For instance, the naturalistic decision making research community emphasizes "the power of intuition." A thesis of the book, "Start with intuition, not with analysis," gets that point across [19, p. 88].

If, at the same time, we consider that humans make a variety of mistakes—some with catastrophic consequences—when relying on intuition, we must ask: Are human beings prone to choosing the wrong tool/approach for attacking different kinds of problems? Obviously, more insights are necessary to understand the role of intuition in different types of decision processes. Moreover, there seem to be intuitive mechanisms which are built into the human brain that provide benefits to the decision maker and beat the limitations with a cognitive-emotional weapon.

So let us face the question: What is intuition?

17.2.1 Defining Intuition

Intuition is what we consider as behavior we cannot observe; in research terms, intuition is a hypothetical construct. A hypothetical construct can be defined as "an abstract concept used in a particular theoretical manner to relate different behaviors according to their underlying features or causes" [14].

In order to grasp the most relevant aspects of this construct, we first survey criteria which are used in different definitions of intuition. A comprehensive definition is given by Webster's Dictionary: "A looking on; a sight or view; but restricted to mental view or perception. Particularly and appropriately, the act by which the mind perceives the agreement or disagreement of two ideas, or the truth of things, immediately, or the moment they are presented, without the intervention of other ideas, or without reasoning and deduction…We know by intuition, that a part is less than the whole." [32].

The so-called gut feeling is closely related to the phenomenon of intuition. The definitions of gut feeling vary in how far the conscious part plays a role.

Le cœur a ses raisons que la raison ne connaît pas. [21, p. 277].

The most well-known reference to "gut feeling" has been made by pascal [21] in the sixteenth century. In this proverb, he describes the heart as the place where information is processed and decisions are made, which are not being understood by the rational part of the information processing system.

Gigerenzer, a cognitive psychologist, defines gut feelings as rules of thumb which "provide knowing without thinking," and uses "the colloquial rule of thumb synonymously with the scientific term heuristic" [11, p. 18]. Gigerenzer presents his doubts on the dogmatic view of rational decision making and portrays an adaptive toolbox as an alternative approach to the "new land of rationality" [11, p. 19].

In order to decompose and further support these definitions, we present a framework that characterizes the nature of intuition and consists of the following dimensions:

- Intuition is related to unconscious and subconscious processes

Although it is widely agreed that intuition is not a conscious process, there is disagreement on the extent it is unconscious or subconscious. However, a person not being able to explain the rationale behind an intuitive decision does not mean that there has not been any reasoning associated with the decision. The assumption that intuition takes place without reasoning and deduction has not been proven and is open to discussion (see Sect. 17.4 for a more detailed discussion).

- Intuition associates to the totality of the situation

Intuition is seen as a synthetic as opposed to an analytical function that apprehends the totality of a situation [30]. This aspect is often addressed within the notion of a holistic approach.

In addition to the un/sub-conscious and the holistic feature, many definitions stress the attribute of affect-relatedness of intuition. For example, Dane and Pratt define intuition as "affectively-charged judgments that arise through rapid, non-conscious, and holistic associations." [5, p. 40] This definition stresses four characteristics of intuition: unconscious, time-related, emotion-related and holistic (see below).

- Intuition is accompanied by affects/feelings/emotions

Many authors also add that the situations in which intuition is used are often accompanied with strong feelings of what will occur—of consequences of pursuing different sets of actions. Vaughan [30], who distinguishes four levels of intuition, physical, emotional, mental, and spiritual, refers also to tension which can be indicated by bodily messages from which further information can be gained and used for coping with the situation at hand. For instance, a sudden bodily discomfort can be interpreted as a source of warning.

- Intuition is fast

Intuitive judgments and decisions surface rapidly. The person suddenly becomes aware of the outcome of the intuitive process, and as mentioned earlier, does not have direct or conscious access to the process itself.

- Intuition uses multi-sensorial stimuli

Intuition can be prompted by different senses such as auditory and tactile input. Especially, expert knowledge is often a combination of haptic, visual, and olfactory knowledge, which only the expert is able to elicit in a specific situation and to gain further information.

- Intuition develops with experience

We can assume that all human beings possess basic behaviors which relate to intuition—primarily linked to survival mechanisms–and that they also learn from experience and internalize knowledge. In other words, it is reasonable to assume that new knowledge is processed into strategies or heuristics in the form of intuition that can provide guidance in the future (see Sect. 17.4 for a more detailed discussion).

- Intuition can stimulate creative solutions

A connection between intuition, creativity, and innovation is implied in the dimensions outlined above. Albert Einstein refers to the outstanding impact of intuition on creativity: "(empirical) knowledge is necessary, too. An intuitive child couldn't accomplish anything without some knowledge. There will come a point in everyone's life, however, where only intuition can make the leap ahead, without ever knowing precisely how. One can never know why, but one must accept intuition as fact" [15, p. 173]. With further enthusiasm for intuition the author finally states: "For it is intuition that improves the world, not just following the trodden path of thought. Intuition makes us look at unrelated facts and then think about them until they can all be brought together under one law. To look for related facts means holding onto what one has instead of searching for new facts. Intuition is the father of new knowledge, while empiricism is nothing but an accumulation of old knowledge. Intuition, not intellect, is the 'open sesame' of yourself" [15, p. 16].

Albert Einstein's position values intuition as a road to new knowledge, and emphasizes its impact on the creative sense-making process.

17.2.2 Theoretical Approaches

There are no theoretical approaches that claim to address intuitive decision making in design. However, in other disciplines, mainly in the area of decision making, there are theoretical approaches which attempt to explain the different facets of human decision making as rational and intuitive. These approaches build on the dual process theory [7, 8, 26, 27], which treats intuition and rational decision making as competitive modes of thought.

These two different modes of thoughts were articulated in detail by James [13] and Freud [10]. Freud distinguishes the associative and unconscious primary system from the conscious and rational secondary system. Similarly, James separates associative and what he called “true reasoning.” According to James, associative knowledge is re-productive as it is built from past experiences, and relies on reasoning processes such as comparison and abstraction. Associative thinking is the “true reasoning” modus as it is also the tool for dealing with unexpected and new situations [13].

About 50 years later, Tversky and Kahneman [22, 23] pursued the idea of two separate systems, which they called the Intuitive System 1 and the Analytical System 2. Embracing an evolutionary view, the intuitive system works fast and automatically, and has an emotional component with focus on survival and reproduction. The person might be conscious about the final product but not about the generating processes. System 2 is slower and is activated when System 1 cannot resolve the issue. System 2 uses conscious judgments and deliberate attitudes to generate new knowledge.

17.2.3 Integrating the Mechanisms of Intuitive (System 1) and Analytical (System 2) Processes

Does the assumption of two different mechanisms of human reasoning deliver an explanation of intuitive behavior? Our position does not assume the existence of two reasoning mechanisms with separate process flows, but integrates both reasoning mechanisms into one process flow instead.

Humans apply different heuristics and strategies when dealing with complex and uncertain problems; these types of problems do not necessitate one single correct solution but follow the criteria of wicked [20] and ill-defined problems [25]. Heuristics are defined as cognitive strategies which are applied when people must decide, and do not have sufficient information, or in cases of information overload.

We postulate that human perception works in a way that in any situation a continuous comparison takes place between expected and actual occurrences with respect to key parameters. When this check results in a significant discrepancy (a mismatch above a threshold), an inquiry is prompted: “What was that?” The analytical system will be activated only in such unexpected or surprising cases. Then, the conscious analysis provides new information, which is integrated into the existing mental model, resulting in the generation of new knowledge. Depending on the outcome, this process might be repeated several times, but after a certain amount of events it will become "routine" and will decrease the conscious attention. Neuroscientists have found that, after a decision, the match between expected and actual occurrences will produce the so called "feel-good" hormone dopamine, whereas in non-expected occurrences the dopamine production is decreased [33].

Moreover, human thinking is based on representations of reality—that can be described as accumulated experience—that are built in order to understand, predict and explain the world. Thus, the identification of a discrepancy and the selection of adequate actions must be based on prior experience within a similar situation. The identification of relevant "pieces of experience" is most likely based on the similarity between the given situation and the schemata stored in memory. Our assumption is that if this similarity-check does not lead to a clear association or fit with an adequate automatism (fixed predetermined response to the cue), then cognitive strategies such as predefined rules or heuristics come into play. Heuristics are built from prior cases of successful experience, but do not guarantee a successful result. In non-routine situations where no suitable schemata are available, new schemata need to be generated. As a consequence, uncertainty surfaces and affects thinking and acting on each level of behavior.

Thus, thinking is primarily steered by (adequate or inadequate) existing schemata, which is very effective and efficient in routine situations and makes the major part of our daily lives more comfortable. We do not need to create a new strategy for each new situation, but can quickly react on the basis of earlier experience. Accordingly, unconscious adaptations and minor modifications frequently occur at the boundary of intuitive and analytical processes.

Imagine a designer whose goal is to redesign a printing machine. In order to assess the current design, he/she will start with scanning the variables he knows being the key parameters of the system [28]. At the same time, the designer will also monitor—not necessarily consciously—other variables whether they are in accordance with his expectations. If there is no discrepancy, further actions in regard to these variables will not become priority of the activities of the designer.

However, new information (can be a specific smell, a noise, or a vibration) might cause him/her to unconsciously recognize a discrepancy. In this example, it is unlikely that he/she can access a successful automatism providing a direct answer because the system encompasses many different components with complex relationships.

Since an automatism was not activated, the designer will be suddenly struck by the intuition—become aware of the issue—that the machine is not operating properly without being able to "put a finger on it." He/she will consciously try to recall other potential relevant information (and rules) from his/her experience. If there is still no pattern match, he has to generate new knowledge in order to develop new solutions to meet his/her goal.

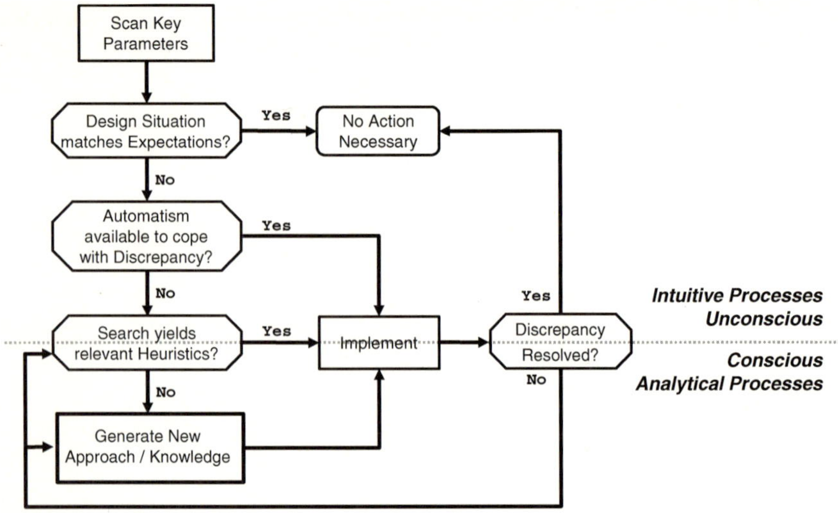

Fig. 17.1 A flow-diagram of conscious and unconscious behavior regulation: Processes integrating the application of automatisms, the search for existing rules, and the generation of new knowledge

In essence, this model assumes that the intuitive and analytical systems are interlinked in a process flow (see Fig. 17.1). They are connected through an intermediate process which is the application of strategies that have been proven to be successful. Also, note that the search of heuristics and implementation (of automatisms, heuristics, and new approaches) are seen as actions that have conscious and unconscious components.

In a similar way, Klein emphasizes the necessity of cooperation between the two systems [19]. Recently, within the "naturalistic decision making" domain, Klein developed the RPD (recognition primed decision) framework [19], which jointly accounts for intuition and analysis. Klein claims that intuition is a pattern-matching process that generates valid solutions to problems; the person quickly identifies the feasible courses of action without a need to compare options. Experts use analysis to consciously and deliberately review the courses of action.

17.3 Implications for Design Methodology

The discussion presented above is primarily a summary of the findings of social and cognitive psychology research. In addition, in the last decade, business and management disciplines have expressed growing interest in non-rational strategies in order to compliment traditional management strategies. However, empirical research on intuition in the design discipline is scarce; the few empirical studies on intuition are mostly conducted in the area of strategic decision making [22].

Although design researchers seem to acknowledge that designers "use" intuition on a daily basis, there is hardly any targeted empirical work in order to understand whether intuition works in designing, and if it does, how and in which phases of the design process it matters the most.

According to the design science approach of Hubka and Eder [16] the designer should work on a knowledge-based level (conscious level), and avoid jumping to conclusions (jumping to solutions) and the use of intuition without thoroughly investigating the problem.

However, if a designer were to absolutely stick to the principle of "not jumping to conclusions," he/she would be paralyzed. "Macroscopic" decisions such as choosing one of the down selected three–five alternatives a designer arrives at after concept selection and testing can lend themselves to such structured analysis fairly well. However, how about the hundreds—maybe even thousands—of "microscopic" decisions that a designer has to make on a daily basis when scoping the project and generating alternatives? It would be almost impossible to strive to attain statistically meaningful information necessary to carry out a structured and explicit process for those decisions; such an approach would bring the design process—especially its front end—to a halt.

In design methodology, the role of intuition as a potential supportive element during the design process has not elaborated. However, Pahl and Beitz state in that intuition might be another reasonable way to arrive at a good solution (1988). In fact, they see both approaches as mutually supportive and claim that a balance would be needed.

Until now this position has not been taken into consideration in the development of design methodology, but there seems to be a trend toward paying more attention to the unstructured and partly unconscious moments of designing. Newer frameworks [31] describe design activities that are opportunistic with hierarchical episodes, from pre-existing plans satisfying constraints, e.g., cognitive economy.

17.4 Design Intuition as a Specific Type of Design Thinking

Cross has reported that professional designers often refer to "intuition" when characterizing their design thinking [3]. He sees that appropriation as a shorthand that describes what really happens in design thinking, and relates it to the notion of abductive reasoning in design. We do not fully follow Cross' argument because, as we stated earlier, we do not see why the development of design intuition has to exclude inductive and deductive reasoning processes, or rather, their outcomes, which may have been internalized as design heuristics. In order to articulate our position, we will focus on and further articulate the experience dimension of the framework we used earlier to characterize intuition.

Expertise in design has been a topic of study for decades. Several studies report that experts, compared to novices, tend to focus less on problem analysis, quickly impose an initial framing of their own on the problem, and generate a relatively small number of solutions [4]. The causes behind this rapid formulation process are unclear, but it is often speculated that past experiences allow the expert to quickly see/recognize/conceptualize the problem in a meaningful way [6].

Our inference is that past experiences do not constitute actionable knowledge by themselves, and, as such, cannot be directly responsible for the reported expert behavior in design. As we argued earlier, there has to be some type of past and current processing connecting experiences to current design situations, which need to be elicited by a situation-driven affordance or by intentions built from the current situation. Our core proposition is that such processing is the basis for design intuition, which does not necessarily deem the use of intuition in design thinking an irrational approach.

In other words, relying on intuition in order to arrive at design decisions might be just as rational as using a systematic and explicit decision making method. One can argue that, given the availability of resources, a systematic decision making method might yield higher quality information, but that does not deem the use of intuition irrational either. Moreover, as we mentioned earlier, resources are often rather limited, so in many design situations, the information an intuitive approach would act on might be of higher quality than the information that can be acquired through a systematic approach under the resource limitations.

Our position is supported by a study carried Ahmed et. al, who also explicitly relate design intuition to experience as a part of an exploratory study (Ahmed et. al [1]). Their preliminary findings, based on interviews of three expert designers, suggest that experts themselves might conceptualize design intuition as *unconsciously* relying on past design experiences while making decisions.

Our formulation of design intuition can be seen as being similar to case based reasoning, but there may some differences—to the extent that we can "observe" design intuition at work. The most significant potential difference stems from the reflective nature of design practice [23], which implies that designers engage in the processing of an experience without waiting to run into a situation that might be relevant to specific past experiences, and that such processing is ongoing, and most likely, cumulative. Another potential difference is the "holistic" nature of intuition (see Sect. 17.2.1), which seems to suggest that intuition can act at a rather abstract level.

To be perfectly clear, our intent in this section is not to judge if design intuition offers more utility than systematic design methods since, based on our discussion, that type of valuation depends mainly on the design situation, experience base of the designer, the ability of the designer to process those experiences, and the availability of resources to collect/generate and process new information. For instance, there is evidence suggesting that novice designers who self-reported higher levels of engineering intuition achieved lower design outcomes than novice designers who self-reported lower levels of engineering intuition [34]. That discrepancy can be attributed to any of these factors. *Our main intent is to differentiate design intuition from irrational—and mysterious—behavior.*

Finally, it is relevant to consider if design intuition is different from intuition as it may be experienced in other domains. In other words, is design intuition a specific type of intuition? We postulate that what differentiates design intuition is the context in which it is exercised: shaping and solving complex ill-defined problems. As we outlined in Sect. 17.2.2.1, in routine situations—provided the problem solver possesses the relevant schemata—the intuitive system will most likely being able to handle the problem without invoking the analytical system. However, most design situations will invoke the analytical system as well since shaping and solving a complex problem is unlikely to be accomplished by relying on existing schemata alone. Moreover, as documented extensively in the design research literature, the generation of a "satisfycing" solution [25] in such a scenario results in iteration, which means that both systems will be invoked repeatedly, and quite possibly, even simultaneously. It can even be argued that this is a fundamental characteristic of goal-oriented creative behavior when tackling complex problems.

17.5 Intuition in Design Practice: Interviews with Designers

In this section, we will put our theoretical framing into context by illustrating it with qualitative data on the role of intuition in design practice.

The data were collected by the master's students in our Design Theory and Methodology course during interviews with professional designers from various domains. The course aims to expose students to a reflective meta-view on the design process and its constituents. More specifically, students learn to study their own design process as well as the design processes of other designers in order to identify key variables and their effects on design outcome. In the last assignment of the course, students plan and carry out interviews with professional designers from different disciplines. The interviews, in combination with the course reader containing literature on fundamental issues in design theory, act as the knowledge base for analyzing the design processes of professional designers. This year, students were asked to pick two topics of interest from the course reader and explore them in the context of their interviews. Therefore, not all students focused on the same topics. The data referenced here were collected during 15 interviews that directly focused on the role of intuition in design practice.

We interpreted and grouped the transcript sections addressing design intuition into factors/themes that are relevant to our theoretical consideration. We will describe and illustrate each theme with excerpts from the transcripts below.

17.5.1 Clients Pay Designers for Their Intuition but also Might Fire Them for Relying on it

While several designers reported that clients hire them because their intuitions are of value to the clients, other designers reported that not being able to present clear rationale for making choices to the client can result in skepticism. Naturally, excess skepticism displayed by the client toward the designer's choices will result in the designer failing to convince the client to go along with the design, which is never a desirable outcome from the designer's perspective.

> I would like to follow my intuition, but most times I have to follow the rational outcomes of research because the rational outcome is more convincing to clients.
>
> Architect with 3 years of experience

The opinion expressed above illustrates this point. However, it also implies that there is a discrepancy between the outcomes of his intuitive and "rational" approaches. Is that because the two approaches fundamentally yield different results, the architect's intuition has not developed sufficiently, or his/her "rational" analysis is flawed? However, the client does not seem to be particularly concerned with this question, and would rather be presented with the "rational" approach.

17.5.2 Novice Designers Should not Rely on Intuition Because They Lack Experience

As we discussed earlier, experience seems to play a critical role in how designers develop intuition in that it is a prerequisite to having sound design intuition but does not necessarily guarantee it. Most of the interviewed designers recognized the link between gaining design experience and developing design intuition. Some went as far as stating that novice designers should not rely on their intuition because of their limited experience.

> The more you experience and know, the greater your intuition would be. Otherwise, if you are a junior designer, choosing the outcome of a rational method is better. Your intuition may mislead you. Therefore, for junior designers, the most important thing is to gain experience.
>
> Architect A with 3 years of experience

However, it can be argued that the lack of experience and well developed intuition might be advantageous since that might cause the designer to be more critical of his/her work. For instance, in the context of the quote below, it is debatable who the better boat designer is: the senior engineer who sizes components based on his/her intuition, or the junior engineer who makes the same type of decisions through analysis?

> I admire the knowledge of experienced guys in the field that are able to make assumptions by using some rules of thumb…The information collected to analyze the design problems is gathered by using the gut feeling… I like to affirm those rules of thumb by calculations, but experienced colleagues almost never fail.
>
> Naval Engineer A with 2 years of experience

The engineer's last sentence is particularly interesting because he/she realizes that the "rules of thumb" approach does indeed fail at times. The consequences of failure can be catastrophic depending on the context of the decision. It is plausible to argue that: an experienced designer who knows how to use his/her intuition well will not use it in critical situations; whereas an experienced designer who does not know how to use his/her intuition will use it in critical situations, and eventually cause a catastrophe?

17.5.3 Understanding the Basis for Your Intuition is Necessary

Not all interviewees who viewed intuition favorably accepted it as is. An artist was supportive of it although she seemed to think that it was not useful unless one knew its basis.

> Intuition often proves to be right, as long as you know where it comes from.
>
> Artist A with unknown duration of experience

> Sometimes gut feeling also plays a role in the decision process… Sometimes this has an influence on the process, but needs argumentation.
>
> Architect B with 5 years of experience

They might have been simply expressing the need to understand the automatism or heuristic at play before feeling comfortable enough to apply it. In other words, listen to your "gut," but do not trust it unless you understand its rationale. Of course, this viewpoint expresses the need for a designer to be open to the possibility that one's intuition might simply be misleading, but it also limits its efficiency. By definition, intuition is efficient because we do not have to think about it rationale before acting on it. Also, by definition, it is highly unlikely that its basis can be fully understood.

17.5.4 Intuition Tells the Designer When a Project Is Finished

Design methodology would advocate that a design project has been successfully concluded when its requirements are met by the solution at hand. That is normally done by applying a set of requirements, metrics and targets that have been made explicit upstream to formally evaluate the performance of the design. Instead, several interviewees reported that they simply rely on their intuition to tell them when the design is "done."

> When you've been working for some years, you know when something is good or not.
> Urban Planner A with unknown duration of experience

> When a song is finished, it's finished. We [the band] just know.
> Musician with 5+ of experience

Designers internalizing the requirements and continuously (and unconsciously) assessing them in real-time is a potential explanation of this viewpoint. However, it has been argued that it is not possible for humans to keep track of a complex list of pros and cons for multiple alternatives to unconsciously arrive at a critical decision [11].

17.5.5 Time Pressure Necessitates the Use of Design Intuition

In according with Cross' view [3], several interviewees saw intuition as a short-cut in decision making, which simplifies and makes the process manageable given the complexity of real-world design situations.

> Sometimes, time pressure is also a very big…under that pressure sometimes you have to be inventive, you have a lot of ideas but you don't know how to deal with it and then you have very little time, and you have to take a decision, to make it simple, because you don't have time to make a complex. Sometimes it's very hard to take all the elements to get them all together, so you have to ignore or find it out later.
> Architect C with 10+ years of experience

In certain situations, ignorance can be beneficial. For instance, Gigerenzer claims that heuristics that drive our intuition might allow us to make certain decisions without a comprehensive information base [11], and that the fact we are ignorant of certain information might actually reveal the irrelevance and insignificance of that information. Of course, there are significant drawbacks to that position as the consequences of ignoring relevant information can be severe.

Project timelines, and thus, the nature of the design work, also have implications for how intuition is perceived with respect to time pressure.

One interesting set of interviews revealed that a flamenco composer with 10 years of experience did not prefer to rely on intuition because the timescale of his projects are rather long, and that he does not feel under pressure to finish a piece and works on it "rationally" until it is done. In contrast, an art director at a large market leading videogame company reported that, under pressure from the competition developing similar products in the marketplace, she almost always relies on intuition to make critical design decision because there is never any time to explore and test the alternatives in a rigorous manner.

17.5.6 When Working in Groups, Relying on Intuition Is Nuanced

Intuition does not seem to always offer a short-cut to making decisions. A product designer found relying on intuition to be problematic in a group setting because groups are not likely to accept the position of individuals unless individuals offer rationale. However, when he reflected on individual work, he said:

> I often have a gut feeling which direction is the right one to go in.
>
> Product Designer A with 5+ years of experience

In contrast, a choreographer perceived a different group effect, which seems to highlight the relevance of the necessary information for making decisions emerging out of social interactions within the group.

> With the preparation I think in a logical way. I think of a theme and make associations on the theme. After that I am reasoning logical steps, the tasks for the dancers. Working with the groups is more feeling and intuition, I see and feel things.
>
> Choreographer A with 25+ years of experience

17.5.7 Intuition Can Lead to Decisions but Does not Tell the Designer how to Implement Them

Several interviewees made a distinction between arriving at a decision based on intuition and implement that decision. They saw intuition being more relevant to the former.

> Intuition tells you what way to go, but the how to go there is something different.
>
> Software Designer A with 10+ years of experience

> I see the solution in one second! But then I need months to build them.
>
> Artist B with 10+ years of experience

This might indicate that some designers are more aware of the role intuition plays in conceptual design thinking. Intuition must come into play during detail design and implementation as well, but the interviews did not perceive its influence downstream in the design process. Conversely, this might be related to the differentiated role analytical thinking might play in the different phases.

17.5.8 Intuition Can Act as a Warning Flag

Intuition does not always tell designers where to go; it can also tell them where not to go. In fact, as we referenced earlier, research has shown that our nervous system unconsciously detects an error before we become aware of it.

When my gut feeling would say that a certain direction is not a good idea, I had no reason to not listen to that. You then would think about changing your principal or system as soon as possible, based on your gut feeling.

You try to get rid of gut feelings...Sometimes you would not get rid of your gut feeling, but you cannot find any alternatives. This usually happens when you get an assignment from the boss and you immediately think that it will fail. Having to design an engine that runs on water, for example.

Machine Designer A with 23 years of experience

17.5.9 *Intuition can Lead to Subjective Thinking and Fuel Creativity*

Although a core criticism against relying on intuition is its subjectivity, that limitation can also be seen as an advantage in terms of creativity.

I think in most cases I rely on my logic thinking more because architecture is mainly about function. The space you design need to be a fit for the users. Therefore, it is all about fitting the regulation and the rules. And the logic thinking as a framework, under the framework you will still using your gut feeling to design. And that's why different designers will design differently under the same regulation and rules.

Architect D with 28 years of experience

Since no two designers are likely to have the same set of experiences and reflections around those experiences when faced with the same design situation, they are likely to experience (slightly or drastically) different intuitions. Naturally, that can be a source of diversity in approaches and solutions, which can be beneficial for creativity.

17.6 Conclusions and Future Work

Intuition has been reported to be an important aspect of judgment and decision making in many domains. In this chapter, we focused on the role of intuition in design thinking and acting, and argued that theoretical integration and empirical investigations that would allow us to understand the mechanisms of design intuition are missing. In response, we offered a detailed theoretical exploration of the concept, and proposed a framework on the nature of design intuition that can lead to a theory of design thinking.

We supported and extended the theoretical framework with empirical findings on how and when practicing designers rely on intuition. More specifically, different situational variables such as time pressure and group and client interactions seem to influence the use of design intuition. Knowing when to trust your intuition surfaced as a critical factor in the use of design intuition.

In conclusion, comprehensive theories of design thinking need to take into account unconscious processes such as intuition. From a methodological perspective, design methods should acknowledge the designer's need to rely on intuition in certain situations—especially under time pressure. At a more advanced level, design methods should support the designer in assessing the limitations and benefits of utilizing intuitive approaches.

There is still much to be learned on the relationships between the use of structured design methods, intuition, and design performance. This work led us to develop the following research questions as the future work:

- In what situations should designers rely on design intuition rather than structured/explicit approaches for better performance? More importantly, how can designers make that decision in real-time during design practice?
- Is the above determination different for novice versus expert designers? Can experts leverage intuition more than novices, and conversely, is it more risky for novices to rely on their intuitions?
- If intuition indeed develops in parallel with experience, and hence is learned to a certain extent, how can design methodology facilitate that learning process?
- How can design methodology, advocating a hierarchically structured design process, integrate design intuition into existing design methods and process models?

Although our exploration deals with these issues and offers preliminary positions on several of them, these are major questions that warrant more detailed and systematic considerations.

Acknowledgments We would like to thank the industrial design engineering master's students of the DTM course at Delft University of Technology, who helped to collect the data referenced in the empirical dimension of this work. We would also like to thank the reviewers for their valuable feedback.

References

1. Ahmed S, Wallace KM, Blessing LT (2003) Understanding the differences between how novice and experienced designers approach design tasks. Res Eng Des 14:1–11
2. Badke-Schaub P, Daalhuizen J, Roozenburg N (2011) Towards a designer-centred methodology: descriptive considerations and prescriptive reflections. In: Birkhofer H (ed) The future of design methodology. Springer, London
3. Cross N (1999) Natural intelligence in design. Des Stud 20(1):25–39
4. Cross N (2004) Expertise in design: an overview. Des Stud 25(5):427–441
5. Dane E, Pratt MG (2007) Exploring intuition and its role in managerial decision making. Acad Manag Rev 32(1):33–54
6. Dreyfus HL, Dreyfus SE (1986) Mind over machine: the power of human intuition and expertise in the era of the computer. Blackwell, Oxford
7. Evans J. St. BT (1984) Heuristic and analytic processing in reasoning. Br J Psychol 75:451–468

8. Evans J (2003) In two minds: dual-process accounts of reasoning. Trends Cognit Sci 7:10
9. Evans J (2012) Questions and challenged for the new psychology of reasoning. Think Reason 18:5–31
10. Freud S (1953) The interpretation of dreams. In: Strachey J (Ed. & Trans.) The standard edition of the complete psychological works of Sigmund Freud, vols 4 and 5. Hogarth (Original work published 1900), London
11. Gigerenzer G (2007) Gut feelings. Penguin Books, London
12. Gunther J, Ehrlenspiel K (1999) Comparing designers from practice and designers with systematic design education. Des Stud 20(5):439–451
13. James W (1950) The principles of psychology. Dover. Original work published 1890), New York
14. Heiman GW (2002) Research methods in psychology, 3rd edn. Houghton Mifflin Company, Boston & New York
15. Hermanns W (1983) Albert Einstein, Einstein and the poet: in search of the cosmic man. Brandon Books, Brookline
16. Hubka V, Eder WE (1988) Theory of technical systems; a total concept theory for engineering design. Springer, Berlin (Translation of V. Hubka, Theorie Technischer Systeme. Berlin: Springer, 1984)
17. Kahneman D (2003) A perspective on judgment and choice. Am Psychol 58:697–720
18. Khatri N, Ng HA (2000) The role of intuition in strategic decision making. Hum Relat 53(1):57–86
19. Klein G (2003) The power of intuition: How to use your gut feelings to make better decisions at work. Random House Digital, Inc
20. Rittel H, Webber M (1973) Dilemmas in a general theory of planning. Policy Sciences 4:155–169 (Elsevier Scientific Publishing Company, Inc, Amsterdam)
21. Pascal B (1958) Pensees, intro. TS Eliot, New York, Dutton
22. Sadler-Smith E (2008) The role of intuition in collective learning and the development of shared meaning. Adv Dev Hum Resour 10(4): 494–508
23. Schön DA (1983) The reflective practitioner. How professionals think in action. Basic Books, New York
24. Simon HA (1972) Theories of bounded rationality. Decision and organization 1:161–176
25. Simon HA (1973) The structure of ill-structured problems. Artif Intell 4:181–201
26. Sloman SA (1996) The empirical case for two systems of reasoning. Psychol Bull 119:3–22
27. Stanovich KE, West RF (2000) Individual difference in reasoning: implications for the rationality debate? Behav Brain Sci 23:645–726
28. Tversky A, Kahneman D (1973) Availability: a heuristic for judging frequency and probability. Cognit Psychol 42:207–232
29. Tversky A, Kahneman D (1974) Judgment under uncertainty: Heuristics and biases. Science 27:1124–1131
30. Vaughan FE (1979) Awakening intuition. Doubleday, New York
31. Visser W (2009) Design: one but in different forms. Des Stud 30:187–222
32. Webster Online (2013). http://www.websters-online-dictionary.org/definitions/intuition
33. Wessel J, Danielmeier C, Morton JB, Ullsperger M (2012) Surprise and error: common neuronal architecture for the processing of errors and novelty. J Neurosci 32:7528–7537
34. Yang MC (2005) A study of prototypes, design activity, and design outcome. Design Studies 26(6): 649–669

Chapter 18
Revisiting Design as an Information Processing Activity

S. J. Culley

18.1 Introduction

Over the years a large number of authors have commented on the importance of information and knowledge manipulation associated with the overall design process, see for example McCloud and Corlett [36], Eder [20] and Ullman [50].

There seem to be two camps here, the knowledge focused camps as exemplified by Vincenzi's [52] seminal work 'What engineers know and how they know it' and the information focussed camps. There is Shears publications in the IEEE [48] 'Engineering design as an information-processing activity'. This paper was published in 1971 and it is surprising how little the paper has been cited. Shear's interpretation of the core process is clearly shown in Fig. 18.1.

There are a number of definitions of engineering design and as Dieter [17] states '.... almost as many definitions as there are designs', such as—'The transformation of ideas and knowledge into a description or artefact, in order to satisfy a set of identified needs; it is the key technical ingredient in producing new products governing the match between product and actual requirements' [13].

Although there is a degree of variation between researchers, engineering designers and other individuals of what is meant by the term 'engineering design', however there are a number of richer and more precise definitions as seen below.

'Engineering design is the use of scientific principles, technical information and imagination in the definition of a mechanical structure, machine or system to perform pre-specified functions with the maximum economy and efficiency' [22].

'Engineering design is a process performed by humans aided by technical means through which information in the form of requirements is converted into information in the form of descriptions of technical systems, such that technical systems meet the needs of mankind' [20].

S. J. Culley (✉)
Department of Mechanical Engineering, Bath University, Bath, UK
e-mail: enssjc@bath.ac.uk

A. Chakrabarti and L. T. M. Blessing (eds.), *An Anthology of Theories and Models of Design*, DOI: 10.1007/978-1-4471-6338-1_18,

Already there are a number of situations where this is formally required and these are discussed below.

18.1.3 Information Outputs from the Design Activity

The areas where a number of specific information outputs are required are listed below, this is a non-exhaustive list but gets across the criticality of the formal information dimension.

1. The Technical Construction File(TCF) This is required as part of CE marking regulations in the EU [7]. This will include such requirements as Internal production control, Conformity to type, Production quality assurance, Product verification, Full quality assurance. This is not an exhaustive list.
2. Information for accreditation systems, such as ISO 9000. ISO 9000 consists of a three-step cycle of planning, controlling and documenting quality in an organisation. It provides minimum requirements needed for an organisation to meet their quality certification standards. Companies will set up procedures and processes that are auditable. These are intended to give confidence to their partners and customers that performance and quality is embedded throughout the company, including the design area. These procedures also dictate workflow patterns [9].
3. Information for formal certification. There are a number of industries, such as Aerospace, Nuclear, defence where there are very strict requirements for certification. There is s very good white paper by Bently [4]. These certification procedures require extensive and specific information output for the inspectors and auditors. The full list of European regulations for the Aerospace industry are shown at the website of EASA [19] in Europe. It is interesting to note the level of detail that are prescribed by these very extensive documents that cover sailplanes, balloons as well as normal passenger aircraft.
4. Information for lifecycle. There is an increasing realisation that the information that is created as part of the design and development activities has a vital role to play in the lifecycle of a product machine or system. This is particularly true of the increasing number of those systems that have 10, 20, 30, 40-year lifespans [40, 46, 54]. The information that is generated will have a vital role to play as maintenance is implemented, upgrades are installed or products are converted from one use to another. There is for example, a large business in converting passenger aeroplanes into freighter aeroplanes.
5. The gated design process is widely used in industry [10]. There are very formal requirements which vary from company to company, but they all will pre-specify certain tasks and certain information sets to be generated before the formal review [49]. This formal review may be undertaken as a paper exercise or as a meeting and face-to-face review activity. Famously Airbus changed the design of the A350 to the A350EWB as a result of a review. The original

specification was found to inappropriate and hence the move to Extra Wide Body—EWB.
6. Information for and Information as part of in-service activities. The full service provision is a vital part of the offering of any organisation [41]. So it is necessary to create information as part of the design process that is supports this activity. Also when design repairs or upgrades, information is generated and archived as part of the auditable process to support the rational for the result [56].

18.1.4 Content of the Chapter

This chapter has three main technical sections on design knowledge, information and engineering design information.

The section on information associated with the design activity establishes three core activities namely: (1) information acquisition for design, (2) information generation as part of the design activity and (3) information storage for design. This is both as a consequence of and as a requirement of the various internal and external demands. The potential use of workflow approaches associated with an information focussed design process is also included. The chapter concludes by highlighting the benefits and the requirements for an information-driven design process.

18.2 Knowledge and Information

The focus of the subsequent sections is on information, however, it is necessary to briefly understand the role and nature of knowledge in engineering design to set the role and nature of engineering information in context, 'information as knowledge' is one of the key ingredients of the first dimensions mentioned above.

Vincenti [52] in his book 'What engineers know and how they know it' argues, through his observations of the historical developments in aeronautical engineering, that engineering knowledge is developed and formalised to meet the needs of engineering designers in a particular domain and that some items of this knowledge are clearly distinguishable, whilst others are not.

However, Ullman [51] has a very pragmatic approach to this topic and proposes three types of knowledge that engineering designers make use of and refer to during their work:

1. *General knowledge,* gained through everyday experiences and general education. The information used in updating this knowledge is that which most people know and apply without regard to the specific domain that they are working in.

Table 18.1 A typology of design knowledge

Knowledge type	Knowledge dimension	Definition	Example
Embedded knowledge	Explicit	Systematic routines, procedures and practices	Company documents on design procedures & sign-off
Encoded knowledge	Explicit	Knowledge represented by signs and symbols in books, manuals and recorded works	Engineering text book on the principles of aerodynamics
Encultured knowledge	A combination of the two	Knowledge from the process of achieving shared understanding.	Personal log-book of experience on design project
Embrained knowledge	Tacit	'Knowledge about'—the ability to work with complex ideas and concepts	Personal experience of a variety of design projects
Embodied knowledge	Tacit	'Knowledge how'—practical thinking; problem solving.	Personal ability to plan and execute a design project

2. *Domain-Specific Knowledge,* gained through study and experience within the specific domain that the designer works in. Information is on the form or function of individual items or groups of items.
3. *Procedural Knowledge,* gained from experience of how to undertake one's tasks within the enterprise concerned. This form of knowledge is often based upon a combination of the previous two.

Many categorisations of knowledge have been proposed, and a number are particularly apposite in the context of design. Ryle [45] distinguished between two different types of knowledge—'know how' and 'know that'. He noted that learning about a subject primarily involves the accumulation of 'know that'—principally data, facts and information. Learning about, however, does not produce the ability to put 'know that' into practical use (i.e. knowledge as some type of competence notion). This, Ryle argued, calls for 'know how', which does not come through the accumulation of information. Learning how to do something can only be carried out in practice, which explains why the same information (e.g. a manual, book, verbal instructions, etc.) directed at different people (with different backgrounds and experiences) does not result in the same knowledge in each—practice and context shapes the assimilation of information by individuals. This distinction can be seen in Blackler's typology of knowledge shown in Table 18.1 with additions by the author [5, 37].

In the context of design, it is suggested that encoded knowledge describes that knowledge and information recorded in books, manuals, codes of practice, specifications and so on, together with recorded information concerning materials, manufacturing processes, machine elements and other components and so on. It

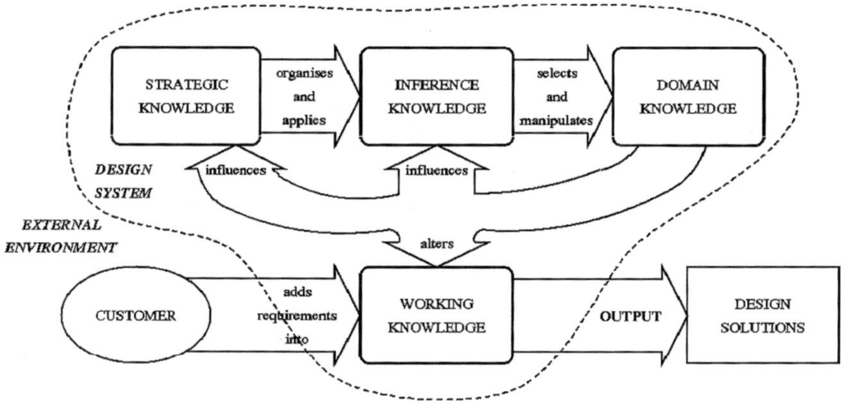

Fig. 18.2 The relationship between the knowledge categories during the design process [16]

becomes clear here where the 'information as thing' is part of this spectrum of design knowledge. This is the embedded and encoded knowledge is concerned with knowledge and information used in design—for example the processes of design analysis and assessment and the formal processes of interaction between the participants in the process. These may be documented in codes of practice, design guides and the like.

Embrained knowledge by contrast describes the implicit or the tacit ability of people to work with complex ideas and concepts, Embodied knowledge is also in general tacit and describes the general problem-solving approaches and attitudes of mind found in design. Embodied knowledge also allows the community to know the limits of its knowledge and where it breaks down. Finally, encultured knowledge may describe the implicit "shared memory" [32] that exists in the design community of practice concerning the shared beliefs and values of the community.

Another view, again introduced as background, is a development of the work on the hierarchical relationships of design knowledge [55]. It is included here to show how a lot of what is being considered is really information and this really is the critical dimension of the overall design sequence.

This work identifies four major critical groups of design knowledge these relationships between the categories that exist in a design system (Fig. 18.2). These categories are still quite loosely defined, and as they say they will vary from domain to domain, and may even vary within a domain when, say, different design strategies are applied, so this description cannot be considered as a generative definition for a design system.

The elements which really are information based are domain knowledge, which when combined with strategic and inference knowledge generates the working knowledge that enables the design to progress. It is with this working knowledge that extensive amounts of 'information as thing' have to be generated and recorded and used. The four groups are summarised below.

Domain Knowledge: This class contains knowledge of the entities which constitute the domain. For configuration design, this group includes knowledge of, the physical elements (and their behaviours) which may constitute a solution. How groups of related elements (up to and including the system level) behave, component parameters, and so on. This category would seem to consist primarily of declarative knowledge—these elements correspond in some way to the external evidence of the design task.

Inference Knowledge: This knowledge is 'reasoning knowledge that allows an abstract element of a design may be made 'more concrete' according to the requirements specified, the intermediate abstractions already formed, design choices made elsewhere, etc.

Strategic Knowledge: This is knowledge of how elements of inference knowledge can be arranged and controlled so as to provide a complete strategy for producing a design. This amounts to a set of high level methodologies for controlling the search for mappings from requirements to solutions. This is procedural knowledge of the design process.

Working Knowledge: This is unique for each design episode and contains the specific requirements, design choices made, knowledge of the reasons for the modifications to a design, feedback from the customer about the application of the designed system, etc. This category represents a 'pool' of knowledge about the current design process, from which elements may be retrieved when they are necessary for invoking or applying elements from the other categories of knowledge.

18.2.1 Overview

Knowledge is an important ingredient in the overall design activity and has been extensively researched. Although the thrust of this overall chapter is about information, covered in subsequent sections. There is a clear blurring of the boundaries between 'information as knowledge' and 'information as thing'. It is also clear that Information outputs from the design activity as listed in Sect. 18.1.4 require a high degree of formalisation.

18.3 Engineering Design Information

It is possible to take the work described above and identify a number of design situations that need documenting and extract the elements formally on knowledge in design. These three areas are listed below:

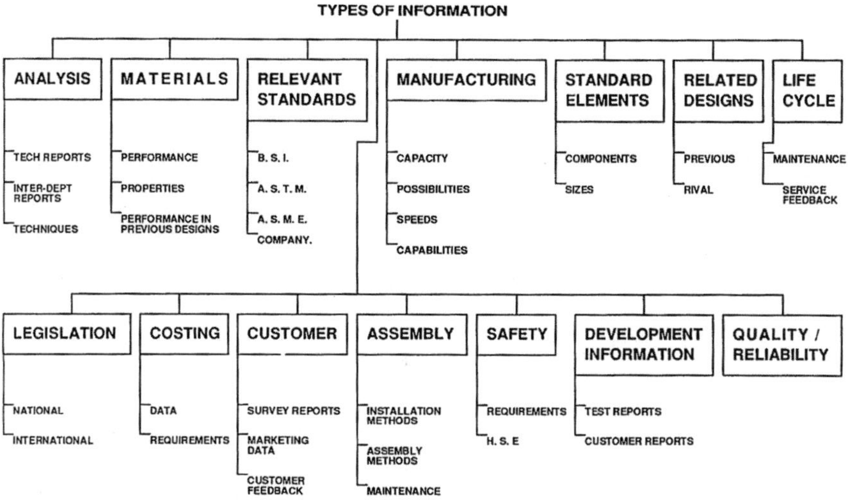

Fig. 18.3 Types of design information [11]

1. Information *acquisitionfor* design
2. Information *generationby* design
3. Information *storage for* design.

These aspects will be discussed in the subsequent sections, but the first section gives some base definitions and restates the differences between Types and Sources of design information, shown in Figs. 18.3 and 18.4.

18.3.1 Basic Definitions

A preliminary list of factors relating to engineering design information can be readily generated from the often vast quantity of information found within an engineering enterprise, which comprises many different types, even in the smallest enterprise. Wall [53] and McLeod and Corlett [36] give excellent and extensive explanations of engineering design information, which is provided in many different formats, consists of many different types of information and may be accessed from many different sources. In this regard it is important to differentiate between an information type and an information source [11].

18.3.1.1 The Types of Engineering Design Information

The following definition of Information Type is proposed [12].

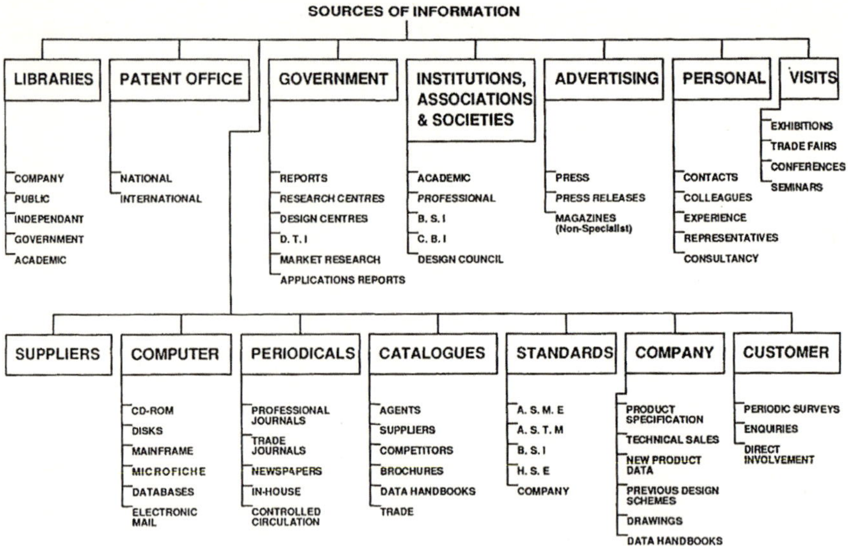

Fig. 18.4 Sources of design information [11]

> ***TYPE*** of information is ***WHAT*** information is required to undertake a particular task. Examples include material strength, production lead time, why a certain design was used in the past or how the design is to be installed in the working environment.

A hierarchical tree has been developed, using this definition as a basis and combining it with the elements of the PDS identified by Pugh [44]. The resulting tree comprises 13 core types of information and 33 sub-types (Fig. 18.3). The tree does not provide an exhaustive set of information types, rather it consists of the essential elements proposed by Pugh and is intended to illustrate what is meant by the term 'type' of information.

18.3.1.2 The Sources of Engineering Design Information

The term 'information source' has also been widely considered by design information researchers, and there is a general consensus of what the main sources of engineering design information should be. The term source of information is sometimes used synonymously with information type, but it is considered important to distinguish between the two. The following definition of 'information source' is proposed [12]: (Fig. 18.4).

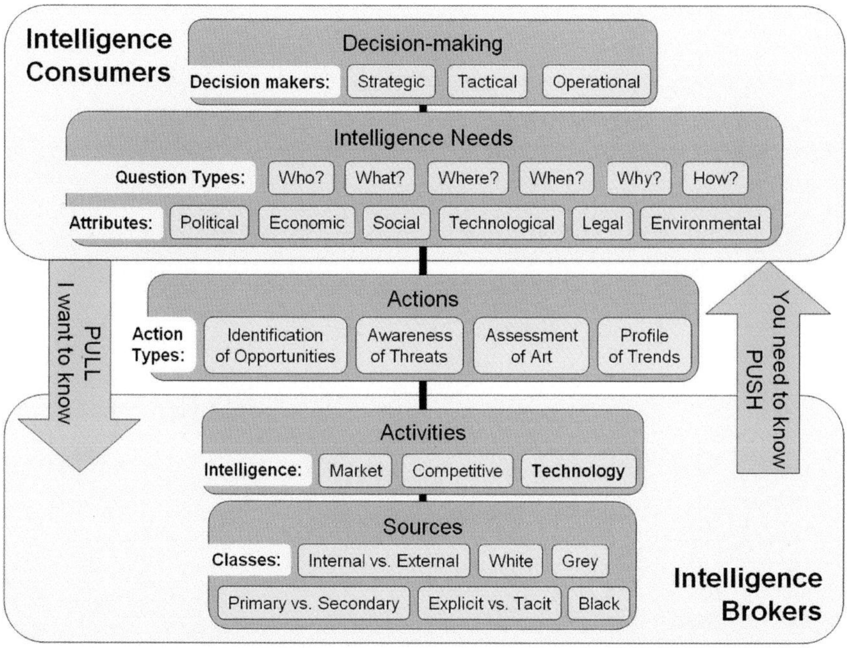

Fig. 18.5 The intelligence framework (adapted from [31])

> ***SOURCE*** is defined as ***WHERE*** such information can be obtained—for example a textbook, a journal, a drawing, a colleague, etc.

18.3.2 Information Acquisition for Design—New Information

It is useful to start with [31] work on technology intelligence, they talk about mining, trawling, targeting and scanning and illustrate these processes clearly as shown in Figs. 18.5 and 18.6.

Figure 18.6 is particularly interesting as it gets across the cyclical nation of the information lifecycle. This lifecycle has been highlighted before [15] but this shows information in the technical domain.

It also shows (in the bottom right) the fact that at various stages that the information will be formally ‘documented’.

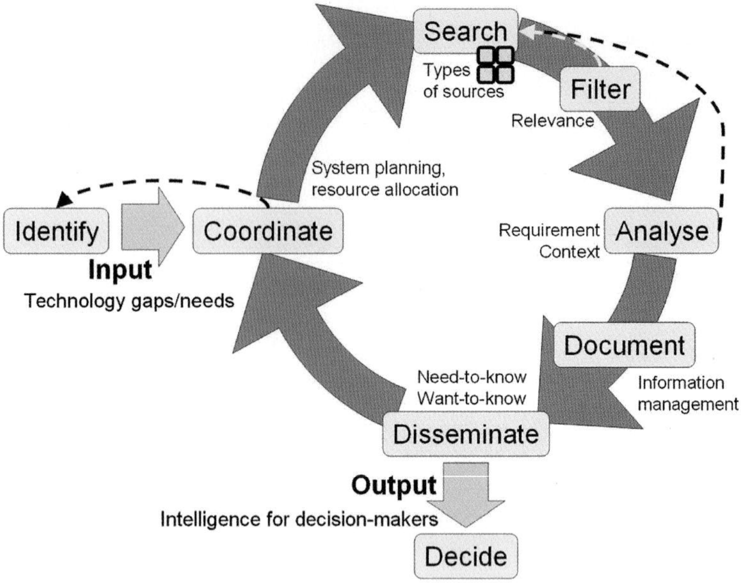

Fig. 18.6 The system operating cycle [31]

Other authors have commented on acquisition for design. These methods for technology transfer (and hence technology tracking) include looking out for opportunities for the company to license technologies, acquire businesses, subcontract, recruit, get involved in joint ventures, or purchase products to incorporate into the company's own product [2]. Also [8] defines two kinds of technology that are essential to a firm's success, yet conflicting in nature. Sustaining technologies are new technologies that improve an existing type of product's performance. Disruptive technologies, on the other hand, are new technologies that initially perform worse than the equivalent sustaining technology, but distinguish themselves by introducing a new method of operation and creating unforeseen market demands that eventually supersede the sustaining technology.

It has also to be appreciated that a lot of engineering information is proprietary and frequently closely controlled within an organisation. This is why the knowledge management aspects discussed in Sect. 18.2 are particularly important.

18.3.3 Information Acquisition for Design: Internal

With the advent of computing networks becoming commonplace, much information previously stored as hardcopy has been transferred into organisations' shared directory structures, document management systems and shared

Table 18.2 Sources of technical intelligence [42]

		Proximity	
		Direct	Indirect
Character	Personal	Personal networks	Gatekeepers
		Sponsored research	Consultants
		Visits	Editors
		Trade shows	Expert panels
		Venture capitalists	Suppliers/vendors
		Universities	Analysts
		Entrepreneurial firms	Retired executives
		…	…
	Impersonal	Patents	Industry surveys
		Patent citations	Trade journals
		Literature searches	Associations
		Reverse engineering	Government records
		Marketing material	UN reports
		Annual reports	
		World wide web	Local newspapers
		Ads for staff	World Wide Web
		…	Buyers' guides

workspaces [38]. With electronic communications and information databases, however, the aspect of context, which is of very high importance in an engineering environment, has not been sufficiently translated. Context is often intrinsically linked to a person possessing a given piece of knowledge or expertise [56, 57].

A typical engineer's first point of contact in searching for information is predominantly a fellow colleague engineer working within the same organisation [47]. The colleague is likely to provide not only the relevant documentation, but an overview of the context of the project, which benefits the seeking engineer greatly in terms of both understanding and time saving [28]. The act of conversation quickly pinpoints the knowledge need and usually results in the relevant literature. Two observations can be made from this: the importance of face-to-face communication, and the inadequacy of catalogue searching.

The speed of understanding, that is, the time it takes for an engineer to convert information of ambiguous usefulness into useful knowledge and understanding through context is a critical consideration. Engineers especially appreciate this because it takes much less time to understand a conversational overview of a topic than it does to delve into an official document, especially since the informed party in the conversation is able to change and tailor the level of technicality in a description during the course of the conversation, whereas a document is static.

Paap [42] created a matrix based on the proximity of the source to the information seeker, and the character of the source based on interpersonal contact (Table 18.2).

Fig. 18.7 Popular supplier communications media compared with resulting information storage levels (adapted from Culley et al. [14]

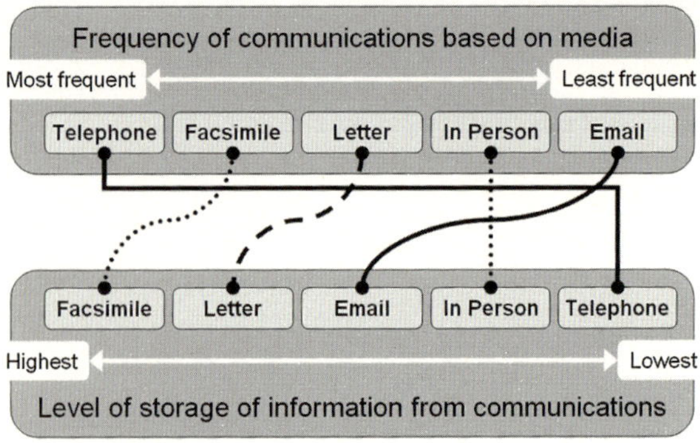

The disadvantage of conversation, however, is that much of the content may be forgotten relatively quickly, necessitating documentary reminders [14, 28]. The point of conversing would be to establish basic knowledge and context as quickly as possible so that the engineer could then deal with the more detailed document in an informed manner.

The discussion above, again, shows the complex relationship between 'Information as knowledge' (the personal interactions) and 'information as thing'. With the requirements to progress through the design process it is clear that a transformation between the two has to take place.

18.3.4 Information Acquisition for Design: External

Information acquisition for design is a complex topic and again the boundary with information generation for design becomes blurred. But the first consideration is that information that is acquired from external sources. The author [14] surveyed the habits of engineers in obtaining information from suppliers, and found the most frequently used media of communications. They also surveyed the level of storage of information gained from these communications. Interestingly, despite the telephone being the most popular form of communication, the information gained from it is the least documented of the media (Fig. 18.7). This implies that a great deal of information gained from suppliers is not being documented, and thus may be lost. It is this information that may be needed for the elements highlighted in Sect. 18.1.4.

Culley et al. [14] discuss the issues surrounding the relationship between suppliers and engineering designers. They highlight the problems incorporating knowledge and information available from suppliers as sources for the engineering design process, the phenomenon of suboptimal supplier selection and inappropriate decision-making on supplier integration, as well as the poor management of information and knowledge available from suppliers.

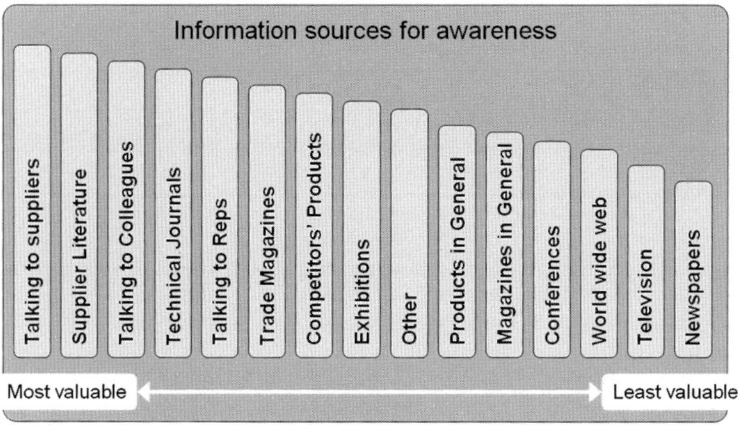

Fig. 18.8 Value of information sources for awareness purposes. Not to scale. (Adapted from Culley et al. [14])

From a survey, they also listed the information sources most valued by engineers for awareness purposes. As seen in Fig. 18.7, suppliers form a very important information source.

18.3.5 Information Generation by Design

The research, design and development activity generates large amounts of data sets as illustrated in Table 18.3 Dodgson & Rothwell [18] describe possible sources of learning which may result in innovation (Table 18.4).

Table 18.4 shows the information sources additional to Table 3 encountered over the course of a project and shows the types of media in which information sources could be presented [18].

A list of sources of design information, shown in Fig. 18.4, Sect. 18.3.1.2. Although extensive, this list is not exhaustive, and serves to give an indication of the variety of sources that could potentially be tracked. Additionally, [36] list a large number of information sources in engineering. (Fig.18.8)

18.3.6 Information Storage for Design

The two previous sections have looked at information acquisition and generation. It is possible to represent this with a simplified model of the total space as shown in Fig. 18.9. It is the elements in the lower boxes that represent the elements that need to be recorded as part of an information based overall design process. There

Table 18.3 R&D activities and outputs [4,18]

Narrowing design space → increasing project cost			
Activities			
Research domain		Advanced engineering domain	Development domain
Basic research	Applied research	Experimental development	Design engineering
Generating new knowledge and options	Generating new knowledge with a practical aim	Demonstrating technical viability	Translating known and demonstrated principles into new products and models
Understanding theory	Developing tools and simulation	Eliminating technical uncertainty	
Developing instrumentation and measurement techniques		Choosing actual technologies and materials	
Tracking and absorbing external knowledge			
Research papers, report, bench-top demonstrators, patents		Demonstrators, know-how	Designs, prototypes
Outputs			

Table 18.4 Innovation as a process of know-how accumulation [18]

Internal learning by…	External learning through…
Developing (R&D)	Suppliers
Testing	Lead Users
Making (production)	Partnerships
Failing	S&T infrastructure
Using in vertically integrated companies	Literature (including Patents)
Cross-project communication	Competitors
	Reverse engineering
	Acquisitions/New personnel
	Customer-based prototype trials
	Servicing/Fault finding

are a number of ways of viewing the issue of storage, first from a quality and future requirement viewpoint and the second is considering the methods and techniques for storage and future reuse.

18.3.6.1 Quality Issues

Once the information has been stored or recorded in the Technical Construction File(TCF) or part of the documented submission for a stage gate review or an ISO 9000 accreditation visit it is also necessary to ensure the quality of the output. This

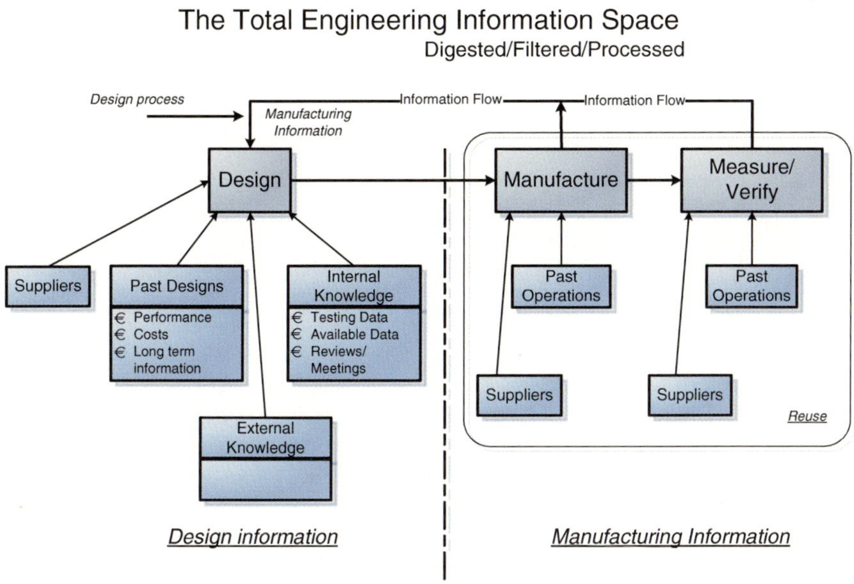

Fig. 18.9 Overview of the design, development, manufacturing environment

is thus an additional dimension to the 'information as thing' paradigm. Pahl and Beitz [43] has defined 'information quality' further and categorised it under nine criteria. These are:

1. *Reliability*, that is, the probability of information being trustworthy and correct.
2. *Sharpness*, the precision and clarity of the information content.
3. *Volume and density*, the number of words, pictures needed for the description.
4. *Value*, the importance to the recipient.
5. *Actuality*, an indication of the point when the information can be used.
6. *Format*, the distinction between graphic and numeric data.
7. *Originality*, the indication of whether or not the original character of the information is preserved.
8. *Complexity*, the structure and similarity between symbols, units, etc.
9. *Degree of refinement*, the quantity of detail in the information.

The important influencing factor of these categories is in the improvement of the match between the quality of the available information to the requirements of the end user. Wall [53] emphasises this point on a number of occasions in his review of information in product design and in particular he states that : '*There is a fundamental need for a comprehensive approach to the subject of data for designers,....which must start with the specification of what designers need*'.

18.3.6.2 Storage Issues

There seem to be a number of core information tasks, for the lifecycle of the project and the lifecycle of the product [1, 39]. These are listed below in the left-hand side of Table 18.5. These summarise a number of the elements that have been discussed above. The right-hand side lists a number of potential approaches for the future developing the ideas of Hicks et al. [29].

18.3.7 Workflow and an Information-Driven Design Process

The argument of this chapter is that to achieve a successful design outcome, 'information as thing' has to be created and recorded. This is at various stages of the design process to enable progression, protection, certification and lifecycle support for products and systems. It may be that workflow approaches could be appropriate to support these activities. These workflow approaches are widely used, in particular in the financial, insurance and other business sectors.

Within an Audit firm there are various processes that must be carried out. On a small audit, one individual from a firm would carry out all these processes, and therefore knows what has been carried out and what needs to be done. However, on a large audit, for a client such as BP, there could be a couple of hundred auditors working all year round. This poses a huge problem in terms of information and knowledge management, which is solved via custom software. Ernst and Young use an online library, in which employees can research any area of audit and find global frameworks, standards, and authoritative literature. The library is known as the Global Accounting & Auditing Information Tool (GAAIT) [21].

The firm like many others have a workflow in place known as the Global Audit Methodology (see [24] for a full description), which is designed to create a sequential consistency to the work produced by the various individuals. This is produced as a piece of custom software known as the GAM-X in which individuals can complete various sections of work, upload the work and then the manager checks it and ticks the box to sign it off. When all the sections are complete and signed off the audit is complete and can be reviewed. Included in this software is an instant messenger, this aids the ability for individuals to work on a client at the same time, such as one person in the office and one at the client site, or a tax specialist advising the auditor.

18.3.7.1 An Information-Driven Design Process

It seems the financial services have embraced the shift to globalisation and developed strong-shared knowledge and work platforms; can this attitude be carried forward to engineering methodologies and design activities? There has been some discussion of this topic, by senior industry figures [23], interestingly, he tends to use the term knowledge.

Table 18.5 Information activities, tasks and approaches

		Core information tasks (CIT)	Potential approaches
Information Activity	Creating/Recording	*1. Creating new information*	*Automatic recognition, recording and visualisation of additional context and links*
		1.Creating new, multi-modal, multi-media information (e.g. sketches, calculations, simulations etc.)	1.Content-centred linking of paper-based info with related digital info
		2.Linking between new information and existing information/knowledge, both physical and digital (e.g. by referencing and pointers)	2.Automatic addition of contextual information (such as author, date, location etc.) to physical documents
		3.Adding contextual information to aid refinding and understanding, and as an aide memoir	3.Domain and activity-specific mark-up/tagging of physical content
		2. Recording new perspectives on existing information	*Effective recording and linking of formal and informal sources*
		1.Summarising of various situations and events—e.g., meetings/teamwork, experiments and test results	1.Augmentation of textual documents with multi-media, 3D models, etc.
			2.Allowing virtual annotation of physical documents
		2.Annotating or supplementing existing information with additional context or understanding	3.Visualisation of links/relations between informal & formal information with formal Document Management Systems (DMS) & Product Lifecycle Management (PLM) systems
	Finding/Refinding	*3. Searching across mixed-mode, distributed information*	*Augmenting information search activities*
		1.Search across physical records and digital records	1.Integrating physical and digital resources from many sources via hyperlinks and AR visualisation
		2.Search across different media, e.g. text, images	
		4. Finding relevant related information	2.Use of visual, multi-faceted searching [32]
		1.E.g. the informal information containing the rationale for the retrieved CAD model or report	3.Overlays of contextual information (author, time, place, activity etc.) on physical content
		5. Recording the results of the search	*Information push*
		1.Recording what information was found—e.g. recording the outcome of a search for material properties in a logbook	1.AR overlay of information based on document recognition and suggestion of related documents, both physical and digital
		2.Recording the process through which the information need was fulfilled—i.e. how it was found and from where?	*Recording the search*
			1.Automatic identification and recording of steps taken, (i.e. the process), as well as the final result

(continued)

Table 18.5 (continued)

	Core information tasks (CIT)	Potential approaches
Reuse/sharing	*6. Making sense of and filtering the information retrieved*	*Novel concepts to support sense-making and reduce information overload*
	1.Ensuring the results of searching multiple sources/stores are recorded in such a way as to allow assessment and comparison	1.AR techniques to effectively visualise large volumes of mixed-mode information
	2.Recording and making visible contextual information, especially the relationships between sources and stores of information	2.AR annotating and hyper linking to aid filtering of paper-based information.
	3.Making a judgement of the value, usefulness, trustworthiness or appropriateness of the collated information, especially in the case of conflicting information	3.Overlays containing info summaries & value/trust measures [33]
	7.Adaption, manipulation and extension of existing information for the new purpose, e.g.	*Techniques to translate information retrieved into formal documents and to aid re-purposing*
	1.Personal notes into formal meeting minutes	1.Automatic recognition of free-hand equations, tables etc.
	2.Sketches transformed into a CAD model	2.'Beautification' of freehand sketches into vector graphics [35]
		3.Integrated annotation of both the digital and physical information
	8. Sharing and communicating the information retrieved	
	1.E.g. Sending to a colleague to continue working on it, or in response to a request	*AR overlay showing the evolution of the information, including:*
		1.The original source of the information
		2.The workflow—i.e. the steps it has been through, including from where it originated, how it was adapted and why—i.e. the rationale for it being shared

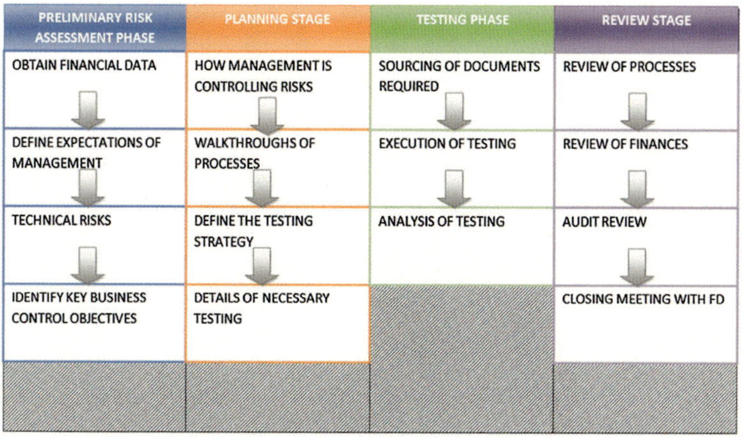

Fig. 18.10 Audit methodology map used for audit management (Ernst and Young [21])

However, the argument put forward in this chapter for a new approach is that the rigour of an information object or 'information as thing' based approach would be what is needed to enable it. This would mean that the generation of particular information objects (replacing tasks completed in workflow terminology) at specific points, both within stages of the design process or within specific gates would be the basis for managing and controlling design activity. The generation of 'super-sets' of information is already frequently undertaken at the end of one phase or as a requirement to pass through a gate, however, this is not universally the case or is not fully formalised. The work of the author in the Aerospace repair design area has seen the critical importance of this approach [56, 57] where it is applied.

As has been highlighted above these information sets are regularly required (see particularly Sect. 18.1.4), the difference here is that they are very formally specified and required by the various actors within the overall design teams, whether from core employees or contractors or system or sub-system suppliers.

18.4 Conclusions

The engineering design process is well established and accepted. The basic steps, however articulated, of brief, specification, concept, embodiment, detail and make are integrated into company processes, national standards and company stage gate-based procedures. The basic proposal of this chapter is that at the heart of these processes is the generation of information, particularly as information objects.

Thus one of the purposes of this chapter has been to review the dimensions of tangible information and tangible knowledge assets as they are required and developed by engineering designers and design teams. This is as part of the creation of a product or system. These are summarised as Information acquisition for design, Information generation by design and Information storage for design.

It always tempting to write 'novel' products and systems in a situation like this. However this temptation has been deliberately avoided. A lot of engineering design is not novel and does not need to be. It will be challenging, difficult and considerable pressure to meet a variety of targets associated with time, weight, size, performance and so on. However these targets will be achieved by using or adapting existing technologies or solutions or optimising and incrementally changing elements or subsystems within existing designs. Thus the information dimension becomes particularly critical.

In addition the steps from embodiment, through detail, to product development and testing and the various stages of prototyping, preproduction runs, production runs and through life developments create information in a formal and regular manner. It is controlling and managing and using these elements that is critical to the success of an overall product or system development programme. Thus the reconceptualisation of engineering design as an information processing activity becomes very important. This is being seen by companies and needs to be understood by engineering design researchers.

In addition, a record of the activities are required for a variety of reasons, controlling IP, ensuring conformance with regulations and performance regimes and to enable designs to be accepted and approved to pass through internal and external gated processes (Sect. 18.1.4). This is to enable a number of next stage activities for product development, such as: make available resource, manpower and finance. Failing to negotiate or pass through a gate with incorrect or insufficient information could potentially have unforeseeable adverse consequences (Fig. 18.10).

Thus the next purpose or underlying consideration emanating from this chapter is to use 'the design as an information-processing activity' view along with the 'information as thing' paradigms and use these to reinforce the requirements to accumulate information objects of the type articulated and described above in a very formal manner. This approach can also be used to manage resource, suppliers and distributed design teams.

This focus or refocus has a number of benefits from an engineering design perspective. It aligns the widely used and accepted design process models and with the transformation of conversion view of the engineering design processes as discussed in the definitions in Sect. 18.1. It also gives substance and aligns with the latest design theories where the concept and knowledge of C-K theory can be formalised, codified, controlled and added to the cumulated information archive knowledge base of an engineering design or in engineering design team.

References

1. Allen RD, Hicks BJ, Culley SJ (2002) Handling electronic information for mechanical systems design. Int J Inf Technol Manage, 1(4):367–385, ISSN 1461–4111
2. Ashworth G, Thornton A (eds) (1986) Technology transfer and innovation. Papers presented to the European conference on technology transfer and innovation held at the University of Salford, London, Taylor Graham, Sept 1986

3. Arnold E, Guy K, Dodgson M (1992) Linking for success: Making the most of collaborative R&D. National Economic Development Office/Institution of Electrical Engineers, London
4. Bently (2010) http://www.ftp2.bentley.com/dist/.../wp_design-engineering_hilmar-retief.pdf
5. Blackler F (1995) Knowledge, knowledge work and organizations: an overview and interpretation. Organ Stud 16(6):1021–1046
6. Buckland MK (1991) Information as thing. J Am Soc Inform Sci 42(5):351–360
7. CE-Mark (2010), http://ec.europa.eu/enterprise/policies/single-market-goods/cemarking/index_en.htm
8. Christensen CM (1997) The innovator's dilemma. Harvard Business School Press, Cambridge, MA
9. Clarkson, Eckert (2005) Design process improvement: A review of current practice [1st Edition]. In: Clarkson, John, Eckert, Claudia (Eds.) vol. XV. Springer, London, p 560
10. Cooper RG, Edgett SJ, Kleinschmidt EJ (2002) Optimizing the stage-gate process: what best-practice companies DoI. Res Technol Manage 45(5):21–27
11. Court AW, Culley SJ, McMahon CA (1993) A survey of information access and storage amongst engineering designers. The University of Bath, UK
12. Court AW, Culley SJ, McMahon CA (1994) The importance of information transfer between the designer and manufacturer. In: Proceedings of the 4th international conference on flexible automation and information Management, Blacksburg, Virginia, USA, pp 352–361
13. Cripps RJ, Smith F (1993) An information system to support the design activity. J Eng Des 4(1):3–10
14. Culley SJ, Boston OP, McMahon CA (1999) Suppliers in new product development: their information and integration. J Eng Des 10(1):59–75
15. Curtis, G. and Cobham, D. Business Information Systems: Analysis, Design and Practice. 6th edn. Harlow: Financial Times/Prentice Hall
16. Darlington MJ, Culley SJ, Potter S (2001) Knowledge and Reasoning:Issues raised in automating the conceptual design of Fluid Power Systems. Int J Fluid Power, 2(2):75–85, ISSN 1439–9776
17. Dieter GE (1983) Engineering Design. McGraw-Hill International Book Company
18. Dodgson M, Rothwell R (eds) (1996) The handbook of industrial innovation, New edn. Edward Elgar, Cheltenham
19. EASA (2010) Regulations. http://easa.europa.eu/agency-measures/certification-specifications.php
20. Eder WE (1989) Information systems for designers. In: Proceedings of the international conference on engineering design, Harrogate, UK, pp 1307–1319
21. Ernst, Young (2011) http://www.ey.com/GL/en/Services/Assurance/Assurance—International-GAAP-Resources
22. Feilden RG (1963) Report on engineering design. Department of scientific and industrial research, London: HMSO
23. Frank D (1999) The importance of knowledge management for BMW. In: Proceedings of the international conference on engineering design, ICED99, (Munich: TU Munich), pp. 33–41
24. Grant Thorton (2012) http://www.gti.org/Transparency-report/strategy-for-robust-audit-delivery/audit-methodology.asp
25. Hales C (1993) Managing engineering design. Longman, UK
26. Hansen MT, Nohria N Tierney T (1999) What's your strategy for managing knowledge? Harvard Business Review, March–April, 106–116
27. Hatchuel A, Weil B (2003) A new approach to innovative design : an introduction to C-K theory. In: Proceedings of ICED conference, Stockholm
28. Hertzum M, Pejtersen AM (2000) The information-seeking practices of engineers: searching for documents as well as for people. Inf Process Manage 36:761–778
29. Hicks BJ, Culley SJ, Allen RD, Mullineux G (2002) A framework for the requirements of capturing, storing and reusing information and knowledge in engineering design. Int J Inf Manage 22:263–280

30. IPO (2012) The website of the UK intellectual property office, http://www.ipo.gov.uk/types.htm
31. Kerr CIV, Mortara L, Phaal R, Probert DR (2006) A conceptual model for technology intelligence. Int J Technol Intell Planning 2(1):73–93
32. Kesselheim AS (2007) Intellectual property policy in the pharmaceutical sciences: the effect of inappropriate patents and market exclusivity extensions on the health care system. AAPS J 9(3) Article 33 (http://www.aapsj.org)
33. Keynote Design (2012) http://www.designconference.org/?menu=185
34. Konda S, Monarch I, Sargent P, Subrahmanian E (1992) Shared Memory in Design: A unifying theme for research and practice, Research in Engineering Design, 4(1):23–42
35. MACHINERY (2009)— Guidance notes on the UK, Regulations http://www.bis.gov.uk/files/file52968.pdf
36. MacLeod RA, Corlett J (eds.) (2006) Information sources in engineering (4th edn) München: KG Saur(ed), pp. 25–39
37. McMahon CA, Lowe A, King A (2002) Managing knowledge to free the designer's mind, Workshop to Honour Herb Simon, INSA, Lyons, March (CD-ROM Publication)
38. McMahon C, Lowe A, Culley SJ (2004) Knowledge management in engineering design: personalisation and codification. J Eng Des 15(4):307–325
39. McMahon CA, Giess MD, Culley SJ (2005) Information management for through life product support: the curation of digital engineering data. J Int J Prod Lifecycle Manage, Inderscience 1(1):26–42
40. MoD (2002) The acquisition handbook: a guide to achieving "faster, cheaper, better". (4th edn), Ministry of Defence
41. Oliva R, Kallenberg R (2003) Managing the transition from products to services. Int J Serve Ind Manage 14(2):160–172
42. Paap JE (1997) Technology management and competitive intelligence: strategies for a changing world. Paap Associates Inc. Available from: http://www.jaypaap.com/articles/CI-TECH-0199.pdf [online]. Accessed April 2007
43. Pahl G, Beitz W (1984) Engineering Design. In: Wallace KM (ed) The design council. London, UK
44. Pugh S (1991) Total design-integrated methods for successful product engineering. Strathclyde, Addison-Wesley Publishers Ltd
45. Ryle G (1949) The concept of the mind. University of Chicago Press, Chicago
46. Saaksvuori A, Immonen A (2003) Product Lifecycle Management, Springer Verlag
47. Schuchman HL (1981) Information transfer in engineering. The Futures Group, Glastonbury, CT
48. Shears (1971) Engineering design as an information-processing activity. In: Shears, Philip A (eds.) Electronics and Power, Pt 1 Volume 27, Issue 5, September 2006, Pages 527–548. Pt 2 Volume: 17, Issue: 3 pp100–103
49. Stagegate (2010) www.feedstockreview.ornl.gov/pdf/stage_gate_management_guide.pdf
50. Ullman DG (1992) The mechanical design process. McGraw-Hill, Inc
51. Ullman DG, Dietterich T (1987) Mechanical design methodology: implications on future developments of computer-aided design and knowledge-based systems. Eng Comput 2:21–29
52. Vincenti WG (1990) What engineers know and how they know it: analytical studies from aeronautical history. The John Hopkins University Press, Baltimore, MD
53. Wall RA (1986) Finding and using product information. Gower, UK
54. Ward M (2005) Information and the engineer (4th ed). In: MacLeod RA, Corlett J (eds) Information sources in engineering. München: KG Saur (Ed), pp.1–24
55. Wielinga B, Schreiber G (1997) Configuration-design problem solving. IEEE Expert Mag, pp. 49–56
56. Xie Y, Culley SJ, Weber F (2011a) Applying context to organize unstructured information in aerospace industry. In: Internal conference on engineering design (ICED'11), Copenhagen, Demark
57. Xie Y, Culley SJ, Weber F (2011b) Opportunities and challenges for context-aware systems in aerospace industry. J Enterp Inf 24(2):118–125

Chapter 19
Constraints and Conditions: Drivers for Design Processes

C. M. Eckert and M. K. Stacey

19.1 Introduction: Factors Shaping Design

Existing theories of design have aimed at understanding design as a unified phenomenon, describing the key steps that all design processes go through or the fundamental elements that all designed products need to have. However, to date, there is no theoretical understanding to explain or predict the differences and similarities that we observe when studying design processes across a range of products and domains. This chapter describes the first steps towards developing a theory of design in terms of the constraints and drivers on the design problem and process, to interpret how a process behaves, and to predict important aspects of the behaviour of new or modified design processes. The view put forward here is neither a theory of design nor a model of design in the sense of a representation of a selected part of the world or a representation of a theory (see [22, 24, 25]). It is an approach to generating partial theories of aspects of design: theory fragments that help to make sense of given situations and can help to predict how design processes might unfold.

The key theoretical concept in this approach is the causal driver: a phenomenon that influences the form or behaviour of a causal system—in our case a design process. The drivers that we are concerned with are characteristics of classes of products or processes or the conditions in which they are created, that enable us to look for and locate important similarities and differences between processes. Thinking about a process in terms of drivers providing causal pushes for design processes can help to make sense of observed behaviour as rational responses to a given situation; and therefore enable an analysis of where the process can be influenced. Particular patterns of constraints shared by similar design problems are among the most important drivers shaping design processes.

C. M. Eckert (✉) · M. K. Stacey
The Open University, Milton Keynes, UK
e-mail: c.m.eckert@open.ac.uk

M. K. Stacey
e-mail: mstacey@dmu.ac.uk

A. Chakrabarti and L. T. M. Blessing (eds.), *An Anthology of Theories and Models of Design*, DOI: 10.1007/978-1-4471-6338-1_19,

In this chapter we argue that incremental theory development is needed, and introduce our perspective of causal drivers of design processes, and the methodology behind our comparative approach to understanding design. We go on to argue that the influence of different types of constraints can be observed across different industries; we discuss some of these, illustrating our argument with two examples.

19.2 Local Incremental Theory Development

Most previous theoretical accounts of design have taken a very broad slice along the dimension of the range of possible types of designing, aiming to cover all design, or the whole of a broad field like mechanical engineering; and have taken a very narrow slice along the dimension of aspects of designing behaviour that they encompass. They tend to come from a discipline perspective, addressing design for example from a social or a cognitive perspective. Theories about the nature of design or how designing is done are typically presented with insufficient consideration of how much of designing they actually cover.

Our research makes a conscious attempt to integrate cognitive, social and cultural perspectives (see [12]). We are primarily interested in why design processes are as they are, and how they could be made to work better, to produce better products, to increase the profitability of companies or produce products faster and with less effort, or involve happier, less stressed, more fulfilled participants. Therefore we are interested in a theory that explains and predicts the behaviour of real design processes at a level that is not trivially true for all processes. The scope of the predictions the theory can make about different processes must therefore be clear from within the theory.

We develop fragments of a theory of design processes by looking at design processes in a range of different domains. This involves looking at the ways in which design processes are similar to those for different products or for similar products in different companies; and at the ways in which they are different. We look at design from a variety of theoretical standpoints, using the methods and conceptual frameworks of a number of different disciplines to make sense of different aspects of design, and trying to integrate the findings and insights they produce into a coherent picture. 'Design' as a human activity—or broad cluster of human activities—is far too complex and too diverse for understanding the whole of design in one step to be feasible, so we need an incremental approach to accumulating understanding. It follows from this that developing design theory involves constructing pieces of theory, assessing their validity, assessing their limits of applicability and progressively stitching them together to make a larger coherent whole.

Applying this approach involves recognising that the theory elements we currently have are at best fragments of a fuller theory of designing, and adding to them, looking at much narrower and more explicitly circumscribed slices of the dimension of the range of possible types of designing, and broader slices of the dimension of aspects of designing behaviour. The theory is provisional not just in that the relationships between drivers and design processes are open to revision, but also the definition of both the drivers and the phenomena they influence are open to revision, to provide better abstractions over concrete cases that yield stronger predictions or more conceptual clarity or better understanding of the differences between processes. Incrementally developing the theory fragments into a more coherent theory of design involves two kinds of operations. The first is to compare pieces of theory with the reality of particular design processes, and explain failures to observe the phenomena the theory fragments predict either in terms of the falsification of the theory, or by elaborating the theory fragments to cover a wider range of causal factors and distinct situations. The second is to connect theory fragments into larger more complete partial theories covering more of the interlocking causal processes shaping how designing is done, by matching and merging the elements of different theory fragments.

19.3 Our Theoretical Perspective: Causal Models of Systems

The conceptual foundations of the approach we are advocating come from systems theory. We view design processes as causal systems comprising a range of human, physical, conceptual and social entities connected by a web of interlocking causal processes operating at different scales. We start with the basic premise that similar influencing factors will tend to cause design processes to be similar in significant ways, and that differences in these influencing factors will tend to cause design processes to be different in significant ways. We want, first, to identify the important causal effects that make design processes operate the way they operate and, second, to describe these influences in a philosophically defensible way that enables us to make predictions about the behaviour of unfamiliar design processes when we observe them, or existing design processes when we try to change them.

This gives us our core theoretical concept: a driver of design behaviour. A driver is a phenomenon that causally influences some other phenomenon or causal system. Descriptions of causal drivers are abstractions of the concrete causal influences that operate in each individual situation—what Weber [46] termed *ideal types*. The scope of applicability of a theory fragment depends on the abstractness and generality of the description of the drivers and other elements and processes it includes.

Causal drivers are closely related to constraints: many, typically most, are direct consequences of the constraints on the product and the process. However, driver is a broader notion than constraint; drivers are theoretical explanatory concepts referring to bigger, more general, more coarse-grained phenomena, whereas constraints are specific, depend on the individual problem, and come in swarms. Some types of constraints, or characteristic patterns of constraints, constitute drivers. Conversely, needs acting as drivers lead to the formulation of particular types of constraints, which then act as external influences on other parts of the design process.

Design processes are influenced not just by the external constraints on product or process, but also by the characteristics of the system itself: the product being designed, the people doing the designing, their organisation and how they conceptualise and structure their work. What is an external causal influence on a system and what is an internal characteristic is a matter of perspective. We can draw a system boundary around the parts of a causal network that constitute (parts of) a system, such as a design process, and think of drivers as factors external to the system boundary. However, we may draw system boundaries in different places according to different conceptions of what the system is, so we do not want to draw any sharp distinction between an external driver of system behaviour and a component of a causal system. The appropriate system boundary for one part of the design process may be very different from the appropriate boundary for another part. Features of the emerging product often strongly restrict the form that other aspects of the design can take: they act as important external constraints on the design processes for those other aspects of the product. The relationship between design processes and their constraints is also bidirectional: identifying constraints, and turning broadly and vaguely formulated needs into exactly formulated requirements, is an important part of many design processes.

19.3.1 Theories and Models

The theory fragments comprise partial *models* of design processes, but with a very different scope and purpose from the process models that are constructed by design practitioners for practical purposes and the prescriptive models proposed by methodologists, which are themselves epistemologically slippery (see [14]). The claim of these theory fragments is that the models *represent* the structure of real, if abstractly described, causal processes (what it means for models to represent reality is a controversial issue in philosophy; see [22], for a discussion). Thus the claims of the theory fragments depend on the relationship between model and reality. Morgan and Morrison [32] summarise van Fraassen [44] as "We assess a theory as being empirically adequate if the empirical structures in the world (those that are actual and observable) can be embedded in some model of the theory, where the relationship between the model and a real system is one of isomorphism". However, Giere [24, 25] also emphasises the non-linguistic and abstract character of models, but argues that scientific theories make claims about *similarity* relationships between model and reality, which do not necessarily require isomorphism. The nature of this similarity relationship is subtle and needs further elucidation, and may differ for different cases; the relationship between scientific models and reality is the subject of extensive debate among philosophers (e.g. [30, 40, 43]). The drivers presented in this chapter are not necessary and sufficient conditions for processes to behave in particular ways. Design processes are determined and affected by far too many factors to ever make such a claim. Nor can the drivers be interpreted in a counterfactual way, meaning that a process would not necessarily have unfolded in a different way, if the driver had not been present. We interpret the drivers as causal pushes, that make a particular situation more likely than another.

19.3.2 Modelling Social Systems

As many design researchers have acknowledged, designing is almost always a collaborative enterprise involving social processes. Understanding design processes as systems involves viewing social processes as causal mechanisms and social facts as elements of causal systems. Whether, when, and how far social structures should be treated as though they have real, objective existence has been fiercely contested in sociology for many decades (see [4]). These arguments are directly relevant to the question of how to understand designing at the level of social and organisational processes, not just for academic purposes but for the application of practical methods for improving work processes and specifying computer tools to support work activities, including designing. On the one hand, how social structures and social processes work at a level broader than individual human thoughts and actions is not only the subject matter of sociology but something we need to think about to live our lives; we all treat them as objectively real and (unless we are hardline interpretivists) believe they are real. Functionalist approaches to social science depend on seeing social structures as systems of interlocking causal entities that are broader and less concrete than individual people (for instance, [34]). On the other hand, it is difficult to see how anything beyond physical objects and individual human thoughts and actions are genuinely objectively real, deserving the ontological status of things; and people can legitimately disagree about what the social structures are and how they work. In the case of the division in sociology between the Durkheimian positivist camp and the Weberian interpretivist camp, we are inclined to be pragmatic: we are in sympathy with the interpretivist view that individuals' differing conceptions of social structures and phenomena are real and primary, and that social structures cannot be said to have objective existence even when there is a great deal of consensus about them, but we take the view that it is frequently both unavoidable and useful to treat social structures as though they were objectively real, and regard doing so as a legitimate pragmatic compromise. Our ideas have been influenced by Soft Systems Methodology [5], which is an approach to understanding and suggesting improvements to work systems that combines a systems theory view of what is going on as comprising interlocking causally active components with the central premise that different views of how the system works are equally valid and there is no objectively correct view. However, Checkland [5] is emphatic that treating any particular view as being objectively correct is not only illegitimate but harmful, and that it is essential to avoid this in applying Soft Systems Methodology.

At a finer grained level, what can exert a causal influence so that it should be regarded as a participant in a causal structure is also debated. One important theoretical approach is Actor-network theory [28, 29], which insists firmly that inanimate objects have properties and behaviour that influence what humans do, and play essential roles in human systems of activity, so should be treated as actors in their own right. Others do not wish to ascribe agency to inanimate objects. As design researchers who are acutely aware of the importance to design communication of the properties of the representations used to convey design information, we favour the view that objects should be treated as participants in causal systems.

19.4 Methodology

This chapter draws on two types of studies carried out by the authors: case studies of design behaviour, developing a detailed understanding of how particular aspects of designing are handled in individual companies, and a research project specifically on comparisons between different design domains. The Across Design project invited 20 designers to present witness accounts of the practices in their fields focusing on one of their own projects (see [3], for a discussion of the methodology; [16], for a summary of the results). The witnesses were design experts with 5–50 years of professional experience from a wide range design fields including engineering design, software design, product design, graphic design, fashion design and food design.

Table 19.1 summarises the empirical case studies, which all followed a similar format of semiformal interviews lasting between 30 and 120 minutes. As far as was possible, these interviews were recorded and transcribed. Otherwise field notes were taken or summaries generated immediately after the interview (in the early knitwear studies). The interviews were conducted to investigate a particular question in each study, such as communication and later inspiration in the case of the knitting studies. However the issues considered in previous studies were revisited with additional questions in later interviews. For example the issue of communication, the subject of the first study, was picked up again in the context of studying planning practice [21]. For the original analyses the transcripts were analysed using a combination of grounded theory [27] and deliberate falsification of current assumptions [37].

This chapter takes a slightly different approach in that it reflects over the insights of the other papers to come up with a higher level perspective on the drivers of the design processes.

We have looked before at the relationship between constraints and how designing is done, focusing on creativity in design. We have argued elsewhere [38] that the main difference in the modes of creativity between engineering and artistic design domains lies in how constrained the design problems are that the designers have to engage with. Engineering designers are usually confronted with difficult, complicated, tightly constrained problems and have to be creative in the way they

Table 19.1 Summary of empirical studies

Domain	Interviews	Companies	Year	Focus
Knitwear	80	25	1992–1998	Communication in teams, inspirations [9, 12, 13]
Engineering	42	2	1999–2003	Engineering change [15]
Engineering	25	2	2000, 2005	Planning [10]
Engineering	15	1	2008	System architecture [17]
Architecture	13	1	2007	Decision making in design
Construction	8	4	2009–2012	Decision making in hospital refurbishment projects [23]
Engineering	11	1	2011–2013	Testing in design processes [41]

reconcile and frame often contradictory constraints to have a well-defined problem; while artistic designers engage in a deliberate constraint-seeking process to narrow the potential design space to again end up with a reasonably well-constrained problem. Design spaces are rarely equally strongly or loosely constrained across all aspects, so that designers can trade off freedom in one area against constraints in another. Following on from this, we argued in [17] that in engineering, designers are often looking for solutions that meet new requirements and constraints but require the fewest changes to existing designs. Creativity therefore often lies in the clever tweak rather than in the radically different solution.

19.5 Constraints

A constraint is a restriction that an action or the solution of a problem must comply with. Designers use constraints on the designs they develop not just to check the viability of design proposals but also to guide the generation of new design ideas. Constraints on designing take a variety of forms: constraints of different types exert different influences on designing (see Stacey and Eckert [38]). They vary in whether the constraint is explicitly stated or tacitly assumed; in whether conformity is binary or a matter of degree; in whether they refer to measurable physical properties or are experiential; in whether conformity can be measured objectively or is a matter of subjective judgement; and in whether they are hard or soft—that is, whether the constraint must be met, or can be relaxed if necessary to reach some solution rather than none.

Explicitly stated constraints play an essential role in more formal approaches to designing. Constraint programming is an important technique in artificial intelligence (for a survey see [36]). It plays an important role in design optimisation [42]; and has been applied successfully to design problems, such as circuit board design, which are well-defined and where any solution can be employed as long as it meets all the constraints [45]. Constraint-based planning is often combined with hierarchical task planning (for instance [2]). The success of constraint programming for some specific design problems notwithstanding, this chapter is not suggesting all design problems are fundamentally search problems that can be resolved if the constraints are known; rather that constraints in the sense of conditions on a solution or its process of creation fundamentally shape the human design process.

Although human reasoning about design problems is radically different, as it employs subtle and powerful pattern recognition and synthesis operations, and cannot employ the extensive combinatoric searches that are easy for computers, humans also depend on understanding or making constraints to contribute elements of solutions and restrict the scope of imagination to possibilities that are likely to be fruitful. Research on creative idea generation, such as Finke's [19] work on pre-inventive forms, where he encouraged people to imagine particular shapes, and then use them in creative tasks, indicates that tasks requiring imagination (but soluble in a wide variety of ways) are made easier by tight constraints that supply elements of solutions to be combined and adapted, and reduce the spaces of possible solutions (see [20]). The nature of the constraints determine what people think the design problem *is*, as well as what appears to be a plausible part of a solution. The central role of the most salient constraints in guiding the conceptualisation of the design problem and the generation of the key elements of the design has been well recognised by design researchers for a long time (see [7]).

Constraints on design have three main sources:

The *problem* that the design must solve or the need that the design must meet; this includes product requirements, manufacturing requirements and constraints stemming from the strategic goals of the company.

The *process* by which this is achieved.

The *emerging solution*—since making certain decisions will rule out or restrict options for other later decisions.

The constraints are different for each design problem. Many constraints are specific to the particular design, but many arise from the individuals that design the product, the product context, the organisations involved in it and the wider market context. Figure 19.3 shows a set of sources of constraints for our case study examples.

In software development, requirements that a system should meet to do its intended job successfully are usually distinguished from constraints, which are restrictions on what the system must be or do separate from user needs. In engineering design where many constraints are expressed explicitly, constraints and requirements are very similar. However, constraints arising from the emerging solution are usually not expressed as requirements. For the purposes of this chapter we can treat requirements as being constraints.

19.5.1 Problem Constraints

Constraints in design are closely linked to requirements but go beyond what is expressed as requirements. Requirements for the product arise from the user needs and desires that are addressed with the product as well as organisational needs for the products, such as cost targets. Many requirements can be immediately expressed as constraints, e.g. the vehicle must carry of a load up to 1,000 kg, while others are far more vague. For example, the aesthetic effect of the sound of a car engine is a vague experiential requirement, but car companies have techniques and procedures for defining the desired acoustic properties and turning them into objective performance requirements. In the early stages of a design processes these requirements are translated into a technical specification. The technical specification defines the characteristic that the product must have in ways that can be assessed. A product is both tested against the technical specification in verification processes and against the requirements in a validation process. In generating a technical specification many of the potential contradictions in the list of requirements are resolved. However design work often starts with a contradictory set of requirements.

Most engineering products are designed by modification from existing products. The products are assessed where they do not meet the existing requirements and are changed accordingly. At the same time engineering companies aim to minimize novelty in a product, and therefore set percentage targets for the degree of reuse or explicitly ring-fence particular components. However, changes have knock-on effects on other parts of the product, which were not intended to be changed [13]. They can also bounce back when a component cannot be changed and a new solution needs to be found. The new design is severely constrained by what can and cannot be changed.

For many products, regulatory frameworks are a major source of constraints. For example, the automotive industry is very severely constrained by emission legislation. Unless they have an engine that meets a particular requirement, they will not have a product to sell. The aerospace industry has to comply with stringent regulations to meet certification requirements. While safety–critical products are most highly regulated, other products are also constrained by legislation, for example around the use of materials.

19.5.2 Process Constraints

All design projects are constrained by time, cost and resources even though the degree of severity of these constraints might vary. The skills of the designers in house and across the supply chain affect the solution itself as well as the process by which it is generated. For example an engineering company with a large in-house control engineering department might develop or adapt their own control software and therefore can make changes to control software throughout the design process. Without this resource the company might subcontract the development of software and be reluctant to ask suppliers for a late change, and deal with an emerging problem in a different way. Time and cost constraints affect the amount of design effort the company can afford. This limits the range of ambition or the degree of innovation. This has to be traded off against other factors such as testing time or cost, or cost and lead time. Organisations typically have standard processes, which govern when and by whom decisions about the product need to be taken. They also have established ways of interacting with their suppliers and dividing the work across the supply chain. Characteristics of the organisation are directly reflected in the process constraints.

19.5.3 Solution Constraints

The way the process is managed and organised constrains the tasks the designers undertake, and the order in which they are undertaken. This is one way in which the emerging solution constrains the further development of the new product. The order in which decisions are taken fundamentally affect the freedom designers have [18]. Some components are given from a previous design and others have to be frozen early because of long lead times. This in turn sets parameters for other components in the product, even though they are not affected directly by external requirements. Typically, companies keep components with shorter lead times open to the end to be able to absorb changes. For this reason many changes in later stages of a project are dealt with by software changes. Lead times are, however, not the only reason to freeze components or parameters. Decisions also need to be taken if multiple teams need to work with set values to reduce iteration. Sometime these products are designed by different designers using preliminary values for each other's decisions to develop a first attempted solution which is then refined during several rounds of iteration until convergence is reached [48]. To avoid this, engineering companies use performance models to derive key parameters from the requirements and cascade them down across the product, making decisions in a planned sequence (see [47]). In the diesel engine case study company, reuse of components is managed by feeding them as constraints into a requirement cascade process.

19.5.4 Meeting Constraints

Engineering designers are often confronted with contradictory constraints; for example weight and performance requirements might be in direct contradiction. The TRIZ methodology for engineering creativity (e.g. [1]) is based on resolving such pair-wise contradictions between requirements by identifying a new solution principle for what might be a well-known class of problem.

Meeting conflicting requirements as well as possible involves two distinct but integrated operations, relaxing constraints and finding solutions. Weak constraints can be relaxed to allow less ideal but feasible designs; conflicting strong constraints, that must be met, make a design solution impossible unless some constraints are relaxed. Design problems are constrained both by explicitly formulated requirements and constraints, and by implicit assumptions about the form of the solution. (Designing is often influenced by *fixation* on the features of previous similar products (see for instance [35]); Altshuller [1], discusses various sources of psychological inertia and how they can be overcome within TRIZ). However, designers often have a choice in how to approach a given design problem. For example the development of a new vehicle might include targets for the percentage of components to be reused, but designers have a choice which components they will try to keep and which they change.

The tightness of the constraints varies enormously between problems, and individual products can be very tightly constrained in some ways and loosely in others. Technical problems can be underconstrained where the requirements are weak or are not yet understood. Sometimes engineering solutions must meet tacit sociocultural expectations, for instance concerning the aesthetic appearance of an engineering product. Similarly, artistic designers are often faced with seemingly impossible constraints from the business context or the technical realisation of the product. In knitwear design it is often novices and outsiders who produce innovations pushing the boundaries of what is possible; Stacey et al. [39] argued that this is precisely because they do not know the tacit and implicit constraints and therefore dare to push for what seems impossible or not worth the fight to the experts.

19.6 Drivers Affecting Product Classes

The constraints discussed in the previous section are individual for each product. However many of the drivers that shape design processes are shared with other products, not necessarily ones in the same domain. This section discusses drivers that shape the design processes of entire product classes. These correspond to broad needs and product characteristics that translate into particular types of constraints for individual products. They also go a long way to explaining differences between processes, for example in explaining why design processes in the aerospace industry are different from those in the automotive industry.

19.6.1 Different Drivers

There are many drivers that affect classes of products. The following ones were particularly pertinent in our case studies. This is by no means a complete list, but illustrates some high-level drivers, which are properties of the product or the context in which they are deployed.

Product complexity affects not only the effort that is involved in a design process, but also much of its execution. Very complex products, such as aircraft, have to be designed by big teams of designers, who have little overview over all of the activities going on the development of the product and therefore need additional layers of management to coordinate. This is for example a marked difference between the development of jet engines and the development of diesel engines. Most of the diesel engine engineers understand how all of the components work at least at a higher level and therefore know the key dependencies. This also enables them to communicate proactively across the design process (see [21]). Product complexity is a strong driver for incremental design and the reuse of product models, to reduce the overall effort and thus the product cost.

Safety criticality is of course a matter of degree, but certain products like aircraft or power plants are subject to the most rigorous safety criteria. These products are very rigorously tested, which adds to the cost of the development. In software safety critical software not only needs to be tested, but needs to be mathematically proven to be correct. Because of the effort involved in testing, designers are also very reluctant to change the product, because every change would require retesting and potentially recertification.

Product lifespan, both in terms of time in production and lifetime of the product, determines many of the design tasks that need to be undertaken. Long lifetime requires very careful user analysis and a system architecture that potentially enables the company to upgrade parts of the product without affecting other parts. This requires a certain degree of redundancy in the product and an awareness of the margins for change of components. Long lifetime products also require spare parts over a long period of time; either these need to be stockpiled or manufacturers need to assure that they can deliver parts. An interesting illustration is how companies handle issues around rare earths. Jet engine manufacturers are aware of long term resourcing issues and are making sure they have the spare parts and look actively for alternative designs. Mobile phones are also affected by shortage of rare earths, but for them it is a cost and an issue of recycling the materials at the end of the product's lifespan.

Volumes of production have a huge effect on how cost-critical all design decisions are. In the automotive industry, volumes for platform components, like car lights, can be in the hundreds of thousands, where every cent saved adds up to a huge amount of money, compared to thousands in the aerospace industry. By contrast building are typically one-off designs. In a one-off design it is possible to negotiate compromises or problem fixes with one client, without needing to be concerned about other users or other contexts. Whereas high volume products have to operate under all the circumstances in which all the potential users might operate them. High volumes also provide companies with enormous power over their suppliers, who will adjust their ways of working to attract volume customers. For one-off products the power often lies with the suppliers who are supplying multiple customers. For products produced in smaller volumes the pattern of supplier relations is more varied. Companies in the aerospace industry usually have a dedicated range of suppliers who make money out of long standing collaborations and the long term spare part contract; however on components where they are competing with the automotive industry, aerospace parts are much more expensive and potentially difficult to source.

Connectivity within product ranges refers to the relation between different products designed in the same company. Some companies offer a range of products or product families which are themselves not very well connected; for example a manufacturing tool maker might make different types of tools and for each type offer a product family without much connection between the product families in terms of functionality or components. Other companies work very hard to develop a product platform, where parts are shared across a wide range of products. Changes to any platform part are expensive, because they not only affect the cost of that particular product but have knock-on effects on the profitability of other products built on the platform, either because the standard part becomes more expensive or because it is produced in smaller volumes. This makes these organisations both reluctant to make changes and bureaucratic in carrying out the changes. Large platforms also almost inevitably mean that the components are better suited to be used in one product than in others. There is an inherent conflict between an optimal platform and an optimal part. In those respects the designers of one-off products have much greater freedom.

Interactions with users vary enormously between different products. Products that are sold directly to the end user require direct engagement with the users' requirements and issues of usability. Components for other products designed without direct contact with the end user still have to comply with requirements from the user, but concerns about human interactions focus on manufacturing and maintenance where the personnel can be trained and the interaction can to some extent be controlled. Products for end users have to engage with issues like ergonomics and inclusive design. For products with direct user interaction aesthetic criteria are a major concern. These products must appeal to the users and not just meet requirements. Products that are sold to professional customers, like production machinery, still benefit from an aesthetic appeal, but are mainly sold on functionality. Aesthetics is not an issue for components inside other products, like hydraulic pumps.

Fashion plays a huge part in the design of many products that are sold to end consumers. For products like garments or many consumer products the functionality remains relatively stable, but the form depends on a context set by other products. Consumers select cars to some extent on their visual appearance, and a wrong call on the car styling can jeopardise the commercial success of a technically very good product. Fashion has a certain built in obsolescence; this affects the lifetime of products and thus sets many technical constraints on a product, e.g. targets for durability.

19.6.2 Profile of Drivers

The factors are not independent, but act as causal drivers for each other. High product volumes for example encourage companies to consider the development of product platforms, because it increases their opportunity to make savings. Many of

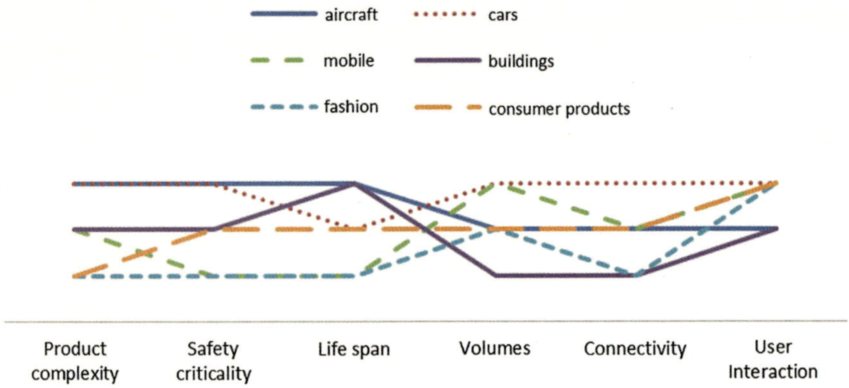

Fig. 19.1 Profiles of drivers

the highly complex products also have long lifespans, because users and clients cannot afford to replace them all that often.

Figure 19.1 shows indicative profiles of drivers for different industry sectors on a simple high, medium and low scale for each driver. The product classes reflect products that are often studied in the design literature. While there is of course an enormous variability within each of these classes, as the graph illustrates the spectrum is completely covered and most combinations of these drivers do exist in some product or other.

19.7 Objective or Subjective Constraints

Products are assessed against the product requirements, but projects are also assessed against whether they have met the other constraints placed on the product. For example, an excellent product that has run over time in the development process can cost its company a lot of money and mean that it will lose market share. Companies are assessed against the profit that they make, but also against the reputation they have built up through their products.

Many of the constraints are explicitly expressed in product specifications as ranges, absolute values or clear descriptions of the target behaviour or function. Others are expressed explicitly elsewhere in the organisation, but might not be spelled out clearly for a particular product. For example a company might aim at a certain level of innovation across its range of products. It does not necessarily parcel out an innovation target for each project, but designers are aware that innovation is an issue. However, constraints may be unstated for a number of reasons. Some are almost too obvious to state. For example an item of clothing needs to be safe to use: this is not included in any specification for a specific product, yet designers are conscious of it and consider it for example in the context of flammability of materials. Experiential constraints work rather differently: some factors that are evaluated perceptually are recognised to be important and govern conscious choices. Tacit constraints influence the procedures designers follow and the choices they make without the designers necessarily being fully aware of them. Brand image in general is very difficult to express explicitly, yet designers are aware of what constitutes a product in the brand style. Jonathan Cagan and his colleagues have explored the use of shape grammars as a means for making perceptually understood brand styling explicit, which is not a trivial task (see for instance [31]). Knitwear designer often comment that they can't describe what their brand's products look like, but they recognise them when they see them. Aesthetic characteristics are also often tacit. They are not defined in product specifications or even on a higher level for an organisation, but it is understood that products need to fit into current fashions and styles. Even if a style or theme is given, what this entails can rarely be expressed explicitly. However, as we have argued elsewhere [11] it can be expressed indirectly by providing indications of a legitimate solution space in terms of other objects that already express this space.

This points to another important distinction, that between constraints that can be objectively assessed and those that cannot. Many engineering constraints can be assessed objectively when they are measured during tests. For example a testing regime reveals fuel consumption of engines under many different circumstances. However many of the perceptual requirements and constraints are very hard to assess objectively. While it is possible to express some aesthetic considerations in explicit rules like the golden ratio and assess them objectively, many are completely tacit. Assessments that require an awareness of a rich context are often very subjective. For example, the assessment of styling of textiles or cars require a deep understanding of other products, and whether the product would appeal to the target customers given the other products that the customers are likely to be familiar with. While other designers or users might concur with these judgements, consensus does not make the assessment objective.

This changes the role that designers play in the design process. With objective product criteria the success of the product and the success of the individual designer can be separated. The product might be late or not sell, but as long as it can be demonstrated that, say, the stress analysis has been correct, individual designers won't be negatively affected by product failure beyond being affected by the fate of the organisation. By contrast, for products which are partially or wholly assessed subjectively, the designer is also the person who makes judgements on how well the design works. The designer becomes the guarantor of success. This is particularly pertinent in fashion products.

Without objective evaluation criteria, companies, designers and customers have to fall back on their own instincts or use reference points to assess a product against. One way of selecting reference points is by looking at the work of established designers or prominent companies. This is extremely important both in the fashion industry and in product design, where famous star designers set trends that are followed by others. In architecture well-known buildings serve as precedents that serve to validate design choices as well as sources of ideas [8, 26, 33]; cf

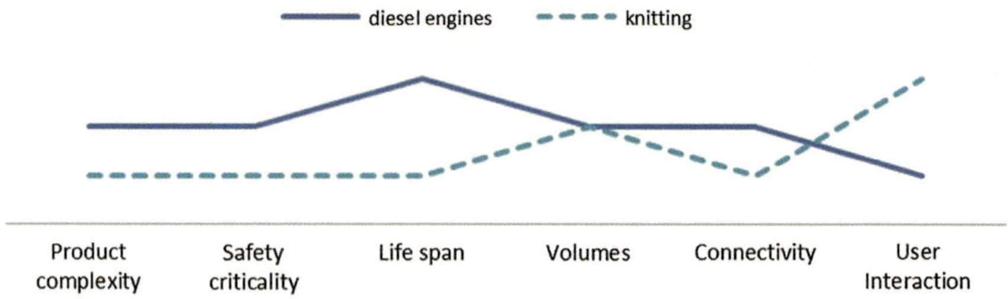

Fig. 19.2 Profiles for diesel engines and knitwear

[6]. The cult of the star designers is a direct result of the inability to assess the quality of designs objectively. There are few famous engineers.

19.8 Examples of Different Behaviour: Diesel Engines and Knitwear

Two processes that we have studied over a long period of time are diesel engine design, in our studies on engineering change, product planning, system architecture and testing; and knitwear design in studies on communication and mechanisms of inspiration. The products and processes are very different in terms of both constraints and drivers. Figure 19.2 shows the profile of drivers for both domains.

Knitwear designs can involve technically tricky detail design but are simple products compared to large-scale engineering (see [9, 13]). The key driver for knitwear design is fashion. For clothing, fashion is still seasonal. While a strict schedule of two season launches every year has been softened to several releases every year in most companies, clothes are still dictated by the weather and seasonal events and designs need to be ready to be launched on time. This makes time one of the greatest constraints on the design process. The designers work on the clothes by season or as groups of garments that are launched together; and start as soon as possible when the previous collection has been passed on to production. Much time is required to get the details of garments right. There are simple characteristics that a garment much have (e.g. fit the body measurements, display the key motifs, etc.), but much is a matter of aesthetics and therefore very subjective. The designers and their technicians work on the designs until they run out of time, and release products even though they know that they could be improved. The textile industry typically produces runs of 100 to 1000 garments and the profit margins on individual garments are very small. Therefore the designs have to be costed very carefully, and development effort even by two or three individuals is a major cost factor. In some cases it is possible to pass the costs on to customers, but usually knitwear designs are sold in price brackets so that they have clear target costs. These costs are balanced across an entire collection. Within a collection there is a balance between successful design features picked up from the previous

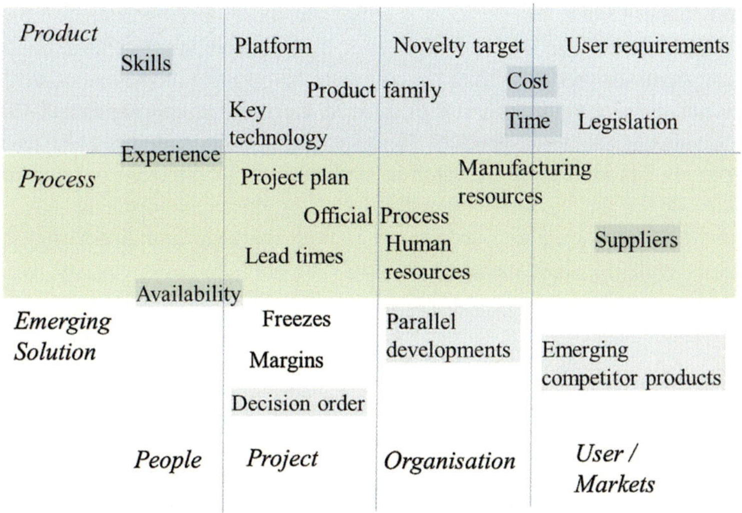

Fig. 19.3 Constraints for knitwear design and diesel engine design

season and new features that are introduced. Designers need to aim for an appropriate degree of novelty in relation both to their own past designs and to fashion trends that are emerging. Knitwear designers do not interact directly with their customers, but they have a sense of the taste of their target market and how it will evolve over time, therefore they know how much novelty they are aiming for in a collection to attract their customers, while not alienating them.

Diesel engines are an established technology that has been refined over the last hundred years. Diesel engines have become increasingly more complex, but have not reached the level of a jet engine or an aircraft. Individual designers have a clear understanding of the core technologies. Diesel engines are very compact, because they typically need to fit into tight spaces. This adds to the challenges posed by their highly interconnected architecture, and keeping track of the product connectivity can be very difficult. Diesel engines are highly regulated in terms of emissions for particular markets. The progress of emissions legislation sets a tight schedule for the development of new products as well as product constraints that need to be met. Off-highway engines are used at nearly peak capacity for most of their lives and have very strenuous requirements in terms of robustness and durability. Reuse of tried and tested components and solution principles is a way to manage that, and the company has stringent novelty targets that must not be exceeded. To meet the strict launch times the company needs to manage its supply chains very carefully and freeze components in time. The frozen components therefore very strongly constrain the rest of the design as it emerges. To enable timely freezes and planning of the design process, decisions are being taken about the design early to enable different teams to work in parallel and follow the targets set by both the official design process and the suppliers. Diesel engines are usually provided to the OEMs for vehicles, so that the diesel engine company has relatively little contact with the end users, but they receive strict requirements from their customers. At the same time they have to be mindful of the use conditions of the product. Key technologies are employed across a range of product families, and innovations on other product families are usually lifted across to a new development. Standardizing as much as possible across different product families is a major driver that generates constraints on product development.

Figure 19.3 shows the key constraints in both knitwear and diesel engines. The constraints with the dark background apply to both domains equally, while the remaining constraints mainly applied to the diesel engines.

19.9 Conclusion

This chapter argues that design processes are influenced by the constraints placed on the product and the process as well as those that arise through the way that the design process unfolds. Many of the constraints are concrete manifestations of factors that apply to classes of products and act to shape their design processes in similar ways. The chapter has discussed a number of these causal drivers and patterns of constraints that were apparent in the authors' case studies, and which explained aspects of how the design processes were organised and the types of problems designers were confronted with.

The view put forward here—modelling design processes as networks of interlocking causal processes influenced by causal drivers—is an approach to building partial theories of aspects of designing that explain why particular design processes are similar or different in particular ways, and help to predict how new or modified design processes will behave. We see the main benefit of the analysis of causal drivers influencing design processes in proving questions that lead to a deeper understanding of an individual design process: Does this causal influence happen here? If so, what form does it take? If not, why not?

This chapter stands at the beginning of a long-term research agenda to understand design processes in terms of drivers, constraints and characteristics, which will need to go through several steps. An ontology of the key concepts, such as driver, constraint and requirement, will need to be defined to enable a clear distinction of concepts, as well as a clearer account of how drivers, constraints, requirements and so on can vary. By looking at the constraints governing specific processes and causal drivers influencing different design domains, typologies of constraints and drivers can be built, and a set of causal maps constructed, which together draw a broad picture of design. This can then be consolidated into larger and fuller causal maps, providing a toolkit for characterising and predicting the behaviour of new design processes.

The future research is required to develop clearer distinctions between different classes of causal drivers, such as requirements, constraints, drivers and conditions or the definition of one clear concept encompassing all—constraints—that is seen in terms of relevance and malleability. To both validate and apply this way of thinking it is necessary to develop a set of categories of drivers and constraints that are applicable to a wide range of processes.

Acknowledgments This chapter draws on a number of industrial case studies by Claudia Eckert and her colleagues at the Cambridge Engineering Design Centre that were supported by the EPSRC. The authors thank their colleagues and industrial informants.

References

1. Altshuller GS (1988) Creativity as an exact science. Gordon and Breach, New York
2. Ambite JL, Knoblock CA, Muslea M, Minton S (2005) Conditional constraint networks for interleaved planning and information gathering. IEEE Intell Syst 20(2):25–33
3. Blackwell AD, Eckert CM, Bucciarelli LL, Earl CF (2009) Witnesses to design: a phenomenology of comparative design. Des Issues 25:36–47
4. Burrell G, Morgan G (1979) Sociological paradigms and organisational analysis. Heinemann, London
5. Checkland P (1981) Systems thinking, systems practice. Wiley, Chichester, UK
6. Clark RH, Pause M (1996) Precedents in architecture. Wiley, New York
7. Darke J (1979) The primary generator and the design process. Des Stud 1:36–44
8. Defazio J (2008) Designing with precedent: a cross-disciplinary inquiry into the design process. PhD thesis, Department of Instructional Systems Technology, Indiana University
9. Eckert CM (2001) The communication bottleneck in knitwear design: analysis and computing solutions. Comput Support Coop Work 10:29–74
10. Eckert CM, Clarkson PJ (2010) Planning development processes for complex products. Res Eng Design 21(3):153–171
11. Eckert CM, Stacey MK (2000) Sources of inspiration: a language of design. Des Stud 21:523–538
12. Eckert CM, Stacey MK (2001) Designing in the context of fashion—designing the fashion context. In: Designing in Context: Proceedings of the 5th design thinking research symposium. Delft University Press, Delft, Netherlands, pp 113–129
13. Eckert CM, Stacey MK (2003) Sources of inspiration in industrial practice: the case of knitwear design. J Des Res, 3(1):article 2
14. Eckert CM, Stacey MK (2010) What is a process model? reflections on the epistemology of process models. In: Clarkson PJ, Vajna S, Heisig P (Eds) Modelling and management of engineering processes. Springer, New York, pp 3–14
15. Eckert CM, Clarkson PJ, Zanker W (2004) Change and customisation in complex engineering domains. Res Eng Des 15:1–21
16. Eckert CM, Blackwell AD, Bucciarelli L, Earl CF (2010) Shared conversations across design. Des Issues 26(3)
17. Eckert CM, Stacey MK, Garthwaite P, Wyatt DF (2012) Change as little as possible: creativity in design by modification. J Eng Des 23(4):337–360
18. Eger T, Eckert CM, Clarkson PJ (2005) The role of design freeze during product development. In: Proceedings of the 15th international conference on engineering design, The Design Society, Melbourne, Australia

19. Finke RA (1990) Creative imagery: discoveries and inventions in visualization. Lawrence Erlbaum, Hillsdale
20. Finke RA, Ward TB, Smith SM (1992) Creative cognition: theory, research and applications. MIT Press, Cambridge
21. Flanagan T, Eckert CM, Clarkson PJ (2007) Externalising tacit overview knowledge: a model-based approach to supporting design teams. Artif Intell Eng Des Anal Manuf 21(3):227–242
22. Frigg R (2003) Re-representing scientific representation. PhD thesis, Department of Philosophy Logic and Scientific Method. London School of Economics
23. Garthwaite PM, Eckert CM (2012) Multilayered change: engineering change in building refurbishment. DESIGN 2012, Dubrovnik, Croatia, May 21–24, 2012
24. Giere R (1988) Explaining science: a cognitive approach. University of Chicago Press, Chicago
25. Giere R (2004) How models are used to represent reality. Philos Sci 71(5):742–752
26. Goldschmidt G (1998) Creative architectural design: reference versus precedence. J Architect Plan Res 15:258–270
27. Goulding C (2002) Grounded theory: a practical guide for management. Business and Market Researchers, Sage, London
28. Latour B (1987) Science in action: how to follow scientists and engineers through society. Harvard University Press, Cambridge
29. Latour B (2005) Reassembling the social: an introduction to Actor–network theory. Oxford University Press, Oxford
30. Mäki U (2011) Models and the locus of their truth. Synthese 180:47–63
31. McCormack JP, Cagan J, Vogel CM (2004) Speaking the buick language: capturing understanding, and exploring brand identity with shape grammars. Des Stud 25:1–29
32. Morgan MS, Morrison M (1999) Introduction. In: Morgan MS, Morrison M (eds) Models as mediators: perspectives on natural and social science. Cambridge University Press, Cambridge, pp 1–9
33. Oxman R (1994) Precedents in design: a computational model for the organization of precedent knowledge. Des Stud 15:141–157
34. Parsons T (1951) The social system. Free Press, New York
35. Purcell T, Gero JS (1996) Design and other types of fixation. Des Stud 17:363–383
36. Rossi F, van Beek P, Walsh T (2006) Handbook of constraint programming. Elsevier, Amsterdam
37. Stacey MK, Eckert CM (1999) An ethnographic methodology for design process analysis. In: 12th International conference on engineering design, Technical University of Munich, Munich, vol 3, pp 1565–1570
38. Stacey MK, Eckert CM (2010) Reshaping the box: creative designing as constraint management. Int J Product Dev 11:241–255
39. Stacey MK, Eckert CM, Wiley J (2002) Expertise and creativity in knitwear design. Int J New Product Dev Innov Manag 4(1):49–64
40. Suárez M (2003) Scientific representation: against similarity and isomorphism. Int Stud Philos Sci 17:225–244
41. Tahera K, Earl CF, Eckert CM (2012) The role of testing in the engineering product development process, theory and methods of competitive engineering 2012. Karlsruhe, Germany
42. Talbi E-G (2009) Metaheuristics: from design to implementation. Wiley, Chichester
43. Teller P (2001) Twilight of the perfect model model. Erkenntnis 55:393–415
44. van Fraassen BC (1980) The scientific image. Oxford University Press, Oxford

45. Van Hentenryck P (1989) Constraint satisfaction in logic programming. Logic Programming Series, MIT Press, Cambridge
46. Weber M (1904/1949) "Objectivity" in social science and social policy. In: Shils EA, Finch HA (Eds) Max Weber's methodology of the social sciences. Free Press, New York
47. Wyatt DF, Eckert CM, Clarkson PJ (2009) Design of product architectures in incrementally developed complex products. In: Proceedings of the 17th International Conference on Engineering Design (ICED'09), Design Society, Stanford, CA, USA, vol 4, pp 167–178
48. Wynn D, Eckert CM, Clarkson PJ (2007) Modelling iteration in engineering design. In: Proceedings of the 16th International Conference on Engineering Design (ICED'07), Design Society, Paris

Chapter 20
Theories, Models, Programs, and Tools of Design: Views from Artificial Intelligence, Cognitive Science, and Human-Centered Computing

Ashok K. Goel and Michael E. Helms

20.1 Introduction

Research on design adopts many perspectives ranging from anthropology to neurobiology to philosophy. The various research paradigms produce not only different theories and models of different aspects of design, but also different types of theories and models. For a quarter of a century, our research laboratory has explored design from the perspectives of artificial intelligence, cognitive science, and human-centered computing. Design research in these paradigms produces information-processing theories and computational models of aspects of design, as well as computer programs that implement and test the theories and models. These products in turn often form the basis for the development of interactive technologies for supporting aspects of design practice as well as pedagogical techniques for teaching elements of design theory and methods.

We have three main goals in this chapter. First, we want to briefly describe the perspectives of knowledge-based artificial intelligence, computational cognitive science, and human-centered computing, and in particular, the types of theories, models, programs, and tools they produce. Second, we want to illustrate some of the methods and artifacts of our research through a case study of problem–solution coevolution in biologically inspired design. Starting with the extant Structure-Behavior-Function model for expressing knowledge of technological and biological systems, we develop a knowledge model of design problems called SR.BID that is grounded in empirical data about biologically inspired design practice. Third, we want to present the SR.BID model that captures problem descriptions as well as problem–solution relationships in biologically inspired design. SR.BID forms the basis for ongoing development of new interactive tools for supporting biologically inspired design practice as well as new pedagogical techniques for learning about problem formulation.

A. K. Goel (✉) · M. E. Helms
Design and Intelligence Laboratory, School of Interactive Computing,
Georgia Institute of Technology, Atlanta, GA 30308, USA

A. Chakrabarti and L. T. M. Blessing (eds.), *An Anthology of Theories and Models of Design*, DOI: 10.1007/978-1-4471-6338-1_20,

20.2 Artificial Intelligence, Cognitive Science, and Human-Centered Computing

Artificial intelligence has several research paradigms. In this work, we are interested in the paradigm of knowledge-based artificial intelligence that has twin goals [22, 31]: to computationally understand human intelligence and to build intelligent systems with human-level intelligence. Theories and models in knowledge-based artificial intelligence typically use knowledge constructs to unify memory, reasoning, and learning processes, and thus address issues concerning the content, representation, organization, use, and acquisition of knowledge.

We are interested in computational cognitive science that seeks to computationally understand animal cognition [43]. A classical paradigm in computational cognitive science is human information processing that seeks to understand human behavior in terms of information processing in the human mind [37]. Another paradigm popular in modern cognitive science is situated cognition [6, 10] that seeks to understand human behavior in terms of interaction with the physical and social worlds.

Human-centered computing is an emerging interdiscipline within modern computing [29]. Human-centered computing takes human experience and its sociocultural context into consideration in the design of computational artifacts. In practical terms, human-centered computing is the next stage in the evolution of human–computer interaction as a discipline. As Fig. 20.1 shows, we are interested in human-centered computing at the intersection of artificial intelligence, cognitive science, and human–computer interaction. In particular, we are interested in research on artificial intelligence and cognitive science that produces interactive tool for supporting human designers in their work. Although not shown in Fig. 20.1, we are also interested in research on artificial intelligence, cognitive science, and human-centered computing that results in pedagogical techniques for teaching and learning design theory and methods.

20.3 Information-Processing Theories, Computational Models, Computer Programs, and Interactive Tools

We use the terms "theory" and "model" here in the sense of a scientific theory and a scientific model [11, 12, 32, 36]. A scientific theory is (i) based on testable hypotheses and makes falsifiable predictions, (ii) internally consistent and compatible with extant theories, (iii) supported by evidence, and (iv) modifiable as new evidence is collected. An important cognitive feature of a scientific theory is that it suggests a process or method for building, evaluating, revising, and accepting (or abandoning) a theory.

As indicated above, we are interested in information-processing theories of design. As an example, for a quarter of century the design research community has

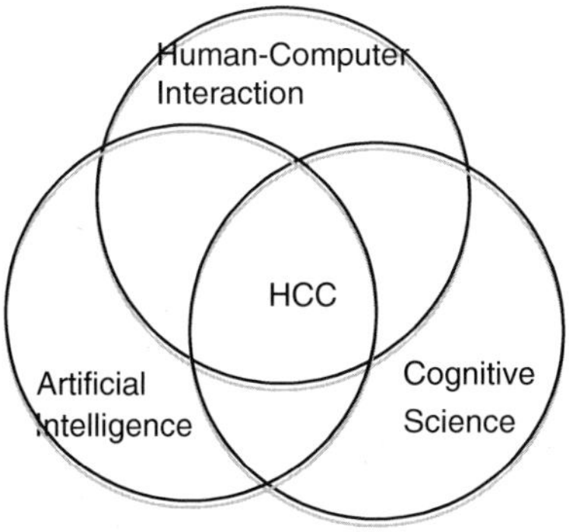

Fig. 20.1 Human-centered computing (HCC) at the intersection of artificial intelligence, cognitive science, and human–computer interaction

been developing Information-processing theories of analogical design (e.g., [17, 21, 35, 49]). In our own earlier work on analogical design, we have developed normative artificial intelligence theories, techniques, and tools for analogical design ranging from case-based design [18, 19] to cross-domain analogies [3, 20]. These theories are based on testable hypotheses about case-based design and cross-domain analogies in design, respectively, and some of their predictions have been evaluated through computational and experimentation.

A scientific model is an interpretation of a target system, process, or phenomenon that proposes or elaborates on the processes and mechanisms that underlie it. Like scientific theories, scientific models too have important cognitive features. First, models are abstractions of reality. They productively constrain reasoning by simplifying complex problems and thus suggest a course of analysis. Second, models are cognitive tools for generating explanations. They serve as tools both for specifying and organizing the current understanding of a system and for using that understanding for explanation and communication.

We are interested in two closely related kinds of models in design. First, we are interested in knowledge models. A knowledge model in design provides an ontology (i.e., a vocabulary) for representing the knowledge and a structure for organizing the knowledge in a design domain. For example, in our work on case-based design, we developed the Structure-Behavior-Function (SBF) knowledge model of the working of technological systems. The SBF knowledge model provides an ontology for expressing the knowledge of the system and a schema for organizing the knowledge [23, 24, 38]. The SBF model enables retrieval, adaptation, evaluation, and storage of design cases in addressing new design problems [19]. Similarly, in our work on cross-domain analogies in design, we developed Behavior-Function (BF) abstractions of SBF models that provide a vocabulary for representing teleological design patterns. The BF design patterns enable cross-domain analogies in designing new technological systems [3, 20].

Second, we are interested in computational models of design. While an information-processing theory of design is based on testable hypotheses and makes

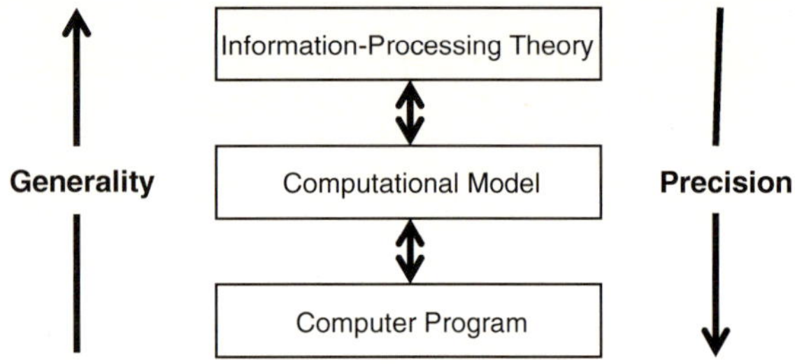

Fig. 20.2 Relationship among information-processing theories, computational models, and computer programs of design

falsifiable predictions, a computational model of design provides architectures, algorithms, and knowledge models for the theory. As Fig. 20.2 shows, computational models are more detailed and precise than information-processing theories. Thus, our computational model of cross-domain analogies in design [3, 20] provides an architecture that integrates memory, reasoning, and learning processes, SBF knowledge models of technological systems and BF knowledge models of design patterns, as well as algorithms for accessing, using, learning, and storing the design patterns.

The artificial intelligence paradigm also develops computer programs. A computer program is an experiment that implements the computational model and evaluates the information-processing theory. A computer program adds enough detail and precision to the computational model to be executable on a computer, as shown in Fig. 20.2. Thus, the Kritik [18, 19] and the Ideal [3, 20] computer systems implement our computational models and evaluated the information-processing theories of case-based design and cross-domain analogies, respectively.

The paradigm of human-centered computing also develops interactive technologies for supporting design practice. Indeed, interactive technologies have revolutionized design practice over the last generation, and insofar as we can see into the future, this trend likely will continue.

20.4 Problem–Solution Coevolution in Biologically Inspired Design: An Illustrative Case Study of Knowledge Modeling

The perspectives of artificial intelligence, cognitive science, and human-centered computing on design are mutually compatible. Thus, a design researcher can move from one paradigm to another depending on the research goal and the design context.

Further, knowledge models are common to all three paradigms. However, our discussion of knowledge models so far has been quite general and abstract. We now illustrate knowledge modeling through a case study of problem–solution coevolution in biologically inspired design (also known as biomimicry, biomimetics, and bioinspiration) [1, 2, 46, 48]. Over the last decade or so, the design research community has been studying biologically inspired design from the perspectives of artificial intelligence, cognitive science, and human-centered computing (e.g., [7, 41]). Our own interest in biologically inspired design spawned in part because it entails cross-domain analogies from biological systems to technological systems and thus provides an arena for further exploration of analogical design.

However, our work on biologically inspired design differs from our earlier work on analogical design in three fundamental ways. First, unlike the earlier normative artificial intelligence theories and models, our new work develops cognitive, descriptive theories, and models of analogical design (e.g., [28, 44]). Second, our work now has the additional goal of using our theories and models to develop interactive technologies (e.g., [25]; http://dilab.cc.gatech.edu/dane/) and pedagogical techniques for aspects of design. Third, our empirical studies have found that biologically inspired design entails not only cross-domain analogies but also problem–solution coevolution [26, 27]. Problem–solution coevolution is a well--known characteristic of creative design [14, 15, 33], but, insofar as we know, biologically inspired design has not been previously studied as entailing problem–solution coevolution. In traditional problem solving, the problem remains fixed even as solutions to the problem are generated. In problem–solution coevolution, the problem evolves as solutions are generated, with the current problem formulation influencing solution generation, and the current candidate solutions influencing problem formulation. Perhaps more interestingly, we found that biological analogies not only help generate solutions to a design problem, but also support inception and evolution of design problems [26, 27].

As much as the scope, focus, and methodology of our work have evolved over the years, our emphasis on grounding design processes in design knowledge has remained constant. The question then becomes what is a good knowledge model that can capture problem–solution coevolution in biologically inspired design? As one might expect, different researchers in biologically inspired design have developed different knowledge models, depending on the goal, scope, focus, and methodology of their work. Thus, Biomimicry 3.8 Institute has developed an ontology of functions of biological systems that purports to support its design model for generating design solutions [4]. Vincent and his colleagues have developed an ontology of biological systems that promises to support a TRIZ-like model of biologically inspired design [45]. Stone, McAdams and their colleagues have proposed the use of the extant function-flow ontology of Functional Basis for the task of concept generation in biologically inspired design [34]. Chakrabarti and his colleagues have developed a detailed SAPPhIRE knowledge model to support biologically inspired design [40]. All these knowledge models are normative, even if some of them are based on notions of best practices in biologically inspired design. Perhaps more importantly, all these models focus on design ideation in conceptual design (and thus do not address problem–solution coevolution).

In contrast, in this work we are interested in developing a knowledge model of design problems that can capture the process of problem–solution coevolution in biologically inspired design. We start with textual data from the practice of biologically inspired design in an educational setting and then derive the knowledge model of design problems called SR.BID. We validate the SR.BID model through comprehensive and repeatable categorization of unstructured textual data collected in the biologically inspired design practice.

20.5 Methodology and Data

Since 2006, we have observed ME/ISyE/MSE/PTFe/BIOL 4740, an interdisciplinary, project-based class taught yearly and jointly by biology and engineering faculty at Georgia Institute of Technology. In this course, mostly senior-level design students work in small interdisciplinary teams of 4–5 on open-ended design projects over the course of a semester. The extended, collaborative design projects typically involve identification of a design problem of interest to the team and conceptualization of a biologically inspired solution to the identified problem. Yen et al. [47] describe the course and the design projects in detail.

We use three data sets collected from observations of the design projects in the biologically inspired design class. The first set of data consisted of the project submissions of one design team in Fall 2008 that focused on capture of solar energy for use in homes. The project was selected as a typical example of biologically inspired design. The data consisted of four individual problem description assignments, a team mid-term presentation, and the team final presentation. We shall refer to this as the 2008 data set. The following is an excerpt from a problem description:

> I think this is a big gap between the static and fragile solar panels that we have so far engineered. So far, most solar panels are set up on a grid basis acting together especially when moving to the sun rather than as individual. Continuing off that tangent I think it would be interesting to have an individual solar panel that can stand alone and still function. The snail shell structure is stand alone and has the ability to passively dissipate heat by using the heat gradient so that it is cooler within the shell than the outside air and ground this would be helpful for allowing the interior of a structure with solar panels to remain cool.

The second set of data consisted of individual assignments given to students in Fall 2010, and collected in the third week of class. This assignment asked students to provide a short 1–2 page design problem description suitable for the biologically inspired design context. A total of 38 assignments were collected (one of which was eliminated as it belonged to a member of our research laboratory who was taking the class at the time). We shall refer to this as the Week 3 2010 data set.

The third set of data consisted of an individual assignment given to students in Fall 2010 and collected during the eighth week of class. This assignment consisted of problem descriptions between one quarter of a page and one full page in length. A total of 32 assignments were collected (the assignment from the member of our laboratory was again eliminated). We shall refer to this as the Week 8 2010 data set.

To analyze these data sets, we used a variation on the methodology of Grounded Theory [16, 42]. In the Grounded Theory methodology, a theory about any phenomenon is derived (solely) from data. In a recent variation, the theory is derived from data but the initial coding scheme is seeded with a predefined ontology [30]. As indicated above, we use the SBF knowledge model as a seed, and then derive the SR.BID model from the data about biologically inspired design.

20.5.1 Brief Review of the SBF Knowledge Model

SBF is a family of knowledge models that includes not only SBF models of biologically and technological systems, but also BF models of design patterns (as well as other models not described here) [19, 38]. Here, we briefly summarize the basic SBF model that consists of three nested high-level schemas, the *structure*, *behavior*, and *function* schemas [23]. The structure schema consists of a set of elements, which may be classified as *elements* such as *substances* or *components*, and *connections* among them. *Elements* may have associated *properties* and *values*, while *connections* express the relationship type (e.g., hinged) between *elements*.

The *behavior* schema consists of *states* and *transitions* between the *states*. *States* consist of a set of *elements*, and a set of *property—value* for the *element*. Each *transition* is annotated by *causal explanations* for the *transition*. Since one kind of *causal explanation* pertains to a *function* of a *component, behaviors* act as indices to *functions* of *components*.

The *function* schema consists of a *given* or *prerequisite state*, and one or more *makes* or *resultant states*. It also specifies one or more external *stimuli*. Also, it specifies the *behavior* that accomplishes the *function*. Thus, *functions* act as indices to *behaviors*. *Functions* can be of several types including *accomplishment, maintenance, prevention,* and *negation*.

20.5.2 Construction of the SR.BID Knowledge Model

We started with a single coder to map the problem description text data in the 2008 data set to concepts in the SBF knowledge model. During initial coding, our goal was to align the SBF ontology with the data and add new conceptual categories as they emerged from the data.

Figure 20.3 shows SR.BID's high-level ontology that emerged from our analysis. The ontology consists of six main concepts: *function, performance criteria, solution, deficiencies/benefits, constraints/specification,* and *operating environment*. *Solution*

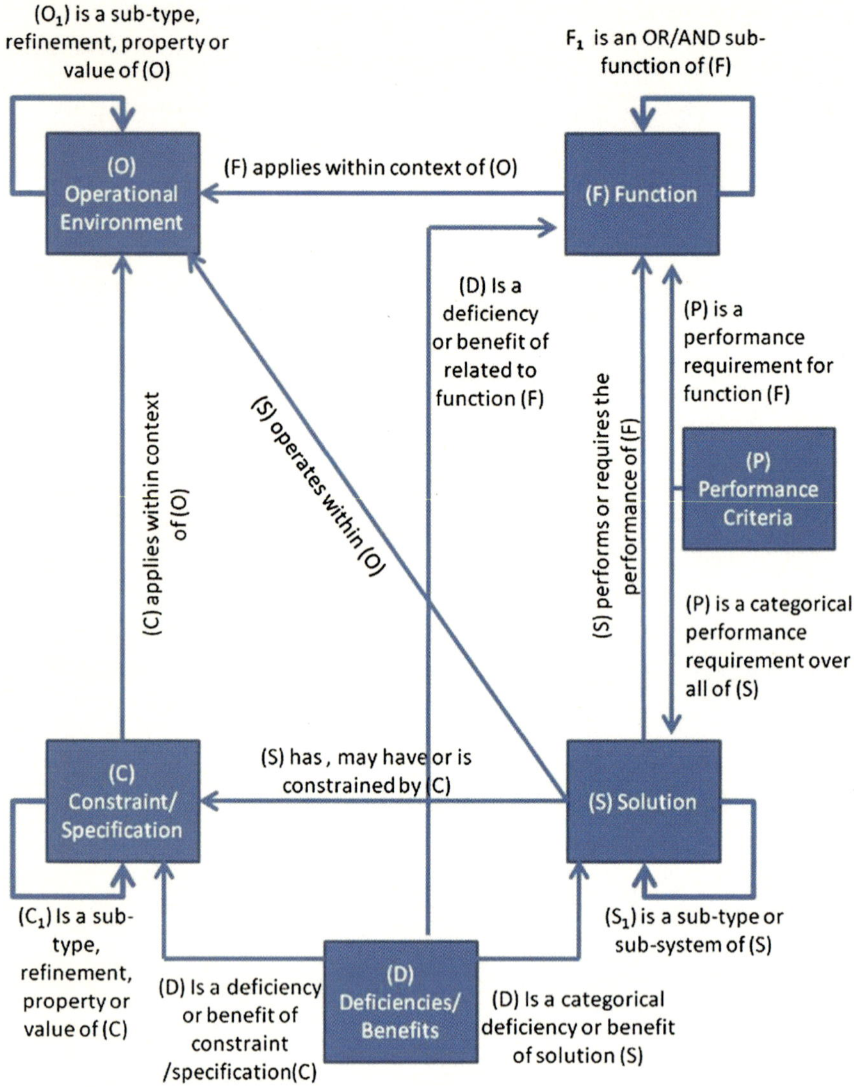

Fig. 20.3 The problem schema in SR.BID including both the main concepts and relationships

here refers to existing systems for achieving the given *function*, and *deficiencies/ benefits* pertain to negative/positive assessments of the *solution. Performance criteria* act as qualifiers on the Function (e.g., dissipate heat *passively*), and *constraints/ specification* describe constraints on the *solution* (e.g., cost).

20.5.3 Refinement of the SR.BID Model

Following the construction of the initial SR.BID model, we used two coders to refine and validate the model using the Week 3 2010 data set, which consisted of 37 design problem statements between one and two pages in length. The first coder was an author on this chapter (Helms) and was well versed with the coding process. The second coder was a third year undergraduate biology student new to the field of biologically inspired design, and without prior background knowledge in design or cognition, SBF, or SR.BID. We allocated half of the data (17 problem statements, selected at random) to training and refinement and used the remaining to draw samples for testing and validation.

This phase led to the identification of relationships among the six concepts in SR.BID's problem model, as shown in Fig. 20.3. This phase also led to identification of additional subcategories of the six categories in the model. Appendix 1 (Detailed Description of the SR.BID Knowledge Model), provides a complete listing and description of each category and subcategory. Note that as required of a knowledge model, the SR.BID model of design problems provides both an ontology for representing knowledge of design problems and a schema for organizing the knowledge, which allows capture of descriptions of specific problems such as the one on page 420.

After two passes on refinement and training, a random sample of five was pulled from the remaining problems to be used for validation. Each coder independently coded each test sample. We found the Cohen's Kappa measure of inter-coder reliability that adjusts for chance agreement to be 0.778. (Generally Cohen's Kappa values close to and above 0.8 are deemed acceptable.) After initial comparison, the two coders entered a negotiation phase, in which they attempted to resolve coding discrepancies. As expected, post-negotiation agreement levels were at significantly higher Cohen's Kappa values: 0.962 of concepts and 0.976 for relationships.

20.5.4 SR.BID Validation

To further test the conceptual soundness and potential usefulness of SR.BID, we applied it to the 2010 Week 8 data set, consisting of 31 brief problem statements. In this test we used a conservative dual-coding strategy over the entire data set. During dual-coding, each of the two coders is present during the session, and while one coder takes the lead, the second coder may question coding decisions leading to discussion and negotiation until a code is agreed upon. This ensures reliability much closer to the post-negotiated numbers shown in the previous test, with the additional cost of requiring two coders to code all documents. Tests of intra-coder reliability, conducted on the recoding of five problem statements selected at random 12 weeks after initial coding, demonstrate an agreement of 0.878 and 0.872 for coding concepts and relationships, respectively.

We found that the *function* concept is pervasive in most problem descriptions, occurring in 72.7 % of all conceptual relationships. The *solution* concept too is quite common, occurring in about half of the relationships. The *function-solution* relationship is the most common relationship, representing about one-fourth of all conceptual relationships in the observed sample. It is noteworthy that nearly 70 % of the *function-solution* relationships in our sample pertained to *existing* solutions rather than conjectured solutions. Understanding the role and influence of existing solutions such as biological analogs is of particular interest in biologically inspired design.

20.6 Discussion

In our perspective, knowledge models in design are intimately connected to information-processing theories and computational models of design tasks. As we study new design tasks, we develop new knowledge models appropriate to the task. Thus, as we studied memory, reasoning, and learning tasks in analogical design a generation ago, the SBF model logically evolved out of Chandrasekaran's Functional Representation [8, 9]: SBF representations supported the inferences required by the memory, reasoning, and learning tasks in analogical design. In a similar manner, as we study problem–solution coevolution in biologically inspired design, the SR.BID model of design problems is evolving out of the SBF model of the working of technological and biological systems.

The SR.BID model allows us to capture problem descriptions more deeply than the SBF model. In the basic SBF schema [23], a system's interaction with its external environment is captured in terms of system's *functions* and *external stimuli* from the environment to the system. Prabhakar and Goel [38] did describe the external and internal environments of a system but those ideas were not fully developed. SR.BID specifies *operational environment* explicitly. Similarly, *performance criteria* establish the metrics against which the *functions* of a design of a system may be evaluated. The frequency of occurrence of the *operating environment* and *performance criteria* concepts in our study seems to highlight their important role of problem formulation: they provide additional information needed to evaluate whether a *solution* satisfies the desired *function*. Dinar et al. [13] provide an alternative schema for representing problem descriptions.

As we noted above, the coded textual descriptions of biologically inspired design frequently refer to biological analogies and other existing solutions. This may have to do with the way in which design problem formulation occurs in biologically inspired design. Given a need, one method for problem formulation is to look to existing solutions that have been used to solve the need, or similar needs, in past. An existing solution provides a base case, a plan, or a pattern from which the designer might abstract key concepts, such as *functions*, which provide the points of traction necessary to begin formulating the design problem. This has deep implications for biologically inspired design because it shows that biological analogies may serve to help (re-)formulate problems as well as solve them.

20.7 Uses of SR.BID

Currently, we are using the SR.BID model in four ways. First, we are using it as a coding scheme to analyze additional data on problem–solution coevolution in biologically inspired design. In particular we are studying the influence of biological analogies on problem formulations and reformulations over time.

Second, we are using SR.BID as part of a pedagogical technique to help students in formulating design problems in the Georgia Tech ME/ISyE/MSE/PTFe/BIOL 4740 course on biologically inspired design. In past, problem formulation has been an extremely difficult task for students in the class [47]. In our pedagogical technique, students define their problems in terms of "four boxes:" *operational environment, function*, *constraints/specifications*, and *performance criteria.*

Third, we are developing an interactive technology for aiding students in evaluating cross-domain analogies in design. Designers in general lack a tool for systematic evaluation for cross-domain analogies. Thus, evaluation of analogies often is *ad-hoc*, and suffers from confirmation bias effects. Our tool uses the same "four-box" method to evaluate analogies in biologically inspired design. Students compare their four-box problem description against a four-box representation constructed for their biological analog, and then use this to frame a discussion of how their analogy is similar and dissimilar.

Finally, where most search engines for biologically inspired design focus on indexing by functions, we are using SR.BID to structure a knowledge base of design problems and biological systems to help facilitate search across the breadth of concepts found in the problem schema shown in Fig. 20.3.

20.8 Conclusions

Methodologies for research in design are receiving much needed attention (e.g., [5]). Our methodology for design research constructs information-processing theories, computational models, and computer programs of design. It also produces knowledge models, interactive tools, and pedagogical techniques for design.

Current information-processing theories of analogical design, including biologically inspired design, typically focus on use of analogy for generation of design ideas, and concepts for a given design problem. However, in tracing collaborative, extended, open-ended episodes of biologically inspired design we found that biological analogies often help not only in generating design ideas for a given formulation of the design problem, but also in (re-)formulating the problem itself. In fact, problem reformulation appears to have been the primary role of some biological analogies since the biological systems were not part of either the preliminary or final design solutions.

Evaluating our information-processing theory of biologically inspired design requires the construction of a computational model that specifies the architecture, algorithms, and knowledge model for problem–solution coevolution, where the knowledge model specifies the ontology and the schema for representing and organizing knowledge of design problems. In this chapter, we focused on the knowledge model. In particular, we used the SBF schema for representing knowledge of biological and technological systems as a seed for developing the SR.BID schema for representing problem descriptions in biologically inspired design. The conceptualization of the SR.BID problem schema was data driven, and grounded in the verbal descriptions designers provided for their designing. As measured by standard tests of coder reliability and coverage, the SR.BID constructs seem to provide comprehensive and reliable encoding of the verbal descriptions of interdisciplinary design teams engaged in biologically inspired design.

The SR.BID problem schema allows us to capture the problem descriptions design teams construct in collaborative, extended, open-ended biologically inspired design; it also enables us to capture the relationships between the problem and the solutions, as well as systematically trace the influence of the problem on the solution and vice versa in problem–solution coevolution in biologically inspired design. The SR.BID problem schema forms the basis of both pedagogical techniques for teaching about problem formulation and interactive tools for assessing cross-domain analogies for addressing a given design problem.

Acknowledgments We are grateful to the instructors and students of the ME/ISyE/MSE/PTFe/BIOL 4740 classes from 2006 through 2012, especially Professor Jeannette Yen, the main instructor and coordinator of the class. We are grateful also for the contributions of Marika Shahid, Swaroop Vattam, and Bryan Wiltgen. We thank the US National Science Foundation for its support of this research through an NSF CreativeIT Grant (0855916, Computational Tools for Enhancing Creativity in Biologically Inspired Engineering Design), and an NSF TUE Grant (1022778, Biologically !nspired Design: A Novel Interdisciplinary Biology-Engineering Curriculum.

Appendix 1: Detailed Description of the SR.BID Knowledge Model

The following tables describe the ontology of the SR.BID knowledge model of design problems that emerged from analyzing problem statements in the Week 3 2010 and Week 8 2010 data sets. These tables refine the high-level ontology of concepts and relationships of Fig. 20.3.

Solution	Description
Primary type	
Biological	The solution is a naturally occurring biological component, organism, or system
Man-made	The designers refer to a system which someone already built or created, or for which they generated prototypes or specifications
New design solution	The designers who are working on the problem are conjecturing a new design (or a design they think is new) to solve the problem
Secondary type	
Sub-solution	A sub-solution consists of many parts that together perform a specific function within the context of a larger solution
Subtype	A subtype solution expresses a "kind-of" relationship with another solution

Function	Description
Primary type	
Accomplishment	The default function type, accomplishment functions change the state of the world in an intended way
Preventative	Preventative functions keep a state OR another function from occurring
Maintenance	Functions that maintain a state are considered maintenance for example "the thermostat regulates temperature" is a maintenance function
Allow	Allow functions enable a state OR another function to occur
Negation	Negative functions are stated as NOT performing another function, for instance this application does not produce light
Secondary type	
Sub-function, AND	When there are multiple sub-function relationships for a given function, AND-type relationships that specify that the related sub-functions must all be accomplished in order to achieve the parent function
Sub-function, OR	When there are multiple sub-function relationships for a given function, the OR-type relationship specifies that one of the functions must be accomplished to achieve the parent function

Operating environment	Description
Primary type	
Location	The places in which the system is intended to operate
Condition-qualitative	Qualitative conditions under which the system is intended to operate
Condition-quantitative	High/low-end values, expected values, or ranges
Time	The time during which the system must operate for example, "at night." Words like "when," "after," "while," "as," and "during" are often used to express a temporal environment
User	The phrase describes an intended user or class of users for the system
Entity	The phrase describes an entity, often biological but sometimes technological, that interacts with the system
System	The phrase describes another system within which the system is intended to work or connect

Constraints and specifications	Description
Primary type	
Material	The material of which one or more components of the design will be composed
Information	Information can be in the form of energetic signals, bits and bytes, or may be encoded in the physical structure of a thing
Energy	Energy can be found throughout a system in many forms; the energy subtype is used when a specified form of energy is discussed within the confines of the system
Time	Includes timeframes not related to the operation of the design
Component	Includes descriptions of specific parts of a solution or design, or groups of parts
Property/value	Concerns the properties of the system as a whole or their values
Shape	Includes the shape of the components or of the design
Spatial orientation	These specify the spatial relationship or orientation between or among one or many components, systems, or subsystems
Structural relationship	Any phrase specifying which components are related by means of connecting joints and contacts points
Cost	Usually in monetary terms, but this could also be in terms of any resource of concern; absolute; or relative
Secondary type	
Limiting	Limiting specifications/constraints are those which require a designer to use a smaller subset of design elements
Enabling	Enabling specifications/constraints offer new possibilities for design elements without enforcing their use
Existing	Existing specifications/constraints discuss the specific properties of an existing design

Performance criteria	Description
Primary type	
Specific	States the specific value or range of the performance criteria
Relative	Uses comparative terms such are "quieter than solution X," without explicitly stating the performance of the compared to solution
Actual	States the performance of an existing solution

Deficiency/ Benefit	Description
Primary type	
Deficiency	Deficiencies can relate to any element of an existing solution or proposed design, highlighting an unfavorable aspect of that element
Benefit	Benefits can relate to any element of an existing solution or proposed design, highlighting a favorable aspect of that element

References

1. Bar-Cohen Y (ed) (2011) Biomimetics: nature-based innovation. CRC Press, Boca Raton
2. Benyus J (1997) Biomimicry: innovation inspired by nature. William Morrow, New York
3. Bhatta S, Goel A (1997) Learning generic mechanisms for innovative strategies in adaptive design. J Learn Sci 6(4):367–396
4. Biomimicry 3.8 Institute (2008) AskNature. http://www.asknature.org/. Accessed on 25 Apr 2011
5. Blessing L, Chakrabarti A (2009) DRM: A design research methodology. Springer, London
6. Brown J, Collins A, Duguid P (1989) Situated cognition and the culture of learning. Educational Researcher 18(1):32–42
7. Chakrabarti A, Shu L (2010) Biologically inspired design. AIEDAM 24:453–454
8. Chandrasekaran B (1994) Functional representation: a brief historical perspective. Appl Artif Intell 8(2):173–197
9. Chandrasekaran B, Goel A, Iwasaki Y (1993) Functional representation as design rationale. IEEE Comput 26:48–56
10. Clancey W (1997) Situated cognition: on human knowledge and computer representations. Cambridge University Press, Cambridge
11. Clement J (2008) Creative model construction in scientists and students: the role of imagery, analogy, and mental simulation. Springer, Dordrecht
12. Darden L (1998) Anomaly-driven theory redesign: computational philosophy of science experiments. In: Bynum T, Moor J (eds) The digital phoenix: how computers are changing philosophy. Blackwell Publishers, New York, pp 62–78
13. Dinar M, Shaj J, Hunt G, Campana E, Langley P (2011) Towards a formal representational model of problem formulation in design. In: Proceedings of ASME 2011 IDETC/CIE Conference, Washington DC
14. Dorst K (2003) The problem of design problems. In: Cross N, Edmonds E (eds) Expertise in design. Creativity and Cognition Studio Press, Sydney
15. Dorst K, Cross N (2001) Creativity in the design process: co-evolution of problem-solution. Des Stud 22(5):425–437
16. Glaser B, Strauss A (1967) The discovery of grounded theory: strategies for qualitative research. Aldine, Chicago
17. Gebhardt F, Voß A, Gräther W, Schmidt-Belz B (1997) Reasoning with complex cases. Kluwer, Norwell, MA
18. Goel A (1992) Representation of design functions in experience-based design. In: Brown D, Waldron M, Yoshikawa H (eds) Intelligent computer aided design. North-Holland, Amsterdam, pp 283–308
19. Goel A, Bhatta S, Stroulia E (1997) Kritik: an early case-based design system. In: Maher M, Pu P (eds) Issues and applications of case-based reasoning in design. Mahwah, Erlbaum, pp 87–132
20. Goel A, Bhatta S (2004) Design patterns: a unit of analogical transfer in creative design. Adv Eng Inform 18(2):85–94
21. Goel A, Chandrasekaran B (1988). Integrating model-based reasoning with case based reasoning for design problem solving. In: Proceedings of AAAI-88 workshop on AI in design, Minneapolis
22. Goel A, Davies J (2011) Artificial Intelligence. In: Sternberg R, Kaufman S (eds) Cambridge handbook of intelligence, 3rd edn. Cambridge University Press, Cambridge
23. Goel A, Rugaber S, Vattam S (2009) Structure, behavior and function of complex systems: the SBF modeling language. Artif Intell Eng Des Anal Manuf 23:23–35
24. Goel A, Stroulia E (1996) Functional Device Models and Model-Based Diagnosis in Adaptive Design. Artif Intell Eng Des Anal Manuf 10:355–370

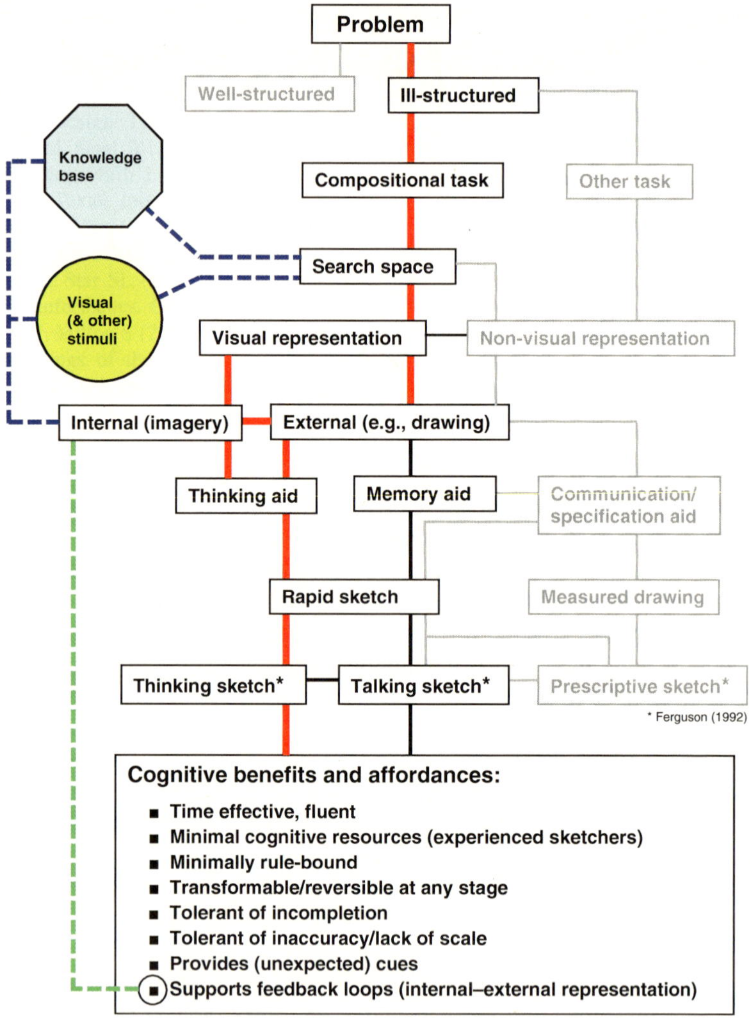

Fig. 21.1 Model: sketching as a mental facilitator in complex, visually mediated tasks (first published in Goldschmidt [22]). *Note* Components and links in *gray* are considered subsidiary

whole—"the larger picture." In design research the purpose of a model is to explicate the process of designing or elements thereof from one or another standpoint.

The paper starts with a brief literature review on sketching in design, to be followed by a presentation of the model developed to explicate the role of sketching (Fig. 21.1). The remainder of the paper presents three studies undertaken to expand and validate some of the model's major tenets.

21.2 Sketching in Design

Numerous studies found that rapid manual sketching is instrumental in the search phase of conceptual design. Visual thinking and reasoning makes use of a large variety of images that are not necessarily formal symbols. It is most useful in reasoning by example (and comparison), as opposed to the complementary mode of reasoning, by rule [41]. Designers infer information from images by attaching meaning to their attributes, including shape and form, color, texture, relationship among components, and more. The latter is particularly important because based on relationships an image may serve as a basis for analogy in the process of designing (e.g., [7]). When visually thinking while ideating, designers utilize images in two ways. The first, which may be called "passive usage," is related to the influence that external images that the designer is exposed to have in the process of idea generating. Such images may act as inspiration [25] or, conversely, they may cause fixation when designers are unable to advance beyond the prompted images that are imprinted in their minds [42]. For our purposes here, the second way in which designers utilize images, namely "active usage," is more relevant. By active usage we mean the self-generation of images which, in the idea-generation phase consists almost exclusively of sketching.

Designers think visually when they generate visual representations either internally in their minds, in which case they use mental imagery, or externally, wherein representations such as drawings, including sketches, are involved. Designers have been sketching on paper while developing their design ideas for centuries, in fact ever since paper of sufficient quality became readily available and affordable, which occurred in Europe in the last quarter of the fifteenth century. This development was the result of the demand for paper following the print revolution that was prompted by Gutenberg's invention of the movable type (Circa 1440; see Eisenstein [12]). Originally produced for book printing, paper changed design practices as well. In the innovative spirit of the renaissance artists and designers (architects and engineers) gained a cultural "license" to explore and come up with less dogmatic designs than had been the rule hitherto, and sketching suited the needs of experimentation and exploration perfectly. In addition, at the same time the newly introduced orthogonal projection system of representation attributed to Raphael [28] that stemmed from the rules of perspective, spread rapidly, and soon became the standard representational mode in design of buildings and artifacts. Thus many sketches, too, utilized orthogonal projections, albeit crude ones. Sketches by Leonardo da Vinci and Michelangelo, to name but the most famous designers of their time, survived and bear evidence of the immersion in this new practice of sketching as of the end of the fifteenth century. The mode of sketching has not changed much since then: rapid freehand strokes on paper, not necessarily precise, complete or to scale.

Arnheim [3], the pioneer researcher of visual thinking, wrote about transformations in visual representation in the course of concept formation, but sketches and sketching were traditionally of interest to researchers primarily from a developmental point of view. Interest in architectural study sketches was first detected in the 1970s. A case in point is the collection of small study sketches by the architect James Stirling, who became one of Britain's leading modern architects. He was in the habit of producing many very small sketches while searching for concepts at the outset of new projects but his sketches were published for the first time only in the 1970s, when competition entries by his office for three German museums that were published by professional magazines (Architectural Review CLX957, 1976; Lotus International 15, 1977), featured for the first time some of the sketches. Years later and after Stirling's death, when asked about earlier sketches, his partner Michael Wilford wrote: "Prior to the German museum projects, the early exploratory material was considered of little value and discarded once the final design had been established. More recent material is stored..." (personal communication 2000). Indeed, sketches are rarely found on the pages of architectural magazines and professional books before the mid-1970s, but in later years it became quite habitual to publish and exhibit early study sketches. Collections of sketches were even published separately in the form of books (e.g., [32, 38]). According to Goldschmidt and Klevitsky [24] there are two reasons for the surge of interest in design sketches. The first is the emerging interest in the design research community in design processes and design thinking, as opposed to design methods that had occupied center-stage in the previous two decades (1950s and 1960s). The second reason is cultural: the 1970s saw the advent of postmodernism, which professed great interest in processes of artistic and design production in all disciplines including, besides design, the visual arts, theater, cinema, literature. Drafts of literary works were published and films about the making of films were produced. Design sketches fitted into this pattern perfectly.

One of the first researchers to write about design sketches was Herbert (e.g., [29]), who studied primarily the structure and composition of sketches and the quality of lines and other graphic marks. Later more researchers turned to the study of sketching; some were interested in better understanding it, primarily from the cognitive perspective, while others hoped to be able to emulate or replace it digitally. An early study is by Fish and Scrivener [15], who developed a theoretical model of the cognitive mechanisms that in their view enable sketching to induce artistic inventiveness. The essence of these mechanisms is the ability to translate propositional information into depictive information and vice versa, possibly leading to new and original descriptions and depictions. A somewhat similar model was proposed by Goldschmidt [17] who claimed that in reasoning designers shift between "seeing as" and "seeing that" arguments. "Seeing as" relates to physical properties of the designed entity and sketching is often instrumental in elucidating them. A series of studies by Suwa and his associates [43–46], based on protocol analysis, further explored how sketching contributes to the formation of design ideas, as designers make discoveries in their own sketches. Goldschmidt studied concrete cases in which designers' interpretations of their own sketches served to solidify design ideas [18, 20, 21], 2003. Purcell and Gero [37] guest-edited an issue of *Design Studies* that was dedicated to design drawing and sketching and in their introductory overview paper they stressed the cognitive aspects of the instrumentality of sketching. They showed how sketching was connected to working memory, imagery and in particular mental synthesis. Some researchers have shown that up to a point, it is possible to rely on visual imagery alone in developing design concepts [4, 5]. However, as Verstijnen et al. [48] have shown, sketching enables more transformations and manipulation of shapes, leading to more complex solutions than are possible without sketching.

The other strand of research is related to CAD. The previous generation of researchers approached CAD from the perspective of the machine's abilities and took it for granted that humans should adapt themselves to the requirements posed by the logic and procedures of the software they were able to develop. The new generation of CAD researchers realized that CAD systems must adapt themselves to human cognition [40]. Ellen Do and Mark Gross are prominent representatives of this group of researchers. Sketching is one of the topics to which they devoted a lot of attention, claiming that since it is vital in the design process it is important to find ways to produce sketches digitally that would have the same affordances as manual sketches (e.g. [9, 10, 26, 27]). At present this line of research focuses on human–computer interaction and Do and Gross fit into this trend [30]. A current example of the HCI line of work can be found in a special issue on understanding design thinking in the journal Human Computer Interaction, guest edited by Scott Klemmer and John Carroll, to appear in 2014.

Contemporary design technology proponents tend to propose the elimination of manual sketching in favor of digital tools only. Given parametric design, simulation techniques and other advances in design computing some researchers believe that designing no longer needs representation at all, and digital models should replace representational concepts such as typologies, precedents and the like, with formation and materialization paradigms [35]. Although there is no denial that computational tools capture a continually growing place in the process of design, human cognition has not changed and in our view claiming that representation has completed its historical role in the process of designing means throwing the baby out with the bath water. We firmly believe that because of its considerable cognitive advantages, sketching is and will remain a useful—not to say essential—strategy in design idea generation, regardless of the computational tools used to develop a design and bring it to fruition. This is reason enough to propose a model of the role of sketching in the (conceptual) design process. Section 21.3 introduces such a model.

21.3 Model of Sketching as Design Search Facilitator

The ultimate purpose of the process of designing is the representation and specification of a design entity at an agreed level of detail and accuracy. Where the design entity is tangible its properties include shape and form and possibly other properties such as texture, color, and so on. These properties are best represented visually and therefore design is a visually mediated mental task. Visual representations are appropriate formats not only for the final product but also for interim and partial representations to be constructed in the course of the design search at the outset of a design task. The sketching model proposed here is based on the underlying notion that to date, sketching is the most productive strategy for the construction of preliminary visual representations at that early phase of designing, at least in the hands of experienced sketchers.

We shall now briefly describe the model, which is shown in Fig. 21.1. The next sections will describe empirical studies that support selected elements of the model.

The problem. Design tasks are initiated with a problem statement in the form of a brief or a program. Such statements are issued with significant variations in their level of comprehensiveness and coherence. As a rule the initial information is insufficient and more information, relevant to the problem and to the designer's disposition, must be sought.

Well-structured problem. In rare cases the problem is on the simple side, and only one satisfactory solution is envisioned. A problem of this kind is well-structured (and usually also well-defined), which stipulates that there is a known algorithm or procedure for arriving at the solution. Since no search is necessary to solve such problems, they are irrelevant to the sketching model.

Ill-structured problem. More often than not, design problems are ill-structured (and usually also ill-defined). This signifies that there can be numerous appropriate solutions to the problem and there is no known algorithm that would lead to a solution. Ill-structured problems present various challenges to designers including, in the case of tangible entities to be designed, a composition of forms and/or shapes, which is to be constructed as a result of a search ('ideation').

Search space. The search for one or more solutions takes place in a design space. Instead of problem space and solution space the term used is "search space," combining both spaces in design problem-solving. For a comprehensive treatment of the design space see Woodbury and Burrow [49] and for a combined problem–solution space see, e.g., Dorst and Cross [11]. In essence, the search space is where partial solutions are elicited, reasoned about and assessed, combined and taken apart, transformed and extended. In addition to externally provided data and directives, information that fuels the search acts is derived from the designer's *knowledge-base* that is stored in memory and from the perception of *stimuli* that direct retrieval of information by eliciting associations. The search space is a platform for representations that may be visual or nonvisual; many of the representations are visual and the model continues to treat only those visual representations.

Internal and external representations. The search space is virtual of course, and the visual representations within its bounds may be of two complimentary types: internal or external. Internal representations are those generated in mental visual imagery, where the formed images may be subjected to various transformations. However, mental images are short-lived and to prevent them from fading away considerable energy must be invested in refreshing them. External images are those created in the physical world. There are two main reasons to produce external representations in the design search: to decrease the load on memory and to assist in explorative thinking. Since designing is always complex and a large number of memory items are involved, creating what may be called an "external memory" that is instantly accessible any time helps reduce the cognitive load of trying to retrieve and maintain items from long- and short-term memory. However, this model is interested in the other function of representations, that is, as thinking aids.

Rapid sketch. The fastest and easiest to generate external visual images are rapid sketches, which differ from measured drawings in that they are freehand drawings or even "doodles" without prior preparation. Ferguson [13] distinguishes among three types of sketches according to their purpose: The *prescriptive sketch*, the *talking sketch* and the *thinking sketch*. The prescriptive sketch specifies the dimensions and other properties of the artifact in question, and is close to the measured drawing, although the latter is not a freehand drawing. The talking sketch is meant primarily for communication: with team-mates and other stake holders (clients, teachers, reviewers). Both of these will be left out of the discussion in this paper, wherein the focus is on the thinking sketch which is produced by a designer for his or her own purposes. It may be reduced to a bare minimum, or be overloaded with a mesh of lines on top of previous lines, such that nobody except the creator of the sketch actually understands its meaning. In Schön's terms [39], the sketch is a medium for the designer's conversation with his or her materials—and with him or herself.

Cognitive benefits and affordances. This is the most significant component of the model. It lists the main cognitive advantages that sketching affords in the idea generation phase, or search, of the design process.

Time effective, fluent. Most sketches are basically orthogonal projections, although not precise ones. An experienced designer is a fluent sketcher who does not need to think about the rules of production of orthogonal projections; these rules are so deeply engrained that they become practically automated (in some fields, like industrial design, many sketches tend to be three-dimensional). Therefore sketching can be executed very fast. To date, the speed with which experienced designers are able to generate sketches cannot be emulated by any other representational medium.

Minimal cognitive resources. Because the designer does not need to worry about production rules, and because the resultant sketch is for his or her own use and does not need to satisfy any external requirement, committing lines and shapes to paper is a very low-cost process in terms of cognitive resources.

Minimally rule-bound. Despite the fact that most sketches roughly follow the norms of orthogonal projections, sketchers can take any liberties they wish since the sketches are meant for their private use. This makes for a process with practically no constraints, including the lack of stop rules, i.e., one can stop whenever the designer feels satisfied or wishes to move on to the next sketch, related or unrelated to the current one.

Transformable/reversible. Sketches "talk back" to the designer [21] and lead to further development. The designer may want to backtrack and go in a different direction, or he or she may wish to transform what one sees on paper: add or

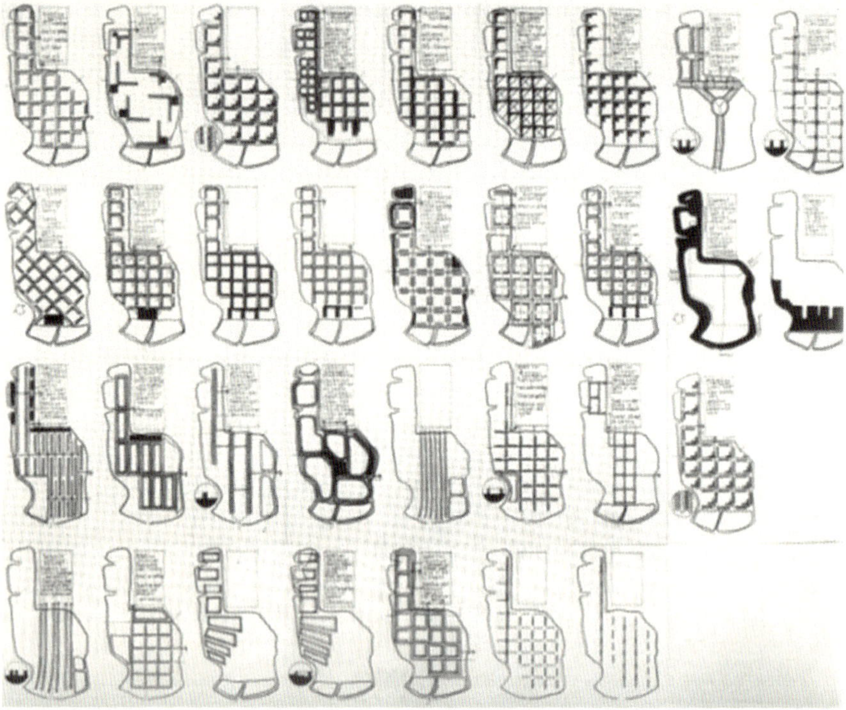

Fig. 21.2 Sketch of alternative urban design patterns for Runcorn New Town by Stirling 1967

delete, make changes of all kinds, fill in details, and so on. A common technique in freehand sketching is overlaying, using translucent "sketch paper" that allows selective preservation (and tracing) of previous notations, and changing others. Tracing is a preferred method for the production of alternative solutions to a design problem (see Fig. 21.2).

Tolerant to incompletion. The thinking sketch does not have to be complete. It can be left incomplete, just as long as the designer has captured in it that which was of interest when the sketch was made. Gestalt principles of perception teach us that we are able to complete in our minds images that are incomplete as long as the information necessary for such completion is stored in our minds. The missing elements that are not represented on paper may well be present in imagery: internal and external representations are well coordinated. Figure 21.3 shows one of a large number of serial sketches made by an architect in which one half of a symmetrical plan was left incomplete.

Tollerant to inaccuracy/lack of scale. Because thinking sketches are explorative studies, they can focus on one aspect of the entity they represent and if desired, that aspect may be exaggerated or imprecise for a better grip on one or another feature that one wishes to focus on. Therefore, neither scale nor accuracy are obstacles and just like completion—they are not required. Figure 21.4 is an

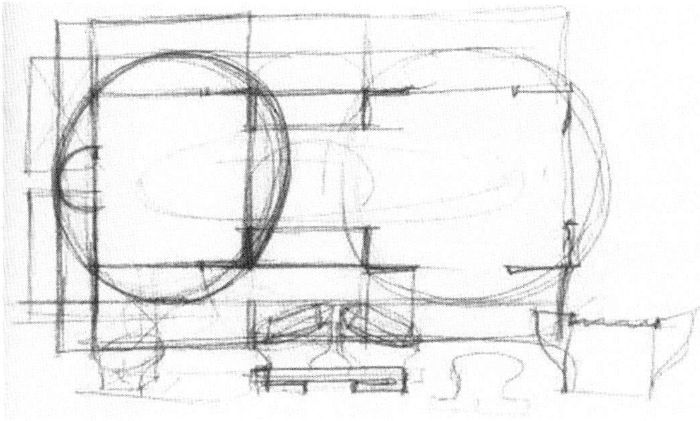

Fig. 21.3 Sketch for Cymbalista Synagogue by Botta, 1996. Only one of two cylindrical structures is fully outlined

Fig. 21.4 Sketch for Lambeth Community Center by Cullinan, 1982. Human figures are out of scale

example of an architectural sketch which knowingly does not respect scale consistency.

Provides unexpected cues. Putting marks down on paper is not always a conscious act that follows explicit and well worked out intentions. Nor are sketches necessarily "downloads" of images that were first constructed and scrutinized in imagery. We claim that sketching is an act of thinking; marks already on paper guide the next acts of sketching. In this process the emerging sketch is not always premeditated and may lead to unexpected cues when contemplated. These cues may trigger further development. This was well understood by the first generation of sketchers, the artists and designers of the late fifteenth century, whose term for sketches was "pensieri," the word for "thought" in old Italian [34].

Supports feedback loops. As mentioned above, internal and external representation feed each other and act in tandem. Mental images may be externally expressed in a sketch, and in turn the contemplation of the sketch, which can never be a precise duplicate of the mental image, gives rise to further associative thoughts that act on the mental image and transform it. This cyclic "dialog" between the mind and the hand holding a pencil and the eyes that perceive the marks on paper is most fruitful and we may talk about a feedback loop between them, in the cybernetic sense.

Having briefly outlined the components of the sketching model we turn to a number of examples of empirical studies and case studies aimed at instantiating and supporting some of the model's claims.

21.4 Sketching Studies

In this section three studies are reported, which illustrate and substantiate the sketching model and in particular the cognitive benefits of sketching and their relationship to visual imagery and existing knowledge.

21.4.1 Unexpected Cues in Sketches (Based on Goldschmidt [18])

The previous section presented sketching as generically activating personal associations that lead to idiosyncratic new interpretations of the same sketch.

The case study in this section presents a process in which sketching activated personal associations that led to idiosyncratic new interpretations of a sketch. In this process an architecture student, Larry, who decided to abandon the solution he had been working on for a kindergarten design, started anew with a different concept. It is based on a long interview with Larry immediately following the event in question. At the outset he had no idea what to do, and therefore he engaged in an activity he was in the habit of busying himself with when he felt a little lost: he scribbled his signature, numerous times. He liked his signature and jotting it down on paper gave him pleasure. In aimlessly sketching while searching for a design idea this young student resembled the prominent architect Alvar Aalto [1], who testified that when confronted with a new task the complexity is so great and the number of requirements so high that no "rational" design method works for him. Instead he starts producing what he calls "abstract art"—lines and shapes that are generated instinctively. Aalto testified that usually these marks on paper became a "starting point" for an idea that led the design development from that moment on. Larry underwent a similar process.

The sheet in front of Larry became full of samples of his signature and at some point when contemplating them he realized that the lines could be interpreted as enclosing spaces. He was already familiar with the site and the program, which

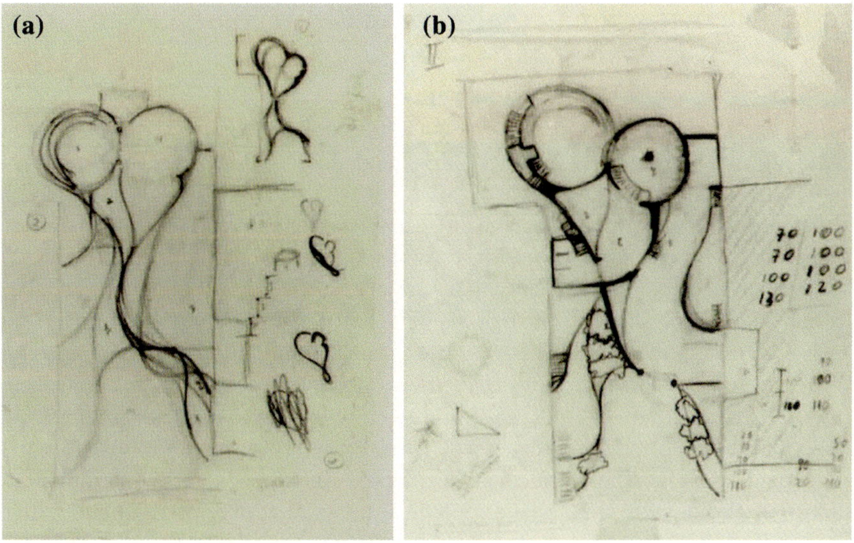

Fig. 21.5 Larry's kindergarten project; plan derived from his signature. **a** First sketch **b** subsequent plan sketch. *Note* on the left edge of the first sheet (**a**), the signature is drawn four times with a very light pencil that is quite difficult to discern

called for three activity spaces, and Larry could see in his mind's eye three spaces within the signature, if it is interpreted as a plan outline. From that moment things happened quickly: Larry drew another version of the signature, this time a little larger and with a plan in mind, and drew the site boundaries around it. He repeated the exercise at a yet larger scale in the middle of the same sheet of paper (interestingly the previous small signatures were drawn along the right and left edges of the sheet). In this sketch the signature became secondary; the outline was already a plan of the kindergarten with several added features that are not part of the signature. Figure 21.5 reproduces the sheet with the signatures and preliminary plan in the middle, and a subsequent development sketch.

This case is a first-hand example of discoveries the designer is able to make in his own sketches, which become stimuli that harbor cues and provide feedback. It is important to remember that most probably Larry's signature would have meant nothing to any other designer engaged in searching for a concept for a kindergarten plan. When contemplating one's own sketches, there is always information in memory to which the sketch may be associated. This information may be strictly personal, as in this case, and not shared by others. Therefore, for discoveries to be made in one's sketches, the personal context may be of the highest relevance and this is why we as observers, and even the designer him or herself, are often surprised by the associations that are formed and that lead to a design solution.

21.4.2 Internal and External Representation (Based on Verstijnen et al. [47, 48])

Verstijnen was interested in work on 'mental synthesis' that was pioneered by Finke and his associates (e.g. [14]) and extended by Anderson and Helstrup [2]. Finke was a mental visual imagery researcher who wanted to show that imagery was a powerful cognitive mechanism, capable of manipulating images in surprising ways. He devised a series of experiments in which participants were shown 15 simple shapes, some geometrical (e.g., square, half-sphere) and other standard objects (e.g., hook, bracket). After memorizing the shapes and their names the images were removed and the participants were blindfolded. A random selection of titles of three of the objects was called out, along with a category of object (e.g., toys). The participants were given 2 min to combine the three objects into a new object that fits into the given category. Then the eye folds were removed and the participants sketched their objects and explained what they were. Finke found that almost all participants were able to complete the task; furthermore, in debriefings most of them said they first endeavored to combine the shapes and only when an interesting combination was successfully found, decided on a particular purpose for the new creation (as opposed to the other way round). Finke called the resultant objects "preinventions." They were scored by naïve judges for creativity by giving separate scores to originality and to practicality. The percentage of objects that reached pre-established threshold scores that qualified them as creative or highly creative varied, contingent on the experimental conditions. Anderson and Helstrup repeated the experiment with one difference: they assigned the same task to a control group that was not blindfolded and that was allowed to sketch during the 2 min in which they endeavored to synthesize a new object. The results were scored as in Finke's experiment and in comparing the experimental and control groups no differences were found in the average creativity scores.

Verstijnen et al. [47, 48], who was skeptical when she saw this result, carried out an additional study which resembled the one by Anderson and Helstrup, with an additional experimental condition that pertained to the participants. Since Verstijnen believed that sketching should impact the results of the mental synthesis exercise, she recruited two groups in each experimental condition (with and without sketching): half the participants were law students and the other half were industrial design students (third year and up). Her assumption was that like in the previous studies where the participants were psychology students, law students would not reach higher creativity scores when allowed to sketch. Conversely, the design students, who were considered expert sketchers, were expected to score higher on creativity in the sketching condition. The results confirmed this expectation. Verstijenen went deeper into the sketches since she wanted to know in what way the high-scoring results with sketching differed from the imagery only results and the sketching with no experience. She found that the objects obtained through expert sketching displayed higher complexity, measured by a number of distinct transformations. A scoring method was developed to measure these transformations, which showed that sketchers obtained significantly higher scores, provided they were experienced, that is, design rather than law students.

The application of transformations was termed "restructuring," which stands in contrast to "combining," which is what participants were engaged in when they performed the mental synthesis task without transformations. The results obtained in the experiments prove that imagery is a perfectly adequate representation strategy to achieve a combined result. However, for restructuring, which tends to lead to bolder and more creative objects, imagery is not sufficient and sketching is required as the representation mode. In terms of our model these results exemplify the path from the display of stimuli as internal representations, through to rapid sketches in which most of their affording advantages can be exploited.

21.4.3 Analogy and Transformation (Based on Goldschmidt [19, 23])

Sketching is beneficial because it supports visual thinking. Visual thinking is a preferred cognitive strategy in design because it is useful to work with visual representations when endeavoring to arrive at the creation of a tangible entity that must by definition have distinct spatial/visual properties. Therefore, designers like to avail themselves of visual stimuli that have the potential of becoming triggers in the search for design ideas and concepts. It is easier for designers to react to an initial image and go from there, than to start from a blank sheet: an image that represents an idea can be reasoned about and thus lead to further progress. When stimuli are external (that is, not mental images that are retrieved from memory), one way to use them is to engage them as sources in a process of analogical reasoning. Analogical reasoning is considered to be typical of creative thinking and problem solving (e.g., [31]) and therefore visual analogy in design is of interest to us.

Stimuli can be just about everything: man-made or natural physical objects, photographs, pictures and drawings of various kinds. In research projects stimuli are often shown to designers at the outset of a design session, with or without instructions to use them as analogy sources. Of course, in real life designers often use stimuli spontaneously, sometimes without being aware of such usage. The use of stimuli almost always has an impact on the designer's activities; this impact may be positive or negative. It is positive if the designer is not only successful in transferring and mapping relevant information from the source stimulus to the target problem he or she is trying to solve, but if this information is abstracted, such that it may acquire new meaning in the context of the target situation. This operation is easier when the analogy is deep, or structural, in which case the information that is transferred pertains to relationships among components of the source stimulus, and not its surface properties [16]. Most people are not very good at spontaneously eliciting deep analogies; the process of identifying relationships among components and abstracting them is cognitively sophisticated and research has shown that without some help in the form of cues or specific instructions to use analogy, most people do not perceive the analogical similarity between a source and a target, visual or otherwise. For this reason mapping and transfer of surface properties, with little or no abstraction, are more common. However, this may not necessarily have a positive impact on the problem solving process, in the sense that the properties that are mapped and transferred do not acquire new meaning and in this case the target solution may have (too) many points of resemblance to the source. In design, wherein a solution is expected to be novel and at best creative and innovative, this is a negative effect which is referred to as fixation (e.g., [6, 23, 33, 36]).

The question, then, is under which conditions can visual stimuli be expected to induce useful analogical reasoning and not fixation? As already mentioned prompting, that is, asking people explicitly to draw analogies from stimuli is usually helpful. Another research finding is that the use of between-domains stimuli is more helpful than within-domain ones (e.g. [23]). Within-domain sources have been used in much of the empirical research on design fixation; many of the stimuli in question were in fact examples of solutions to the problem the designers were asked to solve in the experiments. It is easy to see why it would be difficult for a problem-solver to break away from a given example which, because it is operational, does not encourage the widening of the design space around it. Stimuli from other domains encourage abstraction because they cannot be used as solutions as is. In addition to mapping and transfer, another important condition for successful analogical reasoning in design is transformation.

Analogical reasoning is said to involve mapping and transfer from source to target. Mapping from one situation to another is a precondition for successful transfer (of properties or relations). The more abstraction takes place when information is transferred from source to target, the more options for its interpretation are open in the context of the target. In design, such interpretations and reinterpretations require that the information be not only transferred, but also transformed. This is where sketching becomes instrumental. Since the analogy is visual, that is, images are involved, to play a constructive role in the development of a solution to a design problem the information must be represented visually. To abstract in the source situation and to de-abstract or concretize in the target situation, sketching is a most convenient tool. Abstraction can be achieved by omitting details and creating schematic representations of the essentials, mostly relations. Then the converse process allows the designer to manipulate the abstract sketch, fill in new details, add and change proportions and other properties, such that a new concept may emerge. The key idea, then, is that in visual analogy sketching supports transformation and transformation is a prerequisite for a meaningful analogically derived design idea generation process. Figure 21.6 diagrams such a process. The designer, an architect, was asked to design a library based on a "footprint" outline with a given, problematic, entry point. He created his own visual stimuli by sketching other libraries he knew, notably Mont Angel Library in Oregon by Alvar Aalto, which became an analogy source for his design.

In addition to a sketch of the given footprint (T0) the designer made two sketches of the Aalto library: S1 is a section diagram showing how light penetrates into the building, and S2 is a floor plan. These sketches were in front of him when he made two

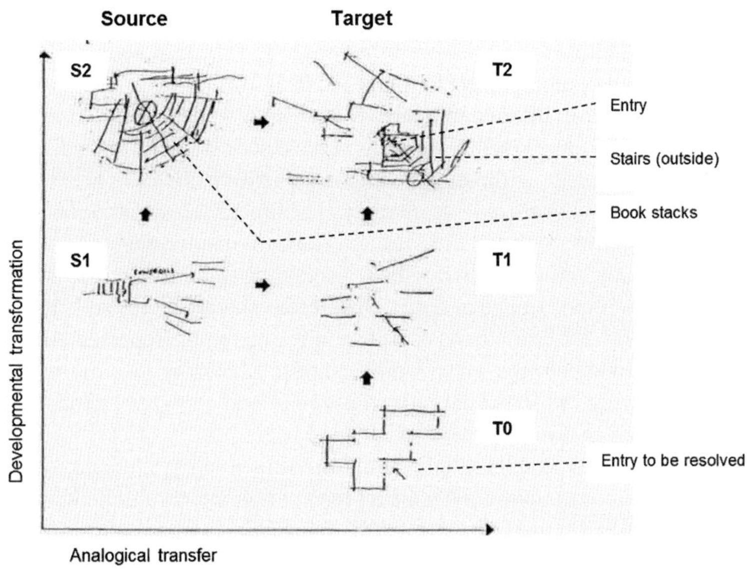

Fig. 21.6 Transfer and transformation with analogical mapping, library design. *S1* Diagram of light penetration *S2* Aalto library plan *T0* given "footprint" *T1* footprint transformed *T2* solution

drawings pertaining to the footprint library design. T1 is a transformed representation of the given outline, with horizontal lines only, one of which somewhat tilted. It is impossible to not notice a resemblance between this sketch and the diagram in S1. Finally, the designer drew the decisive sketch T2, in which the library entry problem was resolved. This sketch again represented the footprint, this time with more tilted horizontal lines, pointing in the opposite direction (compared to T1). In front of the entry there are a few L-shaped steps leading to the actual entrance. When comparing T2 with S2, a striking similarity is discovered in the pictorial structure of the two images: both have longer radial (tilted) lines, and shorter peripheral concentric lines. In S2 the latter represent the library's stacks. In T2 they represent stairs. The designer transferred the pictorial structure from the source to the target, but at the same time transformed it and made sure it also fitted with his previous explorations of the target, that is, the given footprint outline. By abstracting the composition, he was able to transform the stacks into stairs in the new context of the target problem, and he did so by sketching. This case study illustrates the primacy of transformation in successful analogical reasoning in design, which is well served by sketching.

21.5 Conclusion

The model of sketching presented in this paper emphasizes the empowerment that sketching affords the designer, by allowing fast and cognitively cost-free representations to "talk back" to the designer, thus creating loops of representation and feedback. Sketching is a perfect method for the enacting of transformations, which benefit from the ability to draw on both external and internal sources. The tolerance for inaccuracy, incompletion and lack of scale further contributes to the ease and speed of sketching. In a mental activity that involves much visual thinking or even depends on it, this kind of a facile representational tool is indispensable. Designing can take place without sketching, especially if the designer is endowed with strong imagery capacities. But imagery is a limited representational tool; it is less flexible, requires a costly cognitive effort in maintaining and refreshing images that cannot be consulted while producing additional representations; imagery allows only one image at a time. Therefore sketching has advantages over imagery, especially in a complex task that requires numerous representations and a succession of revisions to the same representation. Sketching also has advantages over digital drawing, which is largely rule-bound and requires attention to production rules even at the hands of experienced users.

The model presents sketching as a counterpart of imagery. Indeed, it is not an "either or" situation; the two modes of representation are complimentary and can maintain a fruitful "dialog" by providing feedback to one another. The mind images; the hand interprets the mental image on paper, giving the mind cause to 'revise' the image by reacting, sometimes assessing, the image on paper. This process is central to a search in the design space and also helps to extend it and push its boundaries.

Because sketching has the above advantages only if the designer is sufficiently experienced and fluent to not have to think of the production procedure, it is important to train designers in freehand sketching [8]. Needless to say, the goal is not the production of "beautiful" or artistic sketches, but useful ones that designers can think with. Fluency is an absolute precondition without which sketching cannot fulfill the potential role of a major thinking aid. Understanding the significant advantages of sketching in the process of design ideation should hopefully encourage designers, and design educators and students, to fully exploit this tool rather than dismiss it as a "thing of the past."

References

1. Aalto A (1957) Abstract art and architecture. In: Hoesli B (ed) Alvar Aalto: synopsis. BirkenhäuserVerlag, Basel, pp 223–225
2. Anderson RE, Helstrup T (1993) Visual discovery in mind and on paper. Mem Cogn 21:283–293
3. Arnheim R (1986) A plea for visual thinking.In New essays on the psychology of art. University of California Press, Brkeley, pp 135–152
4. Athavankar U (1996) Mental imagery as a design tool. In: Trappl R (ed) Proceedings of the thirteenth European meeting on cybernetics and systems. Austrian Society of Cybernetics Studies and University of Vienna, Austria, 19–12 April, 1996, Vol II, pp 382–387
5. Bilda Z, Gero JS, Purcell AT (2006) To sketch or not to sketch: that is the question. Des Stud 27(5):587–613

6. Cardoso C, Badke-Schaub P (2009) Give design a break? The role of incubation periods during idea generation. In: Bergendahl M, Grimheden M, Leifer L, Skogstad P, Lindemann U (eds) Proceedings of the 17th international conference on engineering design (ICED'09), vol 2. Design Theory and Research Methodology. The Design Society, Stanford, pp 383–394
7. Casakin H, Goldschmidt G (1999) Expertise and the use of analogy and visual displays: implications for design education. Des Stud 20(2):153–175
8. deVere I, Kapoor A, Melles G (2011) Developing a drawing culture: new directions in engineering education. In: Culley SJ, Hicks BJ, McAloone TC, Howard TJ, Dong A (eds) Proceedings of the 18th ICED. The Technical University of Denmark, Copenhagen, vol 8 paper 426, 151–160
9. Do EY-L (2002) Drawing Marks, Acts and Reacts: toward a computational sketching for architectural design. AIEDAM 16(3):149–171
10. Do EY-L (2005) Design Sketches and Sketch design tools. Knowl Based Syst 18:383–405
11. Dorst K, Cross N (2006) Creativity in the design process: coevolution of problem-solution. Des Stud 22(5):425–437
12. Eisenstein EL (1983) The printing revolution in early modern Europe. Cambridge university Press, Cambridge
13. Ferguson ES (1992) Engineering and the mind's eye. The MIT Press, Cambridge
14. Finke R (1990) Creative imagery: discoveries and inventions in visualization. Erlbaum, Hillsdale
15. Fish J, Scrivener S (1990) Amplifying the mind's eye; sketching and visual cognition. Leonardo 23:117–126
16. Gentner D (1983) Structure-mapping: a theoretical framework for analogy. Cogn Sci 7(2):155–170
17. Goldschmidt G (1991) The dialectics of sketching. Creativity Res J 4(2):123–143
18. Goldschmidt G (1994) On visual design thinking: the vis kids of architecture. Des Stud 15(2):158–174
19. Goldschmidt G (2001) Visual analogy—a strategy for design reasoning and learning. In: Eastman C, Newsletter W, McCracken M (eds) Design knowing and learning: cognition in design education. Elsevier, New York, pp 199–219
20. Goldschmidt G (2002) Read-write acts of drawing. TRACEY (Internet Journal dedicated to contemporary drawing issues); issue on syntax of mark and gesture, University, UK. http://www.lboro.ac.uk/departments/ac/tracey/somag/gabi.html Loughborough
21. Goldschmidt G (2003) The backtalk of self-generated sketches. Design Issues 19(1):72–88
22. Goldschmidt G (2008) Sketching is alive and well in this digital age. In: Poelman W, Keyson D (eds) Design processes: What architects and industrial designers can teach each other about managing the design process. IOS Press, Amsterdam, pp 29–43
23. Goldschmidt G (2011) Avoiding design fixation: transformation and abstraction in mapping from source to target. J Creative Behav 45(2):92–100
24. Goldschmidt G, Klevitsky E (2004) Graphic representation as reconstructive memory: Stirling's German museum projects. In: Goldschmidt G, Porter WL (eds) Design representation. Springer, London, pp 37–61
25. Goldschmidt G, Smolkov M (2006) Variances in the impact of visual stimuli on design problem-solving performance. Des Stud 27(5):549–569
26. Gross MD (1996) The Electronic Cocktail Napkin—a computational environment for working with design diagrams. Des Stud 17(1):53–69
27. Gross MD, Do EY-L (2000) Drawing on the back of an envelope. Comput Graph 24(6):835–849
28. Heidenrich L, Lotz W (1974) Architecture in Italy 1400–1600. Penguine Books, Middlesex
29. Herbert DM (1988) Study drawings in architectural design: their properties as a graphic medium. J Architect Educ 41:26–38
30. Johnson G, Gross MD, Hong J, Do EY-L (2009) Computational support for sketching in design: a review. Found Trends Hum–Comput Interact 2(1)1:1–93

31. Johnson-Laird PN (1989) Analogy and the exercise of creativity. In: Vosniadou S, Ortony A (eds) Similarity and analogical reasoning. Cambridge University Press, Cambridge, pp 313–331
32. Lacy B (1991) 100 contemporary architects: drawings & sketches. Harry N. Abrams, New York
33. Linsey J, Tseng I, Fu K, Cagan J, Wood KL (2009) Reducing and perceiving design fixation: initial results from an NSF-sponsored workshop. In: Bergendahl M, Grimheden M, Leifer L, Skogstad P, Lindemann U (eds) Proceedings of the 17th international conference on engineering design (ICED'09), vol 2. Design theory and research methodology. The Design Society, Stanford, pp 233–44
34. Olszweski EJ (1981) The draughtsman's eye: late renaissance schools and styles. Cleveland Museum of Art/Indianna University Press, Cleveland
35. Oxman R (2008) Digital architecture as a challenge for design pedagogy: theory, knowledge, models and medium. Des Stud 29(2):99–120
36. Purcell AT, Gero JS (1996) Design and other types of fixation. Des Stud 17(4):363–383
37. Purcell AT, Gero JS (1998) Drawings and the design process. Des Stud 19(4):389–430
38. Schank Smith K (2005) Architects' drawings: a selection of sketches by world famous architects through history. Architectural Press, Oxford
39. Schön D, Wiggins G (1992) Kinds of seeing and their functions in designing. Des Stud 13(2):135–156
40. Shapir O, Goldschmidt G, Yezioro A (2007) Conceptual design: an operational prescription for a computer support system. In: Banissi E, Sarfraz M, Dejdumrong N (eds) Computer graphics, imaging and visualization: new advances. 4th CGIV07 international conference. Bangkok, 15–17 Aug. IEEE & Computer Society, London, pp 513–521
41. Sloman SA (1996) The empirical case for two systems of reasoning. Psychol Bull 119(1):3–22
42. Smith SM, Linsey JS, Kerne A (2010) Using evolved analogies to overcome creative design fixation. In: Taura T, Nagai Y (eds) Design creativity 2010. Springer, London, pp 35–39
43. Suwa M, Tversky B (1997) What do architects and students perceive in their design sketches? A protocol analysis. Des Stud 18(4):385–403
44. Suwa M, Tversky B (2001) How do designers shift their focus of attention in their own sketches? In: Anderson M, Meyer B, Olivier P (eds) Diagrammatic representation and reasoning. Springer, London, pp 241–254
45. Suwa M, Tversky B (2002) External representations contribute to the dynamic construction of ideas. In: Hegarty M, Meyer B, Narayanan NH (eds) Diagrammatic representation and inference, proceedings of diagrams 2002. Lecture notes in artificial intelligence series. Springer, London, pp 341–343
46. Suwa M, Tversky B, Gero JS, Purcell T (2001) Seeing into sketches: regrouping parts encourages new interpretations. In: Gero JS, Tversky B, Purcell T (eds) Proceedings of visual and spatial reasoning in design II. Key Centre of Design Computing and Cognition, University of Sydney, pp 207–220
47. Verstijnen IM, Hennessey JM, van Leeuwen C, Hamel R, Goldschmidt G (1998) Sketching and creative discovery. Des Stud 19(4):519–546
48. Verstijnen IM, van Leeuwen C, Goldschmidt G, Hamel R, Hennessey JM (1998) Creative discovery in imagery and perception: combining is relatively easy, restructuring takes a sketch. Acta Psychol 99:177–200
49. Woodbury RF, Burrow AI (2006) Whither design space? AIEDAM 20:63–82

Index

A. Chakrabarti and L. T. M. Blessing (eds.), *An Anthology of Theories and Models of Design*, DOI: 10.1007/978-1-4471-6338-1,